# DATE DUE

| | | | |
|---|---|---|---|
| | | | |
| | | | |
| | | | |
| | | | |
| | | | |
| | | | |
| | | | |
| | | | |
| | | | |
| | | | |
| | | | |
| | | | |
| | | | |
| | | | |
| | | | |
| | | | |
| | | | |

DEMCO 38-296

STP 1063

# Masonry: Components to Assemblages

*John H. Matthys, editor*

ASTM
1916 Race St.
Philadelphia, PA 19103

Copyright © by AMERICAN SOCIETY FOR TESTING AND MATERIALS 1990

## NOTE

The Society is not responsible, as a body,
for the statements and opinions
advanced in this publication.

### Peer Review Policy

Each paper published in this volume was evaluated by three peer reviewers. The authors addressed all of the reviewers' comments to the satisfaction of both the technical editor(s) and the ASTM Committee on Publications.

The quality of the papers in this publication reflects not only the obvious efforts of the authors and the technical editor(s), but also the work of these peer reviewers. The ASTM Committee on Publications acknowledges with appreciation their dedication and contribution of time and effort on behalf of ASTM.

Printed in Chelsea, MI
July 1990

# Alan H. Yorkdale

## 1931–1987

# Dedication

*Alan H. Yorkdale's passing from this life left his many friends, associates, and colleagues with large personal voids in their lives. Alan's personable, friendly helpfulness at all times will be sorely missed by all.*

*Alan's participation and accomplishments, within and for the Masonry Industry, include Vice-President, Engineering and Research, Brick Institute of America; membership in the American Ceramic Society, American Concrete Institute, American National Standards Institute, American Society for Testing and Materials, Building Officials and Code Administrators of America International, Inc., National Executive Committee of the National Institute of Building Sciences Consultative Council, National Society of Professional Engineers, Society of American Military Engineers, Southern Building Code Congress International, The Masonry Society, U.S. Chamber of Commerce, and many other affiliations.*

*Alan was especially active in ASTM activity, believing that clear standards for masonry testing and materials were vital to the understanding and use of masonry materials in construction. Alan was active at all levels of the*

*Society: Society membership from 1971, membership on ASTM C-12 on Mortars for Unit Masonry, serving as Chairman of ASTM C-12.03 on Specifications during the period 1977 to 1978; membership on ASTM C-15 on Manufactured Masonry Units, serving as Secretary during the period 1971 to 1974, Chairman of ASTM C-15.02 on Clay Brick and Structural Clay Tile since 1976 and 1st Vice-Chairman since 1976; Membership on ASTM E-5 on Fire Standards; Membership on E-6 on Performance of Building Construction, serving as Chairman of E-6.55 on Exterior Building Walls; at the Society Alan served on the Committee on Standards and the Committee on Terminology and the Society Board of Directors from 1982 to 1984. His tireless work and devotion to ASTM resulted in his being honored with the ASTM Award of Merit and the Honorary Title of ASTM Fellow in 1985.*

*Alan was formally educated at Montgomery College, George Washington University, and the University of Virginia. He joined the Brick Institute of America in 1960 and served in various engineering capacities during his long service to the brick industry culminating in his position as Vice-President, Engineering and Research in 1980. His vast knowledge of masonry was always available to those who sought his expertise. His enthusiastic, dynamic presentations and lectures were always well received by his audiences. His clear, concise writings and papers contributed greatly to the masonry industry's store of knowledge.*

*WHEREAS, the members of ASTM Committee C-15 on Manufactured Masonry Units have suffered a great loss of a valued contributor and sincere friend.*

*BE IT RESOLVED that this memorial to Alan H. Yorkdale be included in the minutes of this meeting and copies sent to members of his family with our most sincere sympathy and condolences from all our members.*

*BE IT ALSO RESOLVED that the Alan H. Yorkdale Memorial Award be established for significant papers related to clay or concrete manufactured masonry units and published in the English language. ASTM Committee C-15 on Manufactured Masonry Units will administer the Memorial Award.*

*ASTM Committee C-15*
*December 1987*

# Foreword

This publication, *Masonry: Components to Assemblages*, contains papers presented at the symposium of the same name held in Orlando, Florida on 5 Dec 1989. The symposium was sponsored by ASTM Committees C-7 on Lime, C-12 on Mortars for Unit Masonry, and C-15 on Manufactured Masonry Units. John H. Matthys, Professor of Civil Engineering, The University of Texas at Arlington, presided as symposium chairman and was editor of this publication.

# Foreword

# Contents

**Introduction**                                                                 1

COMPONENTS

**Initial Rate of Absorption of Clay Brick Considering Both Bed Surfaces in the As**     5
**Received Condition and After Outside Exposure**—J. H. MATTHYS,
W. BAILEY, AND J. EDWARDS
Discussion                                                                      22

**ASTM C90 Concrete Masonry Units Using Lightweight Aardelite Aggregates**—      27
E. R. DUNSTAN, JR. AND P. HAY
Discussion                                                                      36

**Recommended Test for Slip Resistance on Brick Surfaces**—B. E. TRIMBLE        38

**The Properties of Masonry Grout in Concrete Masonry**—E. G. HEDSTROM AND       47
M. B. HOGAN

**Effects of Aggregate Gradation on Properties of Masonry Cement and Portland**   63
**Cement-Lime Masonry Mortars**—C. E. BUCHANAN, JR. AND B. M. CALL
Discussion                                                                      81

**Masonry Cements—A Laboratory Investigation**—J. W. RIBAR AND V. S. DUBOVOY     85
Discussion                                                                      96

**A Bentonite Clay Plasticizer in Masonry Mortars**—B. DICKELMAN               108
Discussion                                                                     120

**The Development of Ready Mixed Mortar in the United States**—R. E. GATES,     123
R. L. NELSON, AND M. F. PISTILLI
Discussion                                                                     143

**Quality Control of Mortars: Cubes vs. Cylinders**—S. SCHMIDT, M. L. BROWN, AND   147
R. TATE
Discussion                                                                     162

**Performance of Mortars Produced Under the Proportion Specification of ASTM C 164 270-86a**—J. H. MATTHYS
Discussion 171

**Corrosion of Reinforcing Steel and Wall Ties in Masonry Systems**—C. HAVER, 173
D. KEELING, S. SOMAYAJI, D. JONES, AND R. HEIDERSBACH
Discussion 191

**Experimental Investigation of Mortar Compressive Strength Using 5.08 cm Cubes** 194
**and 7.62 cm x 15.24 cm Cylinders**—J. H. MATTHYS AND R. SINGH

**The Effect of Constituent Proportions on Stress-Strain Characteristics of Portland** 206
**Cement-Lime Mortar and Grout**—W. SRIBOONLUE AND E. M. WALLO

ASSEMBLAGES

**IRA and the Flexural Bond Strength of Clay Brick Masonry**—W. M. McGINLEY 217
Discussion 230

**A Compilation of Flexural Bond Stresses for Solid and Hollow Nonreinforced Clay** 235
**Masonry and Portland-Lime Mortars**—B. GABBY

**Nondestructive Evaluation of Masonry: An Update**—J. NOLAND, G. KINGSLEY, 248
AND R. ATKINSON

**Compressive Tests of Hollow Brick Units and Prisms**—R. H. BROWN AND 263
J. G. BORCHELT

**Anchor Connections of Stone Slabs**—J. E. AMRHEIN, R. H. HATCH, AND 279
M. MERRIGAN
Discussion 292

**A Critical Review of the Field Adapting ASTM E 514 Water Permeability Test** 299
**Method**—M. T. BROWN
Discussion 308

**A Discussion of the Abuse of Some Common Masonry Industry Practices**— 309
R. W. CROOKS AND F. A. HERGET

**Problems and Cures in Masonry**—A. TOMASSETTI 324

**Research of Physical Properties of Masonry Assemblages Using Regional** 339
**Materials**—W. A. LASKA, O. W. OSTRANDER, R. L. NELSON, AND
C. C. MUNRO

**Concrete Masonry Prism and Wall Flexural Bond Strength Using Conventional** 350
**Masonry Mortars**—J. H. MATTHYS
Discussion 362

**The Potential for Traffic Noise Reduction by Thin Masonry Panels**—M. MEHTA     366

**In-Plane Seismic Resistance of Two-Story Coupled Concrete Masonry Walls**—     378
    M. MERRYMAN, G. LEIVA, N. ANTROBUS, AND R. E. KLINGNER

**Masonry Wall Drainage Test—A Proposed Method For Field Evaluation of**     394
    **Masonry Cavity Walls for Resistance to Water Leakage**—
    N. V. KROGSTAD

**Towards Developing a Flexural Strength Design Methodology for Concrete**     403
    **Masonry**—A. A. HAMID, G. F. ASSIS, AND H. G. HARRIS

**Aspects of Blast Resistant Masonry Design**—D. E. VOLKMAN     413

**Summary**     423

**Indexes**     429

# INTRODUCTION

The papers in this ASTM STP 1063 were submitted and successfully peer reviewed for the sixth in a series of masonry symposia sponsored by ASTM Committees C-7 on Lime, C-12 on Mortars For Unit Masonry, and C-15 on Manufactured Masonry Units. Like its predecessors this symposium provided a forum for the dissemination and exchange of information related to masonry components and assemblages. Committee C-15 held the distinction of being the committee in charge of the symposium.

This symposum and its proceedings are dedicated to ASTM Fellow Alan H. Yorkdale, who spent a significant portion of his life both inside and outside of ASTM promulgating masonry information exchange and technical advancement of masonry, particularly as related to manufactured masonry units. His enthusiasm and devotion to his professional efforts and his personable character and individual concern for others gave him a special place in the hearts of those whose lives he touched.

To produce a lasting tribute to a special friend, ASTM Committee C-15 established the ASTM Society Alan H. Yorkdale Memorial Award. In addition ASTM Committees C-7, C-12 and C-15, along with ASTM Headquarters worked diligently to produce this special symposium, "Masonry: Components To Assemblages" and the corresponding proceedings STP 1063. Special thanks go to the 28 authors who submitted and completed the ASTM review process that produced the outstanding array of papers herein presented. Gratitude is extended to the 69 members of ASTM Committees C-7, C-12, and C-15 who peer reviewed the submitted papers.

Thanks are extended to the ASTM staff: Kathy Greene, Dorothy Savini, Barbara Stafford, Therese Pravitz, John Vowell, David Jones, and Gerald Davis for providing the extra efforts and considerations to make this a special symposium and proceeding publication. Finally, recognition goes to George Judd (Chairman C-7), Hugh McDonald (Chairman C-12), Colin Munro (Chairman C-15), J. O'Grady (President ASTM), J. Grogan (Chairman of Alan H. Yorkdale Memorial Award Committee) and D. M. Greason (ASTM Chairman of the Board) for their symposium tributes to Alan H. Yorkdale and the formal presentation to the first recipient of the Alan H. Yorkdale Memorial Award. This symposium and its proceedings were indeed a family affair. Thanks to all for a job well done.

John H. Matthys
Professor Civil Engineering
Director Construction Research Center
University of Texas at Arlington
Symposium Chairman

# Components

William G. Bailey, John H. Matthys, and Joseph E. Edwards

INITIAL RATE OF ABSORPTION OF CLAY BRICK CONSIDERING BOTH BED SURFACES IN THE AS RECEIVED CONDITION AND AFTER OUTSIDE EXPOSURE

---

REFERENCE: Bailey, William G., Matthys, John H., and Edwards, Joseph E., "Initial Rate of Absorption of Clay Brick Considering Both Bed Surfaces in the As Received Condition and After Outside Exposure," Masonry: Components to Assemblages: ASTM STP 1063, J. H. Matthys, Ed., American Society for Testing and Materials, Philadelphia, 1990.

ABSTRACT: The initial rate of absorption test (suction) on a clay brick unit is generally conducted in the laboratory according to ASTM C-67. This test attempts to represent the effect of the unit on pulling water out of the mortar during assemblage construction. In ASTM C-15 committee work, two questions that have been raised are:

1) Is the IRA of a unit the same for both bed surfaces that will be in contact with mortar?

2) When as received units are exposed to typical Texas heat, is the IRA significantly affected?

To address these questions a project was conducted at The University of Texas at Arlington. In addition data was requested and received from various U.S.A. brick manufacturerers. Analysis of this data provided answers to the above questions.

William G. Bailey is Manager of Technical Services for the ACME Brick Co. of Fort Worth, TX. John H. Matthys is Professor of Civil Engineering at The University of Texas at Arlington. Joseph E. Edwards is Vice President of Engineering and Research of General Shale Products Corp. of Johnson City, TN.

KEYWORDS:    Initial Rate of Absorption, solid brick, cored brick, flashed brick, extruded brick, hand molded brick

The pores or small openings in clay products function as capillaries which tend to draw water into the unit. This action in a brick is referred to as its initial rate of absorption or suction [1].

Capillarity or suction has little bearing on water transmission through masonry, but it has an important effect on the adhesion or bond between brick and mortar [1].

Suction of brick is determined by partial immersion of the unit to a depth of 0.32 cm. (1/8 inch) in water for a period of one minute [2]. The initial rate of absorption test conducted according to ASTM C-67 requires that the test brick be oven dried. The bed surface area is determined. In the case of cored brick the net bed surface area has to be determined by deducting the area of the core holes. The initial rate of absorption of a test brick is corrected to 194 cm.$^2$ (30 in.$^2$) of bed surface area. The test results are reported in grams of water absorbed in one minute per 194 cm.$^2$ (30 in.$^2$) of bed surface area. Suction may be calculated by the following formula:

$$S = \frac{(W^1 - W) \times 194 \text{ cm.}^2}{A}$$

Where

$S$  = suction, in grams per minute per 194 cm.$^2$

$W$  = weight of unit prior to partial immersion, in grams

$W^1$ = weight of unit after partial immersion for one minute, in grams

$A$  = net cross-sectional area of surface of unit immersed, in square cm.

Numerous tests of the tensile strength of bond between mortars and brick, and of the permeability of brick walls to water penetration indicate that, other factors remaining constant, maximum bond strength and minimum water penetration are obtained with bricks having suctions not exceeding 20 grams per minute per 194 cm.$^2$ (30 in.$^2$) when laid [1]. However, Robinson and Brown [3] and Gazzola [4] report that IRA appears to have little influence on bond strength. Additional research is needed before most individuals in the

masonry industry accept the fact that bond strength and
IRA are not related in some fashion.

The field term for a high initial rate of absorption
brick is a "Hot Brick". The field correction for high
IRA brick is to prewet the brick prior to laying. This
correction keeps the brick from pulling excessive
amounts of water out of the mortar. Excess water is
required for proper curing. Incorrect mortar curing
results in poor adhesion and water permeable mortar
joints. Note 2 in ASTM C-216 recommends that units
having initial rates of absorption exceeding 30
gms./min./194 cm.$^2$ (30 in.$^2$) should be well wetted prior
to laying [5].

**WHY WAS THIS STUDY CONDUCTED?**

Grimm studied the effect of brick suction on
bricklayer productivity [6]. Productivity was measured
for walls constructed with "vitrified face brick" (low
IRA) and moderate IRA brick. In that study he reported
marked differences in the initial rate of absorption of
one bed surface versus the other bed surface for the
same brick. Grimm's report and additional discussions
at ASTM C-15 meetings sparked the interest in additional
research to examine this phenomena.

**ADDITIONAL QUESTIONS RAISED AS THE STUDY PROGRESSED**

As the project progressed additional questions were
raised. Is the IRA different between bed surfaces for
extruded brick versus molded brick? Is there a
difference between hand molded and machine molded brick?
Are flashed bricks different from non-flashed bricks
when IRA for each bed surface is tested? Will cored
brick versus solid brick be different when IRA for each
bed surface is determined?

Concern about brick walls constructed during a hot
Texas summer brought about another question. Is a
laboratory determined IRA different from the IRA of job
site brick stored in the hot sun? In addition are IRA
test results significantly different for brick oven
dried as prescribed in ASTM C-67 versus brick tested in
the "as received" condition?

**BRICK TESTED**

A total of 344 brick from more than eight different
manufacturers made up the 62 samples that were tested.
Nineteen samples including 103 brick were tested in the
"as received" condition as well as in the oven dried
condition. Fifteen samples including 81 brick were
tested after exposure to hot weather for an extended

period of time.  Table 1 depicts a breakdown of the
samples tested.

TABLE 1 -- Sample Summary

| Process | Cored | Flashed | Samples | Total Brick |
|---------|-------|---------|---------|-------------|
| Extruded | Yes | No | 23 | 136 |
| Extruded | Yes | Yes | 9 | 46 |
| Extruded | No | No | 9 | 45 |
| Extruded | No | Yes | 8 | 40 |
| Hand Molded | No | No | 6 | 31 |
| Machine Molded | No | No | 7 | 46 |
| | | TOTAL: | 62 | 344 |

**TEST PROCEDURE**

All brick were tested for initial rate of absorption
as prescribed in ASTM C-67.  Each brick was tested with
one bed surface placed in the water for one minute and
the absorption was recorded.  The brick was then placed
with the other bed surface in the water and that
absorption was recorded.

Fifteen samples of extruded cored clay brick from
several states were shipped to UTA for investigation.
Upon receipt at UTA specimens were unboxed, inspected,
and appropriately marked.  The IRA test method as
stipulated in ASTM C-67 was conducted on the brick in
the "as received condition" in an environmentally
controlled lab.  Results are shown in Table 5.  The
specimens were returned to their shipping boxes and
stored in the laboratory for approximately one year.
The specimens were then transported to wood pallets
outside to be exposed to "Texas heat" during the summer
months.  The specimens were exposed for a three week
period to temperatures averaging 100° F.  The ASTM C-67
IRA test was conducted on the brick in the "in the sun"
condition.  Results are given in Table 5.  The specimens
were returned to their shipping boxes and stored in the
laboratory for approximately seven months.  The
specimens were then subjected again to the ASTM C-67 IRA
test after being brought to the "oven dried" conditions
as stipulated in ASTM C-67.  Results are listed in Table
5.  The personnel that conducted the "as received" IRA
test was different than the personnel that conducted the
"In the Sun" and "Oven Dried" IRA.

One sample of hand molded brick and three samples of machine molded brick were also tested to determine "as received" versus oven dried IRA comparisons.

## RESULTS

The test data with respect to the IRA on the two opposing bed surfaces of a given brick are given in Table 2 (Extruded Brick), Table 3 (Molded Brick), and Table 4 (Group Data Summary). This same data is plotted graphically in Figures 1 through 8. Table 2 (Extruded Brick) data is presented according to five groups: Group I - All Extruded; Group II - Cored Non-Flashed; Group III - Cored - Flashed; Group IV - Solid Non-Flashed, and Group V - Solid Flashed. Table 3 Molded Brick data is presented by two groups: Group VII Hand Molded and Group VIII Machine Molded. The test data comparison with respect to IRA for "As Received," "In the Sun," and "Oven Dried" conditions are given in Table 5. On the _average_ all of the extruded brick including cored, solid, flashed and nonflashed show no significant difference in the IRA of one bed surface versus the other as indicated in Table 4. However, an individual brick may have a large difference from one bed surface to the other. One extruded brick tested had a difference of 17.4 grams between bed surfaces. The maximum and minimum differences for individual samples are listed in Table 2. These results are shown graphically in Figures 1a through 1d.

There was a small difference (1.4 versus 2.0) for the average difference in bed surface IRA's for brick oven dried as prescribed in ASTM C-67 versus brick tested in the "as received" condition. The cored extruded brick, like the entire group of extruded brick, were not significantly different when one bed surface IRA was compared to the other. See Figures 1 and 2. An individual brick in this group had a large difference as depicted in Group II of Table 2. The results for oven dried versus "as received" values were the same.

The cored extruded flashed brick exhibited the same small differences as observed for the overall group of extruded brick. See Figure 3. Again the oven dried versus the "as received" results were very nearly the same. Additional tests should be conducted for flashed brick. If one bed surface is flashed and the other surface is not flashed, significant difference in IRA may be observed. Time did not permit collecting the samples needed to show true differences for that condition.

Cored extruded nonflashed brick (Figures 2a and 2b) versus solid extruded nonflashed brick (Figure 4) were also not significantly different. On the average the IRA differences were small for the solid brick tested although on individual brick one might see a large difference (12.8) as listed in Table 2, Group IV.

Solid extruded flashed brick (Figure 5) did show a larger average difference in bed surface IRA's (2.7) than the entire group of extruded brick but the differences for the brick tested were small. Additional testing should be conducted on selected samples of flashed brick (one surface flashed and one surface not flashed).

Group VI (Figure 6) included all of the molded brick tested. Molded brick do show large average differences in IRA values from one bed surface to the other (8.8) as shown in Table 4. Individual brick had very large differences in bed surface IRA's. One brick in Group VII (Figure 7) Table 2 had a difference of over 50 grams and the sample averaged over 22 grams difference. The differences between "as received" and oven dried brick were small.

Hand molded brick show slightly larger differences in IRA values than the machine molded brick (9.6 versus 8.2) as shown in Table 4.

Table 5 listed actual IRA values for brick tested in the "as received" condition, in the "in the sun" condition, and in the "oven dried" condition. For the brick tested there are only small differences for the different test conditions.

**CONCLUSIONS**

The average IRA of one bed surface compared to the other bed surface for extruded brick was nearly the same. An individual extruded brick may have a large difference in bed surface IRA's.

Flashed brick may exhibit significantly different IRA values from one bed surface to the other if one surface is flashed and the other surface is not. The flashed brick tested were flashed on both surfaces.

Molded brick definitely have different IRA characteristics between bed surfaces. Hand molded and machine molded brick show about the same differences between bed surfaces and are much larger values than extruded brick.

Brick tested in the "as received" condition versus field sun conditions versus oven dried conditions show very similar IRA values.

## TABLE 2 -- Extruded Brick

| I.D. | Sample Size | As Rec'd. | Oven Dried | Avg.[1] $\triangle$ | $V^3$ % | $\triangle^2$ Max./Min. |
|------|-------------|-----------|------------|---------------------|---------|-------------------------|

Group I = Groups II, III, IV, & V - All Extruded Brick

Group II - Cored Non-Flashed

| I.D. | Sample Size | As Rec'd. | Oven Dried | Avg. | V | Max./Min. |
|------|-------------|-----------|------------|------|-----|-----------|
| 1 | 6 | X | | 2.65 | 139.2 | 10.1/0.3 |
| 1 | 6 | | X | 3.03 | 94.4 | 8.8/1.1 |
| 3 | 4 | X | | 2.47 | 104.4 | 6.1/0.0 |
| 3 | 4 | | X | 1.60 | 80.6 | 3.3/0.3 |
| 5 | 6 | X | | 0.57 | 77.2 | 1.3/0.2 |
| 5 | 6 | | X | 0.45 | 82.2 | 1.1/0.0 |
| 6 | 5 | X | | 0.50 | 92.0 | 1.3/0.1 |
| 6 | 5 | | X | 2.22 | 89.6 | 5.7/0.9 |
| 7 | 6 | X | | 0.48 | 99.8 | 1.3/0.0 |
| 7 | 6 | | X | 0.73 | 126.0 | 2.5/0.0 |
| 8 | 6 | X | | 1.75 | 66.3 | 3.8/0.7 |
| 8 | 6 | | X | 2.20 | 56.8 | 3.8/0.6 |
| 9 | 6 | X | | 1.23 | 138.2 | 4.6/0.0 |
| 9 | 6 | | X | 0.58 | 156.9 | 2.3/0.0 |
| 10 | 4 | X | | 1.40 | 65.7 | 2.3/0.3 |
| 10 | 4 | | X | 0.77 | 107.8 | 1.9/0.1 |
| 11 | 6 | X | | 3.00 | 76.7 | 7.2/0.8 |
| 11 | 6 | | X | 2.48 | 93.1 | 6.1/0.1 |
| 12 | 5 | X | | 1.12 | 112.5 | 2.9/0.0 |
| 12 | 5 | | X | 1.66 | 59.0 | 2.6/0.3 |
| 13 | 5 | X | | 1.37 | 87.6 | 2.5/0.1 |
| 13 | 5 | | X | 1.22 | 45.0 | 2.0/0.7 |
| 14 | 6 | X | | 1.64 | 68.9 | 2.8/0.1 |
| 14 | 6 | | X | 1.78 | 84.8 | 3.7/0.3 |
| 18 | 12 | | X | 1.66 | 89.4 | 5.4/0.4 |
| 19 | 12 | | X | 7.19 | 70.8 | 17.4/0.4 |
| 25 | 5 | | X | 0.86 | 72.1 | 1.6/0.0 |
| 26 | 5 | | X | 1.04 | 119.2 | 3.1/0.1 |
| 27 | 5 | | X | 0.38 | 94.7 | 0.9/0.0 |
| 28 | 6 | | X | 0.67 | 89.6 | 1.5/0.0 |
| 29 | 6 | | X | 0.40 | 72.5 | 0.8/0.1 |
| 30 | 5 | | X | 1.40 | 70.7 | 2.8/0.0 |
| 31 | 5 | | X | 0.94 | 91.5 | 1.6/0.0 |
| 32 | 5 | | X | 2.10 | 108.1 | 6.0/0.0 |
| 33 | 5 | | X | 1.16 | 104.3 | 2.9/0.0 |

[1] $\triangle$ = average of difference in IRA between the bed surfaces of brick for the entire sample

[2] $\triangle$ max./min. = the largest and smallest differences in IRA between the two bed surfaces for the entire sample

[3] V = coefficient of variation

**TABLE 2 -- Extruded Brick Cont'd.**

| I.D. | Sample Size | As Rec'd. | Oven Dried | Avg. $\triangle$ | V % | $\triangle$ Max./Min. |
|------|------|------|------|------|------|------|
| Group III - Cored-Flashed | | | | | | |
| 2 | 6 | X | | 1.88 | 47.9 | 3.1/0.6 |
| 2 | 6 | | X | 1.50 | 92.7 | 4.1/0.0 |
| 4 | 5 | X | | 1.36 | 54.4 | 2.0/0.5 |
| 4 | 5 | | X | 1.02 | 86.4 | 2.4/0.3 |
| 15 | 5 | X | | 0.32 | 112.5 | 0.8/0.0 |
| 15 | 5 | | X | 0.68 | 61.8 | 1.3/0.1 |
| 34 | 5 | | X | 4.42 | 40.0 | 6.5/1.9 |
| 35 | 5 | | X | 2.52 | 25.0 | 2.8/1.4 |
| 36 | 5 | | X | 2.78 | 46.0 | 4.6/1.5 |
| 37 | 5 | | X | 0.60 | 136.7 | 1.5/0.0 |
| 38 | 5 | | X | 2.98 | 34.2 | 3.0/1.5 |
| 39 | 5 | | X | 2.74 | 108.4 | 7.7/0.0 |
| Group IV - Solid Non-Flashed | | | | | | |
| 40 | 5 | | X | 1.24 | 38.7 | 2.1/1.0 |
| 41 | 5 | | X | 3.86 | 30.6 | 4.8/1.9 |
| 42 | 5 | | X | 2.00 | 93.5 | 5.0/0.0 |
| 43 | 5 | | X | 1.98 | 72.2 | 3.3/0.0 |
| 44 | 5 | | X | 1.10 | 140.9 | 3.3/0.0 |
| 45 | 5 | | X | 1.74 | 112.6 | 4.3/0.0 |
| 46 | 5 | | X | 2.22 | 52.7 | 3.4/1.0 |
| 47 | 5 | | X | 4.16 | 57.4 | 6.6/1.1 |
| 48 | 5 | | X | 3.86 | 133.4 | 12.8/0.0 |
| Group V - Solid Flashed | | | | | | |
| 49 | 5 | | X | 2.16 | 117.6 | 6.5/0.0 |
| 50 | 5 | | X | 2.60 | 62.6 | 5.4/1.1 |
| 51 | 5 | | X | 0.72 | 55.3 | 11.7/1.7 |
| 52 | 5 | | X | 0.76 | 103.9 | 1.9/0.0 |
| 53 | 5 | | X | 1.70 | 57.6 | 3.2/1.0 |
| 54 | 5 | | X | 3.92 | 47.4 | 5.2/1.0 |
| 55 | 5 | | X | 3.18 | 57.2 | 6.0/0.9 |
| 56 | 5 | | X | 0.44 | 136.4 | 1.1/0.0 |

## TABLE 3 -- Molded Brick

| I.D. | Sample Size | As Rec'd. | Oven Dried | Avg.[1] $\triangle$ | V[3] % | $\triangle$[2] Max./Min. |
|------|-------------|-----------|------------|----------|--------|-----------|
| **Group VI = Groups VII & VIII - All Molded** | | | | | | |
| **Group VII - Hand Molded** | | | | | | |
| 20 | 5 | | X | 7.68 | 70.7 | 12.8/1.0 |
| 21 | 5 | | X | 4.74 | 46.8 | 8.5/2.9 |
| 22 | 5 | | X | 21.76 | 91.2 | 50.7/0.9 |
| 23 | 5 | | X | 16.12 | 90.4 | 37.6/0.8 |
| 24 | 5 | | X | 3.24 | 34.3 | 4.6/2.3 |
| 61 | 6 | X | | 4.60 | 28.3 | 6.0/2.8 |
| 61 | 6 | | X | 3.90 | 59.0 | 7.7/0.7 |
| **Group VIII - Machine Molded** | | | | | | |
| 16 | 10 | | X | 5.05 | 55.2 | 8.7/0.2 |
| 17 | 10 | | X | 4.40 | 125.9 | 18.7/0.6 |
| 58 | 5 | | X | 9.92 | 69.2 | 20.1/1.2 |
| 59 | 5 | | X | 14.18 | 56.2 | 24.0/2.2 |
| 60 | 7 | X | | 5.20 | 43.3 | 7.6/2.3 |
| 60 | 7 | | X | 4.80 | 59.0 | 8.9/1.0 |
| 62 | 6 | X | | 9.10 | 130.0 | 32.6/0.1 |
| 62 | 6 | | X | 7.30 | 109.6 | 21.8/0.0 |
| 63 | 3 | X | | 12.00 | 40.1 | 15.4/6.5 |
| 63 | 3 | | X | 12.00 | 29.9 | 15.9/8.8 |

[1] $\triangle$ = average of difference in IRA between the bed surfaces of brick for the entire sample
[2] $\triangle$ max./min. = the largest and smallest differences in IRA between the two bed surfaces for the entire sample
[3] V = coefficient of variation

### TABLE 4 -- Group Data Summary

| Group | | As Rec'd. | Oven Dried | Avg. △ | S* | V* % |
|-------|---|-----------|------------|--------|-----|------|
| I - | All Extruded | X | | 1.4 | 0.8 | 57.1 |
| | | | X | 2.0 | 1.5 | 74.3 |
| II - | Cored Extruded Non-Flashed | X | | 1.7 | 1.9 | 111.8 |
| | | | X | 1.6 | 1.4 | 87.5 |
| III - | Cored Extruded Flashed | X | | 1.2 | 0.8 | 66.7 |
| | | | X | 2.1 | 1.2 | 57.1 |
| IV - | Solid Extruded Non-Flashed | X | | -- | -- | -- |
| | | | X | 2.5 | 1.2 | 48.0 |
| V - | Solid Extruded Flashed | X | | -- | -- | -- |
| | | | X | 2.7 | 2.0 | 74.1 |
| VI - | All Molded | X | | 7.7 | 3.5 | 45.4 |
| | | | X | 8.8 | 5.7 | 64.7 |
| VII - | Hand Molded | X | | -- | -- | -- |
| | | | X | 9.6 | 7.6 | 79.2 |
| VIII - | Machine Molded | X | | 8.8 | 3.4 | 38.6 |
| | | | X | 8.2 | 3.9 | 47.6 |

*S = standard deviation; V = coefficient of variation

### TABLE 5 -- IRA*

| Sample # | As Rec'd. | In The Sun | Oven Dried |
|----------|-----------|------------|------------|
| #1 | 26.3 | 24.6 | 24.3 |
| #2 | 21.0 | 19.1 | 19.0 |
| #3 | 17.8 | 16.4 | 16.2 |
| #4 | 10.2 | 8.8 | 10.8 |
| #5 | 7.4 | 5.7 | 5.3 |
| #6 | 15.5 | 13.5 | 13.4 |
| #7 | 14.4 | 13.3 | 12.6 |
| #8 | 11.4 | 11.4 | 11.5 |
| #9 | 7.8 | 8.4 | 8.1 |
| #10 | 15.9 | 15.3 | 14.8 |
| #11 | 20.5 | 20.0 | 19.5 |
| #12 | 16.9 | 13.8 | 14.4 |
| #13 | 29.5 | 26.9 | 24.6 |
| #14 | 29.8 | 29.3 | 25.2 |
| #15 | 4.9 | 4.5 | 3.9 |

*Initial Rate of Absorption in gms./min./194 cm.$^2$

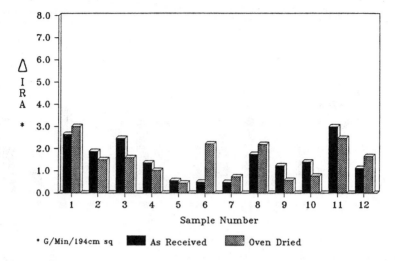

Figure 1a.                    Group I. Extruded Brick
                     Mean IRA Differences of Bed Surfaces

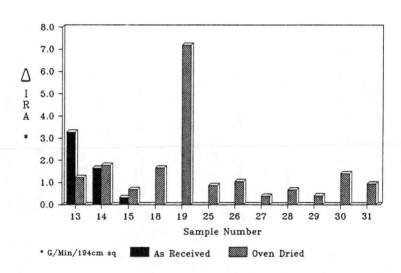

Figure 1b.                    Group I. Extruded Brick
                     Mean IRA Differences of Bed Surfaces

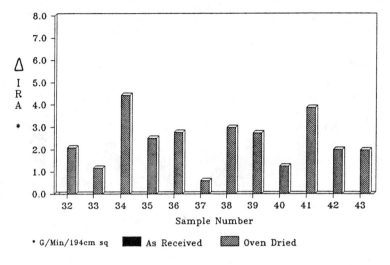

Figure 1c.                Group I. Extruded Brick
               Mean IRA Differences of Bed Surfaces

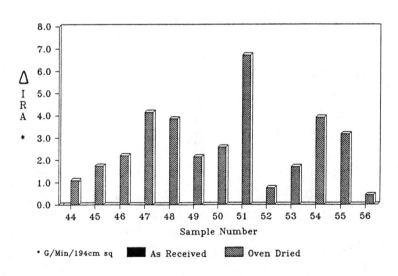

Figure 1d.                Group I. Extruded Brick
               Mean IRA Differences of Bed Surfaces

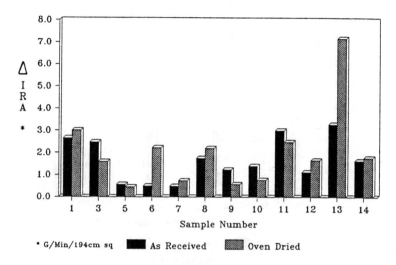

Figure 2a. Group II. Cored Extruded Non-Flashed Brick
Mean IRA Differences of Bed Surfaces

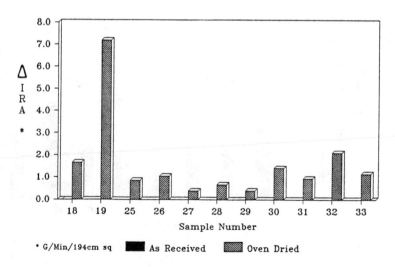

Figure 2b. Group II. Cored Extruded Non-Flashed Brick
Mean IRA Differences of Bed Surfaces

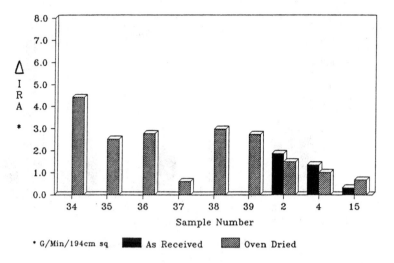

Figure 3.     Group III. Cored Extruded Flashed Brick
              Mean IRA Differences of Bed Surfaces

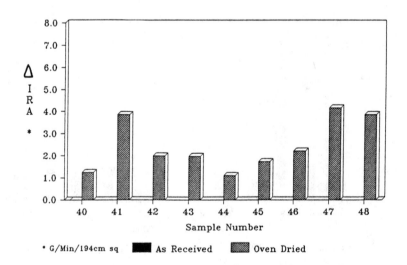

Figure 4.  Group IV. Solid Extruded Non-Flashed Brick
           Mean IRA Differences of Bed Surfaces

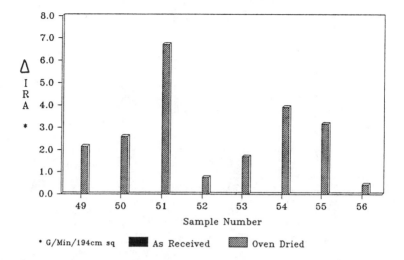

Figure 5.    Group V. Solid Extruded Flashed Brick
Mean IRA Differences of Bed Surfaces

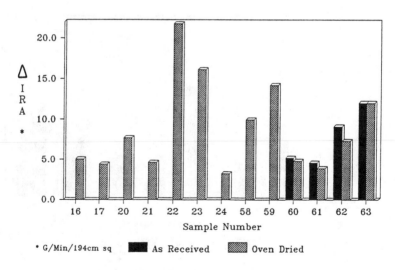

Figure 6.    Group VI. Molded Brick
Mean IRA Differences of Bed Surfaces

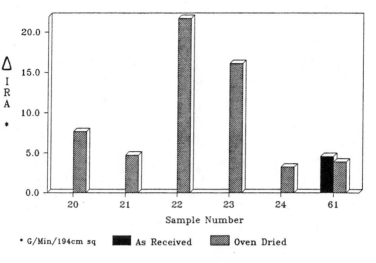

Figure 7.          Group VII. Hand Molded Brick
                Mean IRA Differences of Bed Surfaces

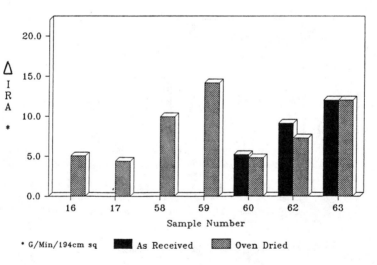

Figure 8.          Group VIII. Machine Molded Brick
                Mean IRA Differences of Bed Surfaces

## ACKNOWLEDGMENTS

The tests were conducted at The University of Texas at Arlington, The General Shale Brick Company, The ACME Brick Company, The Glen-Gery Brick Company Laboratories and others. The authors appreciate the laboratory work on this project by Frederick S. Quick, Farrell Long, Leonard Bliss, and other locations. Contributions of testing and/or brick from General Shale, Boral, Cushwa, Glen-Gery, Old Carolina, ACME, and other brick companies were greatly appreciated.

## REFERENCES

[1]  Plummer, Harry C., Brick and Tile Engineering, Brick Institute of America, 1750 Old Meadow Road, McLean, Virginia  22201, 1962.

[2]  "Standard Methods of Sampling and Testing Brick and Structural Clay Tile," ASTM C-67-87, American Society For Testing and Materials, Philadelphia, PA, 1988.

[3]  Robinson, Gilbert C. and Brown, Russell H., "Inadequacy of Property Specifications in ASTM," Masonry: Materials, Design, Construction, and Maintenance, ASTM STP 992, Harry A. Harris, ed., American Society For Testing and Materials, Philadelphia, PA, 1988, pp. 7-17.

[4]  Gazzola, Edward, Bagnariol, Dino, Toneff, Janie and Drysdale, Robert G., "Influence of Mortar Materials on the Flexural Tensile Bond Strength of Block and Brick Masonry," Masonry: Research, Application, and Problems, ASTM STP 871, J. C. Grogan and J. T. Conway, Eds., American Society For Testing and Materials, Philadelphia, PA, 1985, pp. 15-26.

[5]  "ASTM Specification For Facing Brick (Solid Masonry Units Made From Clay or Shale)", ASTM C-216-87, American Society For Testing and Materials, Philadelphia, PA, 1988.

[6]  Grimm, C. T. and Fowler, D. W., "Effect of Brick Suction on Bricklayer Productivity," Proceedings of the North American Masonry Conference, Arlington, TX, June 1985, pp. 8-1 to 8-18.

Discussion of Paper, "INITIAL RATE OF ABSORPTION OF CLAY BRICK"

by William C. Bailey, John H. Matthys, and Joseph E. Edwards

The discussion is authored by
Lynn R. Lauersdorf and Gilbert C. Robinson

## 1. INTRODUCTION

IRA is considered significant to mortar bonding, but scant attention has been given to the variation of IRA within brick prior to this work by Bailey et al. Instead, emphasis has been placed on average values obtained from a designated sample. Mortar bonds to individual units and not to an average. An acceptable average may include individual brick with unacceptably high or low values and thus produce a segment of poor bonding within a structure.

Test results from our laboratories confirm the results of Bailey et al. In addition, other sources of IRA variation have been examined and these may make an appropriate addendum to assist in interpretation of IRA data.

## 2. SOURCE OF SIDE TO SIDE VARIATION

The major sources of bed surface to bed surface variation are the kiln setting pattern, fuel, and firing schedule. Molding practice is an added source for molded brick.

Two brick may be pushed together with their bed surfaces forming an interface and leaving the opposite faces exposed to the kiln atmosphere. The heating exposure will differ between interface and exposed faces and produce a difference in porosity and IRA. The slower the heating rate, the lower the difference.

Solid fuels may increase the difference because of less uniform heating and localized atmospheres. Flashing may be applied to produce desired colors. It also produces greater fusion and thus lowers the porosity of the brick. The joined interface may be shielded from the time limited flashing environment and thus increase IRA difference. This difference can be observed end to end in a brick as well as bed surface to bed surface. The IRA difference from these sources is usually minimal, but will change from kiln to kiln. It is possible to find brick which will show a high IRA difference particularly in flashed brick but the population of these samples is small.

## 3. VARIATION BRICK TO BRICK

The change in IRA from one brick to another probably is more significant than the side-to-side variation. This is illustrated in Table 1. Two samples of five brick each and one sample of ten brick were obtained from a single shipment. Notice that the individual brick IRA spread from 14.9 to 39.4. A mortar compatible with the 14.9 brick will produce a leaky wall when a few 39.4 IRA brick are included even though the average may be acceptable.

Comparing averages of the three samples showed a change of 28 to 21 grams or a larger change than the side-to-side variation for most extruded brick of the Bailey et al report.   It would appear that the testing of one side is adequate for indicating the spread of IRA values for extruded brick;  however, neither ASTM C-64 or C-67 specify the number of brick to be tested and useful information depends on an adequate sample size and sampling procedure.   Furthermore it is suggested that the spread between maximum and minimum IRA is as significant as the level of IRA in predicting the quality of mortar bonding.

4.   TEST PROCEDURE

The same brick were sequentially tested by three different laboratories with results identified by individual brick numbers.  The results of Table 1 show the maximum of 3 grams difference between mean values from the different laboratories.  Individual values differed by as much as 5.4 grams.  The laboratories used room dried specimens (labeled 'as received').  The time intervals between testing are indicated by the dates of testing listed in Table 2.  Laboratory A repeated the measurement but this time drying the specimen as instructed by C-67.  This increased the difference in IRA from the original measurement.  Then laboratory A repeated the 'as received' measurement and found still higher differences.  The results in Table 2 show differences as high as 9.3 grams with mean values for the three samples of 6.5, 4.6, and 5.8.

A comparison of dried with 'as received' brick was made by laboratory A for the test made one after the other.  This showed a difference of less than 4 grams for all specimens and suggests minimal differences between the two procedures.  However, mortar bonding of brick depends on the remnant IRA at the time of mortar application.  Any moisture pickup at the job site will influence this IRA and so IRA should be determined on brick as conditioned for laying in order to predict mortar bonding.

5.   INFLUENCE OF REPEATED RUNS

It was noticed that each successive laboratory determination produced progressively lower IRAs.  There were three exceptions to this trend among the twenty brick tested by laboratory B and one for laboratory C.  It was suspected that the brick were changing rather than laboratory identity causing the change.  Laboratory A continued the repetition of measurements for three additional weekly trials.  A comparison of the seventh run with the original run shows a marked change in IRA.  The mean difference of the three samples was 12.0, 9.3, and 10.4.  This amounted to a 50% reduction in IRA for many of the samples.  Bailey et al showed a similar trend in their Table 5. Twelve out of 15 brick showed a progressive decrease in IRA with the third test versus the first test.

The reason for the change is not known.  One possibility may be attributed to the presence of soluble salts.  There are small quantities of soluble salts in brick usually in the amount between 0.01 and 0.07%.  Also there are soluble salts in municipal water supplies and distilled water is not specified for the IRA tests,

although it was used in this series. Repeated solution and evaporation may concentrate the salts in the surface pores and reduce IRA.
Another possibility centers on the unusually strong attraction for
water by the smallest pores. The water may not be removed by room
temperature conditioning in a seven day period. The residue of water
would reduce the apparent IRA of the brick. The results suggest that
IRA changes with repeated wetting and aging.

6.  SURFACE TEXTURE

The exposed faces of extruded brick may be decorated with sand
coatings or may be textured. The coatings or textures may cause
pronounced changes in IRA. The influence of sand coating on a
commercial machine molded brick was evaluated by determining the IRA
with the coating in place and then redetermining the IRA after sawing
off the coated surfaces. The IRA was 61 before and 4 after removal of
the coated surfaces. A less dramatic, but still significant change,
occurs in extruded brick. A coated brick may show a high IRA around
the perimeter of the bed surface and low IRA over most of the
interior. This gives differential bonding between a narrow band
around the perimeter compared to the interior surface.

7.  MOLDED BRICK

Molded brick are shaped in such a way as to encourage difference
in IRA between bedding surfaces. One bed surface may be sanded while
the opposite face may be struck to produce a shaggy surface. Again
the other surface may be smooth. The difference in texture can
produce large differences in IRA. This is shown by the work of Bailey
et al Tables 3 and 4. The average difference for all molded brick was
7.7 to 8.8 between bed surfaces. This compares to a value below 3 for
extruded brick. Examination of the results for machine molded brick
in their Table 3 shows an average of 17.4 for the maximum differences.
It would have been helpful for the authors to have listed the IRA
level of the brick since a difference of 17.4 grams in a 60 gram IRA
brick would be of little concern. The same difference in a 20 gram
brick would be of major significance. The results do suggest that it
is important to identify the IRA of both bed surfaces of molded brick.

Answer (J. H. Matthys, University of Texas at Arlington):

The authors appreciate the discussion of Mr. Lauersdorf and Professor Robinson that not
only supports the findings of the authors' paper, but also contributes to the understanding
of the behavior observed. Your perceptive comments supported by your data add
significantly to the value of this paper for the masonry industry with respect to the IRA
performance of clay brick.

TABLE   I

A COMPARISON OF IRA RESEARCH
FROM LABORATORIES A, B AND C

| SAMPLE | LAB A | | | LAB B | | | LAB C | | |
|--------|-------|---|---|-------|---|---|-------|---|---|
| No. | x | s | v | x | s | v | x | s | v |
| A | 25.8 | 5.9 | 23 | 23.9 | 5.7 | 24 | 22.8 | 4.2 | 18 |
| B | 28.2 | 7.9 | 28 | 28.2 | 8.3 | 29 | 25.8 | 7.5 | 29 |
| C | 20.7 | 4.6 | 22 | 19.7 | 4.2 | 21 | 17.7 | 4.2 | 24 |

x = Mean IRA, g/min. 194cm$^2$

s = Standard Deviation,  g/min  194 cm$^2$

v = Coefficient of variation, %

TABLE 2
THE INFLUENCE OF LABORATORY SELECTION
AND REPEATED RUNS ON IRA

| SAMPLE LOT | IRA* AS RCVD LAB A | DIFFERENCE IN IRA* FROM LABORATORY A | | | | DIFFERENCE IN IRA* AT LAB A | |
|---|---|---|---|---|---|---|---|
| | | AS RCVD LAB B | AS RCVD LAB C | AS RCVD REPEAT RUN LAB A | DRIED LAB A | AS RCVD FROM DRIED | 7TH RUN FROM FIRST |
| | 8/22 9/27 | 10/17 | 11/7 | 11/15 | 11/13 | 11/15 11/13 | 12/4 |
| A | 26.1 | -1.7 | -3.1 | -7.4 | -5/5 | -1.9 | -13.0 |
| | 16.6 | -1.3 | -1.1 | -3.3 | -3.7 | -0.2 | - 7.8 |
| | 26.5 | -3.3 | -4.6 | -6.1 | -4.9 | -1.2 | -12.3 |
| | 33.2 | -1.9 | -3.2 | -6.4 | -5.3 | -1.1 | -11.6 |
| | 26.7 | -1.5 | -3.2 | -9.3 | -5.4 | -3.9 | -15.3 |
| MEAN | 25.82 | 1.94 | 3.04 | 6.50 | 4.96 | 1.66 | 12.00 |
| B | 25.4 | -2.0 | -3.9 | -6.3 | -4.8 | -1.5 | -11.8 |
| | 29.1 | 3.7 | 2.3 | -2.5 | -0.2 | -2.3 | - 6.6 |
| | 29.6 | -0.8 | -3.2 | -4.5 | -3.2 | -1.2 | - 7.7 |
| | 17.6 | -0.2 | -2.1 | -3.7 | -2.6 | -1.1 | - 8.8 |
| | 39.4 | -0.6 | -5.4 | -5.9 | -3.8 | -2.1 | -11.4 |
| MEAN | 28.22 | 1.46 | 3.38 | 4.58 | 2.92 | 1.66 | 9.26 |
| C | 23.7 | -0.7 | 3.3 | 7.0 | 4.9 | 2.1 | 10.2 |
| | 21.2 | -2.9 | 3.9 | 7.3 | 4.9 | 1.4 | 10.4 |
| | 31.7 | -1.6 | 4.7 | 5.1 | 3.8 | 1.3 | 8.6 |
| | 19.0 | -1.8 | 2.9 | 4.3 | 4.1 | 0.2 | 9.6 |
| | 19.7 | 0.6 | 2.6 | 4.7 | 3.7 | 1.0 | 8.6 |
| | 20.0 | -0.6 | 2.4 | 7.2 | 4.0 | 3.2 | 11.6 |
| | 21.5 | -1.4 | 3.7 | 7.5 | 4.4 | 3.1 | 13.7 |
| | 16.9 | -1.1 | 4.3 | 6.4 | 4.2 | 2.2 | 11.5 |
| | 14.9 | 1.1 | 1.7 | 3.6 | 2.4 | 1.2 | 7.2 |
| | 18.2 | -0.6 | ---- | 4.9 | 4.1 | 0.8 | 12.9 |
| MEAN | 20.68 | 1.24 | 3.24 | 5.80 | 4.05 | 1.65 | 10.43 |

* INITIAL RATE OF ABSORPTION IN g/min/ 194 $cm^2$

EXTRUDED 3 HOLE BRICK FLASHED

Edwin R. Dunstan, Jr., and Pete Hay

ASTM C90 CONCRETE MASONRY UNITS USING LIGHTWEIGHT
AARDELITE AGGREGATE

---

REFERENCE: Dunstan, E. R. Jr., and Hay, P.,
"ASTM C90 Concrete Masonry Units Using Light-
weight Aardelite Aggregate," Masonry:
Components to Assemblege, ASTM STP1063, American
Society for Testing and Materials, Philadelphia,
1989.

ABSTRACT: This report gives test results for
ASTM C90 block produced throughout Central
Florida using a new lightweight aggregate. In
May 1988, Florida Mining and Materials started
producing medium and light weight block using
Aardelite. Aardelite consists of pelletized
particles produced from a mixture of hydrated
lime and fly ash. The Aardelite plant is
located at Florida Power's Energy Center near
Crystal River, Florida.

KEYWORDS: lightweight aggregate, masonry units

INTRODUCTION

In May 1988 Progress Materials, Inc. began producing
a new lightweight aggregate - Aardelite. The Aardelite
plant is located at the fly ash silo for Florida Power
Corp.'s coal fired units 1 & 2 at Crystal River, Florida.

Aardelite is a Dutch process which produces
lightweight aggregate from fly ash conforming to ASTM
C618, Class F and quicklime with a minimum of 94%
available Ca0. The quicklime is first hydrated with
excess water to form a slurry. The lime slurry and fly
ash are introduced into a high speed, high efficiency
mixer, after which it is fed onto a rotary pelletizer
pan. A small amount of additional water is sprayed onto

Edwin R. Dunstan, Jr. is Engineering Manager for the
Fly Ash Division at Florida Mining & Materials, 13228 N.
Central Avenue, Tampa, Florida; Pete Hay is Manager of
Aardelite Operations, Progress Materials, Inc., 235 3rd
Street South, St. Petersburg, Florida 33701.

the pelletizer.  After the pellets leave the pelletizer,
they are dusted with dry fly ash to prevent the pellets
from sticking together.  The pellets are then moved to a
silo and steam cured at 170°F.  Following a 24-hour cure
the pellets are screened.  The undersize is returned to
the pelletizer.  The oversize is crushed and rescreened.
The properly sized pellets are now "Aardelite", a
lightweight aggregate for use in concrete masonry units.

CONFORMANCE WITH ASTM C331

Aardelite is a lightweight processed fly ash
aggregate in that air is entrapped in the pelletizing
process.  Aardelite is chemically similar to ancient
Roman concretes.  The calcium from the hydrated lime
reacts with silica and alumina in the fly ash to form
calcium-silicate-hydrates, calcium-alumina-silicate-
hydrates, calcium-alumina-hydrates and other hydrates
similar to those of portland cement.  These hydrates are
very stable as demonstrated by Roman concretes still
intact after 1900 years.

Aardelite has been tested for compliance to ASTM
C331, "Lightweight Aggregates for Concrete Masonry
Units."  The test results are shown in table 1.  When
first produced, Aardelite met the requirements of ASTM
C331 as a combined fine and coarse aggregate.  The
gradation was too coarse and it has been modified over
the first year to produce the texture of block required
for the Central Florida market.  Aardelite is now a fine
aggregate.  The initial and current production gradations
are shown in Table 1.  (ASTM C331 Gradation Waiver Sec.
5.1.4)

ASTM C90 - MEDIUM WEIGHT MASONRY UNITS

The full output of the Crystal River plant is
purchased by Florida Mining & Materials to produce medium
and light weight concrete masonry units.  Aardelite is
purchased based on a shipped weight of 60 lb/ft$^3$,.  Daily
production tests at the plant have averaged 61 lb/ft$^3$
(see table 2) and recent changes produce a lighter
material.

To insure a strong aggregate to produce ASTM C90
structural block the pellet strength is tested each day.
The pellet strength is controlled based on a  test
developed by the Dutch (Appendix A).  There is no ASTM
standard, however, our in-house standard minimum strength
is 3.0 Newtons/mm$^2$.  The pellet strength has averaged 4.0
Newtons/mm$^2$.

Florida Mining & Materials produces about 1.5
million 8-inch equivalents each month using over 8,000

tons of Aardelite aggregate.  The aggregate is used in
9-block plants in Central and Northern Florida, and
Southern Georgia.  Randomly sampled medium weight blocks
were tested between September 6, 1988 through February 2,
1989.  The average test results are shown in Table 3.
The block mix proportions vary slightly in each plant
with a basic medium weight block mix being that shown in
Table 4.

The fire ratings, shown in Table 3, are based on
fire tests made on Aardelite aggregate block at 90 lb/ft$^3$
and 100 lb/ft$^3$ unit weights which match fire ratings
versus concrete density given in the Standard Building
Code or South Florida Building Code.  The fire ratings
shown in Table 3 are calculated based on a concrete
density of 115 lb/ft$^3$.

ASTM C90 LIGHT WEIGHT MASONRY UNITS

During the first year, Aardelite was used mostly in
medium weight block.  In recent months Aardelite has been
used to produce light weight block.  Typical mix
proportions are shown in Table 4.  These proportions
produce a fairly open texture block.  For a finer texture
light weight block, or for 100% lightweight aggregate
block, we use a blend of Aardelite and a finer gradation
lightweight aggregate from another producer.


CONCLUSION

Aardelite has proven to be a very economical and
technically sound aggregate for masonry units.  Its' use
will continue to increase in Central Florida.

TABLE 1

ASTM C331 Test Results for Aardelite

| | | |
|---|---|---|
| Soundness | ASTM C151 | No Popoputs |
| Staining | ASTM C641 | Very light<br>Stain Index 20 |
| Loss on Ignition | ASTM C114 | 10% (Max 12%) |
| Gradation | ASTM C136 | |

| Sieve Size | % Passing | Limits | Current Grading<br>(Sec 5.1.4)<br>% Passing |
|---|---|---|---|
| 1/2 in. | 100 | 95-100 | |
| 3/8 in. | 96 | | 100 |
| No. 4 | 60 | 50-80 | 76 |
| No. 8 | 25 | | 34 |
| No. 16 | | | 4 |
| No. 50 | 5 | 5-20 | |

| | | |
|---|---|---|
| Unit Weight Dry | ASTM C29 | 55 lb/ft$^3$ |
| Shrinkage | ASTM C157 | 0.07% (Max 0.10%) |
| Freezing & Thawing | ASTM C666 | N/A - Florida |

TABLE 2

Production Tests - Aardelite

| | |
|---|---|
| Ash Shipped Unit Weight | 61 lb/ft$^3$ |
| *Pellets Crushing Strength | 4.0 $^n$/mm$^2$ |
| *See Appendix A | |

TABLE 3

ASTM C90 Testing of Aardelite Medium Weight Block

* Independent Testing September 6, 1988 -
February 2, 1989

| TEST | STANDARD BLOCK | NO. OF TESTS (3 units/ test) | AVERAGE RESULT |
|------|------|------|------|
| Strength (lb/in$^2$) | 4-inch | 1 | 1510 |
| | 6-inch | 1 | 1100 |
| | 8 inch | 13 | 1140 |
| | 12-inch | 1 | 1070 |
| Weight (as received) (lbs) | 4-inch | 1 | 20.5 |
| | 6-inch | 1 | 25.2 |
| | 8-inch | 13 | 31.5 |
| | 12-inch | 1 | 46.4 |
| Oven-Dry Concrete (lb/ft$^3$) | 4-inch | 1 | 112.4 |
| | 6-inch | 1 | 112.6 |
| | 8-inch | 13 | 115.3 |
| | 12-inch | 1 | 112.5 |
| Absorption (%) | 4-inch | 1 | 10.6 |
| | 6-inch | 1 | 11.1 |
| | 8-inch | 13 | 10.6 |
| | 12-inch | 1 | 10.9 |
| Absorption (lbs/ft$^3$) | 4-inch | 1 | 12.0 |
| | 6-inch | 1 | 12.4 |
| | 8-inch | 13 | 12.1 |
| | 12-inch | 1 | 12.3 |
| Moisture (%) of Absorption | 8-inch | 13 | 33.5 |
| Equivalent Thickness (inches) | 4-inch | 1 | 2.6 |
| | 6-inch | 1 | 3.1 |
| | 8-inch | 13 | 4.0 |
| | 12-inch | 1 | 5.7 |
| Shrinkage (%) | 8-inch | 7 | 0.051 |
| ** Fire Rating (Hrs) | 4-inch | | 0.9 |
| | 6-inch | | 1.3 |
| | 8-inch | | 2.2 |
| | 12-inch | | 4.5 |

* Block randomly sampled from Florida Mining & Materials
block plants located at Auburndale, Orlando,
Brooksville, Tampa, Sharpes, Daytona and Ocala.

** Medium weight blocks oven-dry concrete weight 115 lb/ft$^3$.  Calculated fire resistance for concrete and concrete masonry based on the Standard Building Code and the South Florida Building Code.

TABLE 4

Mix Proportions - Medium Weight Aardelite Block

|  | lbs | #/Blk |
|---|---|---|
| Cement | 335 | 3.28 |
| Aardelite | 1002 | 9.82 |
| Sand | 1827 | 17.91 |
| TOTAL | 3110 | 31.01 |

Yield      102
Cmt/Blk    3.28
Dry Density (lb/ft$^3$) 113

MIX PROPORTIONS - LIGHT WEIGHT AARDELITE BLOCK

|  | Mix 1 (lbs) | Mix 2 (lbs) |
|---|---|---|
| Cement | 380 | 380 |
| Aardelite | 1680 | 1890 |
| Sand | 810 | 495 |
| TOTAL | 2870 | 2765 |

| | | |
|---|---|---|
| Yield | 102+ | 102+ |
| Cmt/Blk | 3.75 | 3.75 |
| Dry Density (lb/ft$^3$) | 102 | 98 |

APPENDIX A

AARDING TEST METHOD I

COLLAPSE STRESS OF THE PELLETS

The collapse stress of the pellets is determined as
follows:

C1 - From a random closely-sized sample of Aardelite®
     aggregate (consisting of at least 20 pellets) pellet
     strength is determined by compressing individual
     pellets between parallel metal plates until the
     pellet fails by fracturing. The force (F) to
     fracture the pellets is averaged and expressed in
     Newtons (4.4482 N = 1 lb$_f$).

C2 - For a given size fraction the average collapse
     stress ($\bar{o}$) is calculated as the average force ($\bar{F}$) to
     cause fracture divided by the average (as defined in
     this method) central surface area ($\bar{A}$) of the pellets
     (usually expressed in mm$^2$).

$$\bar{o} \; = \; \frac{\bar{F}}{\bar{A}} \; = \; \frac{\bar{F}}{\pi\, r_{min} r_{max}} \; = \; \frac{\bar{F}}{\pi\, \frac{dmin}{2}\, \frac{dmax}{2}}$$

C3 - The standard deviation for the individual pellet
     strengths is calculated and substituted into the
     collapse stress formula to yield a standard
     deviation for collapse stress. This value is then
     subtracted from the mean collapse stress to give an
     indication of pellet strength consistency. See
     example in C-5.

C4 - The requirement for Aardelite® aggregate is that the
     mean value of the collapse stress be greater than
     3.0 Nmm$^2$ and that the mean value of the collapse
     stress minus one standard deviation be greater than
     2.4, N/mm$^2$.

C5 - Example

     - Select 20 (n) pellets retained on a No. 6 (3.36 mm,
       .132 in.) U.S.
     - Standard Sieve and passing a No. 4 (4.76 mm,
       .187 in.) sieve.
     - Fracture test yields the following Force (F)
       results:

```
            45.3  Newtons
            39.8     "
            37.3     "
            34.2     "
            21.2     "
            33.5     "
            38.7     "
            39.8     "
            40.5     "
            38.3     "
            39.6     "
            34.3     "
            38.2     "
            36.1     "
            44.1     "
            39.7     "
            36.9     "
            35.1     "
            38.3     "
            50.7     "
           _____
∑x  =      761.6
```

For these F values

$\overline{x}_f$ = 38.1 N = $\overline{F}$
$s_f$ = 5.69 N

The mean collapse stress ($\overline{\sigma}$) equals:

$$\overline{\sigma} = \frac{\overline{F}}{\overline{A}} = \frac{38.1\ N}{\pi\ \dfrac{3.36\ mm}{2}\ \dfrac{4.76\ mm}{2}} = \frac{38.1\ N}{12.6\ mm^2} = 3.02\ \frac{N}{mm^2}$$

The mean collapse stress ($\overline{\sigma}$) minus one
standard deviation ($s_f$) equals:

$$\overline{\sigma} - \frac{s_f}{\overline{A}} = 3.02 \ \frac{N}{mm^2} - \frac{5.69 \ N}{12.6 \ mm^2} = 3.02 - .452 = 2.57 \ \frac{N}{mm^2}$$

Since the mean collapse stress ($\overline{\sigma}$) is greater than 3.0 N/mm$^2$, and the
mean collapse stress minus one standard deviation is greater than 2.4
N/mm$^2$ this sample would meet the acceptance criteria of this test
method.

NOTE:

As used in this method: Arithmetic mean $= \overline{x}_f = \frac{\leq x}{n}$

$$\text{Standard Deviation } s_f = \sqrt{\frac{M}{n^2}}$$

Where $M = n\leq x^2 - (\leq x)^2$
n = Number of Specimens
x = Individual Test Values

DISCUSSION


"ASTM C90 Lightweight and Medium Weight Masonry Units Using Aardelite Aggregate" - E. R. Dunstan, Jr., R. Keck, P. Hay.

Discussion (T. A. Holm, Solite Corporation, Richmond, VA)

According to the definitions presented in the INTRON Institute report, "PROPERTIES OF CONCRETE MADE WITH THREE TYPES OF ARTIFICIAL PFA COARSE AGGREGATES", by G.J.L. van der Wegen and J.M.J.M. Bijen, Netherlands, August 1985, the Aardelite aggregate presently produced in Florida is a "cold bonded" fly ash particle type. The INTRON report divides fly ash aggregates into three types based upon process temperatures reached during formation of the particle microstructure.

1) sintering (temperature T > 900 deg. C)
2) hydrothermal treatment (100 deg. C < T < 250 deg. C)
3) cold bonding (10 deg. C < T < 100 deg. C)

Whereas sintered lightweight aggregates have a ceramic microstructure, both hydrothermal and cold bonding aggregates are composed of cementitiously bound fly ash particles that bear a closer resemblance to the paste matrix than the aggregate fraction. The Aardelite aggregates used in the INTRON report were autoclaved (hydrothermal) prior to use in the concretes tested. Concretes composed of autoclaved coarse Aardelite and natural sand fine aggregates developed the highest shrinkage of the three fly ash concrete types tested with a strain of .07% reported at 100 days.

Commenting upon the higher shrinkage the authors stated, "Apart from the smaller restraining effect of the Aardelite particles (because of the lower modulus) the higher shrinkage of Aardelite concrete is probably due to the shrinkage of the Aardelite aggregate itself. It is known that autoclaved lime-silica materials show drying shrinkage whereas sintered ceramic materials such as Lytag and dense natural aggregates do not."

The data presented at the symposium suggested that the cold bonded fly ash Aardelite concrete (presumably entirely Aardelite aggregate) developed drying shrinkage strains of .07% when measured in accordance with ASTM C331. The same drying shrinkage strain (.07%) was reported in the INTRON paper, despite fundamental differences between the Aardelite aggregate type used in the INTRON investigation when compared with that produced in Florida. In addition to being autoclaved the Aardelite aggregate tested by INTRON incorporated 45% quartz sand within the pellet in addition to the fly ash. Furthermore, the INTRON Aardelite concretes contained natural river sand as a fine aggregate. Equal drying shrinkage strains reported on the autoclaved, sanded Aardelite concrete and the Florida Aardelite (presumably all lightweight aggregate concrete incorporating only cold bonded fly ash particles) contradicts usual experience when drying strains of autoclaved and non-autoclaved concrete products are compared.

Qualifications of a lightweight aggregate, according to ASTM C331, require lightweight aggregate concretes (composed entirely of lightweight aggregates) to develop drying shrinkage less than .1% (ASTM C331 paragraph 5.2.3). It is essential that aggregates used in masonry units meet ASTM C33, or C331 specifications to insure dimensional stability of concrete products in keeping with the expectations of design professionals.

The influence of shrinking aggregates with dimensional stability characteristics closer to the cementitious matrix than the inert aggregate fraction may be analyzed by the Hansen-Nielsen report "Influence of Aggregate Properties on Concrete Shrinkage" ACI Journal, July 1965.

Answer               Your question has addressed the shrinkage properties of Aardelite aggregate. You have referenced work by INTRON Institute on studies of Aardelite manufactured in Holland. As you pointed out, the Aardelite from Holland is "different" than Florida Aardelite.

First, our formula uses much less lime; about 1 part lime to 20 parts fly ash whereas the Holland material is 1 to 10. They use autoclave; we steam at 175°F for 16 plus hours. The Holland material is a coarse aggregate whereas our material is a fine aggregate. It is difficult to make direct comparisons.

The shrinkage test procedures in the INTRON report are different than that of ASTM C157. The aggregate in the INTRON report is soaked for 30 minutes before the specimens are cast. We do not soak any lightweight used in our block plants and we thus make our specimens from Aardelite at the shipping weight and moisture content. The INTRON Aardelite shrinkage tests requires soaking the bars in water for 7 days before drying whereas ASTM C157 is storage in a fog room. The INTRON concrete slump was almost 5-inches, whereas ASTM C331 specifies 2-3 inch slump. The shrinkage bars are different in the INTRON test and ASTM C157.

The differences and how they effect shrinkage could be discussed in depth. Where would it lead? The Florida experience is that Aardelite was tested by ASTM C157 procedures and found to have a shrinkage of 0.07%.

Brian E. Trimble

RECOMMENDED TEST FOR SLIP RESISTANCE ON BRICK SURFACES

REFERENCE: Trimble, B.E., "Recommended Test for Slip Resistance on Brick Surfaces", Masonry: Components to Assemblages, ASTM STP 1063, John H. Matthys, Editor, American Society for Testing and Materials, Philadelphia, 1990.

ABSTRACT: This paper examines current slip resistance test procedures and recommends an appropriate test method for use on brick surfaces. The current ASTM standard for paving brick, ASTM C 902-89a, Standard Specification for Pedestrian and Light Traffic Paving Brick acknowledges the need for considering skid/slip resistance; yet there is not a recommended test procedure to use. Other countries have tests and values for slip resistance on brick surfaces, as well as this country for materials similar to brick. Various test methods that are available for measuring slip resistance are discussed including advantages/ disadvantages of each test in regard to brick surfaces. This paper also recommends an action to implement the test into appropriate ASTM standards. This paper addresses slip resistance only.

KEYWORDS: brick, coefficient of friction, slip resistance, test methods.

INTRODUCTION

At the current time, there are several methods to measure the slip resistance of flooring materials. The aim of this paper is to determine which method is best suited for measuring slip resistance on brick surfaces. The current ASTM standard for paving brick, ASTM C 902, states a need for developing a test method for slip resistance. There has been some discussion from ASTM subcommittee C 15.02 which has jurisdiction over ASTM C 902, but no actions have been taken.

Brian E. Trimble is a staff engineer at the Brick Institute of America, 11490 Commerce Park Drive, Reston, VA 22091.

Slip resistance is related to heel/flooring friction and is applicable to pedestrian traffic.  Skid resistance is related to tire/roadway friction and is applicable to vehicular traffic.  Even though ASTM C 902 acknowledges a need for considering both slip and skid resistance, it is difficult to discuss the two together.  Therefore, this paper will address slip resistance only.

In the past it was felt that it was unneccessary to consider the slip resistance of a brick surface.  The coefficient of friction is regarded to be very high for brick (0.75).  Even when wet, most brick tend to have an adequate coefficient of friction (0.50)[1,2].  So why is it necessary to have a requirement for testing slip resistance?

One of the main reasons for a test is to give designers and/or consumers a guide in determining the safety of a particular flooring surface.  National awareness, as well as the number of litigations, has increased over the years necessitating the need for a recommended and applicable test method.  The Consumer Product Safety Commission (CPSC) collects data relating to slips and falls and has indicated the need for better testing methods [3].

Even though there are many slip resistance testers, this paper will only consider those tests which already have ASTM approval.  This will reduce the amount of discussion involved in deciding how to apply a specific test method.


TEST METHODS


There have been numerous attempts at developing a testing machine to simulate the mechanisms of slip.  Each type of testing machine meets specific needs.

There are three types of slip resistance testers.  These are described very well in Brungraber's Research [4] and repeated here.

1)  <u>Drag type meters</u> - This method consists of a weight, faced with a shoe sole or heel material, which is pulled across a material. Both static friction and dynamic friction can be measured.

2)  <u>Pendulum type meters</u> - This method consists of a pendulum, faced with a shoe sole or heel material, which sweeps a path across a material.  The loss of energy of the pendulum is measured and is related to dynamic friction.

3)  <u>Articulated strut devices</u> - A known vertical force is applied to a shoe or heel material.  An increasing lateral force is then applied until slip occurs.

There exist four ASTM standards of these three types.  They are:
ASTM C 1028-84  Test Method for Evaluating the Static Coefficient of Friction of Ceramic Tile, and other Like Surfaces by the Horizontal Dynamometer Pull Meter Method - Type 1

ASTM D 2047-82   Test Method for Static Coefficient of Friction of
(1988)           Polish-Coated Floor Surfaces as measured by the
                 James Machine (also ASTM F 489) - Type 3

ASTM E 303-83    Method for Measuring Surface Frictional
                 Properties Using the British Pendulum Tester -
                 Type 2

ASTM F 609-79    Test Method for Static Slip Resistance of
(1984)           Footwear, Sole, Heel, or Related Materials by
                 Horizontal Pull Slipmeter (HPS) - Type 1

A review of the literature on slip resistance reveals that each
method has its own advantages and disadvantages.


DISCUSSION


By reviewing the amount of literature on slip resistance and
coefficient of friction testing, it becomes apparent that it is very
difficult to simulate all that occurs during slipping.  There have
been numerous papers on the mechanisms of walking.  As was reported by
Adler and Pierman [5], data indicates that the static coefficient,
rather than the dynamic coefficient of friction, is the important
variable for walking.  This has a decided effect on which test method
is to be used.

A basic understanding of slip resistance terminology is important
to comprehend the differences in friction.

Friction is defined as the force which resists the relative
movement of two surfaces in contact with each other.

Coefficient of Friction is the ratio of the force required to move
one surface over another to the total force pressing the two
surfaces together.
$$\mu = \frac{\text{Horizontal Force}}{\text{Vertical Force}}$$

Static Coefficient of Friction is the ratio of forces at the
instant of motion.

Dynamic Coefficient of Friction is the ratio of forces when
movement occurs at constant velocity.

Each ASTM standard has a scope of when the test method is
applicable and other features.  The advantages or drawbacks of each
test will be discussed.


ASTM C 1028

The Horizontal Dynamometer Pull Meter (HDPM) is a drag type device
(see Figure 1).  A dynamometer is used to determine the force

necessary to cause a 50 lb (22 kg) sled assembly with a Neolite heel
material to slip continuously across a surface. A friction index is
then determined by dividing the force necessary to move the assembly
by the weight of the sled. The meter can theoretically measure both
static friction and dynamic friction. The dynamometer is hand
operated.

As stated in the scope of the standard, this test method evaluates
ceramic tile surfaces. Since ceramic tile and brick are similar
materials, this test method is applicable. The major disadvantage to
this test is that the dynamometer is hand operated which introduces an
undesirable variable.

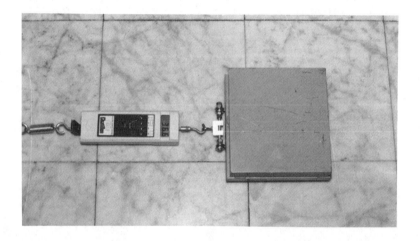

FIGURE 1 -- ASTM C 1028 Test Assembly

## ASTM D 2047

The James Machine is an articulated strut device used to measure
the static coefficient of friction (see Fig. 2). A known load is
applied to a pivoted vertical strut with a sole material attached to a
plate. The test specimen rests on a flat table which moves with a
constant velocity. As the table moves, the strut rotates until slip
occurs. The point of slip is recorded on a chart which can be easily
read.

As stated in the scope of the test method "The apparatus is not
suitable for use on wet, rough or corrugated surfaces. The apparatus
is suitable for laboratory testing." Both of these statements are
disadvantages. Typically, brick have a good slip resistance value
when dry, whereas the important friction value is when the brick are
wet. Also, it is important to have a portable testing machine to
measure values in the field.

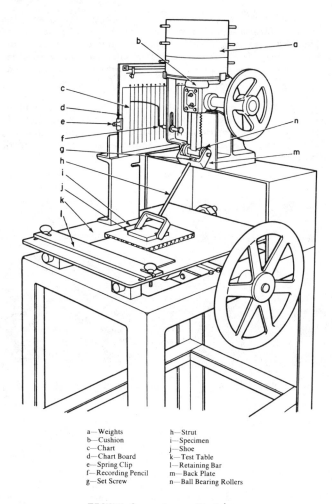

| a—Weights | h—Strut |
| c—Chart | i—Specimen |
| d—Chart Board | j—Shoe |
| e—Spring Clip | k—Test Table |
| f—Recording Pencil | l—Retaining Bar |
| g—Set Screw | m—Back Plate |
| | n—Ball Bearing Rollers |
| b—Cushion | |

FIGURE 2 -- James Machine

ASTM E 303

The British Pendulum tester is a pendulum type device (see Fig. 3). A rubber slider is attached to the end of a pendulum. The slider is positioned to barely come in contact with the test surface when the pendulum swings. A pointer records the "British Pendulum (Tester) Number (BPN)". This procedure measures the loss of energy which is not directly related to coefficient of friction. The usefulness of this method is that it can rank different materials in regard to relative slip resistance. The British Pendulum Tester measures the dynamic coefficient of friction. As stated before, static friction is more applicable to pedestrian walking. The British Pendulum Tester (BPT) is an excellent test method for measuring skid resistance. Many foreign standards use the BPT to measure roadway frictional characteristics and polished stone values.

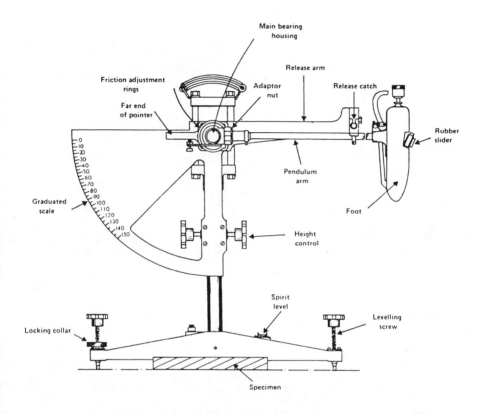

FIGURE 3 -- British Pendulum Tester

## ASTM F 609

The Horizontal Pull Slipmeter is a drag type device (see Fig. 4). It is similar to the HDPM except that a sled is attached to a fixed speed motor by a flexible but stiff cord. A 6 lb (3 kg) sled is then pulled across the testing surface. The Horizontal Pull Slipmeter (HPS) is designed to test footwear materials, but can also be used to test floor surfaces when faced with a standard rubber heel material. This test method is useful for field testing as well as laboratory use.

The advantage of this test method is that it is similar to the HDPM but eliminates the operator variable by the use of a fixed speed motor.

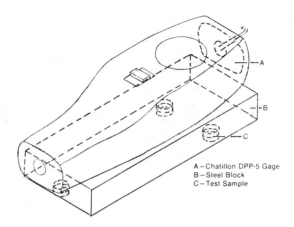

A — Chatillon DPP-5 Gage
B — Steel Block
C — Test Sample

FIGURE 4 -- Horizontal Pull Slipmeter

RECOMMENDED TEST

As stated in numerous papers, it is practically impossible to
design a test method to accurately measure friction because of the
complexity involved.  At best, a test can be used to measure relative
slip resistance, and predict how resistant a surface will be.

Based on the previous Discussion, two of the tests can be
eliminated.  The British Pendulum Tester measures dynamic friction.
Since static friction is more important, this test is not applicable
for slip resistance.  The James Machine also has a flaw in that it is
a laboratory test only and does not measure specimens that are wet.

Most brick are relatively slip resistant when dry.  Brick that
fall into the marginal range are typically smoother surfaced brick
with a ceramic tile-like surface.  The Horizontal Dynamometer Pull
Meter is used to measure ceramic tile surfaces.  This would be an
excellent test to propose if not for the operator variable.

The best method of the four ASTM test methods is the Horizontal
Pull Slipmeter (ASTM F 609-79 (1984)).  This method is similar to the
HDPM but does not have problems with operator variation.  Again, no
test is perfect for measuring friction, but the HPS is the best to
date.  Other good test methods exist, such as the NBS-Brungraber
tester, but they have yet to be approved as an ASTM standard.  Future
developments may warrant a review of those standards.

RECOMMENDATIONS

One of the purposes of this paper is to recommend to ASTM Subcommittee C 15.02 the inclusion of the test method in ASTM C 902. It is already stated in the standard that there is a need for a "skid/ slip resistance" test. Since most paving brick units have a rough surface and are fairly resistant to slip, it should not be a requirement to have a test run. It most definitely should be included as a guide to designers so that they may choose an appropriate test without having to become familiar with the topic of slip resistance. Also, it is important that brick manufacturers use the same test procedure so that designers may compare results equitably between paving units.

The test method and an explanation of how to review results from the test can be included in the "Explanatory Notes" section. The wording can be as follows:

"Slip Resistance - The coefficient of friction or slip resistance of brick is typically high and there is not a need to test many paving units. There are some units though that have smooth surfaces or those units in combination with water or soapy water that could possibly lead to slips and falls. These units should be tested in accordance with ASTM F 609 'Test Method for Static Slip Resistance of Footwear, Sole, Heel or Related Materials by Horizontal Pull Slipmeter (HPS)'. It is the responsibility of the designer to request test results."

Further research should be conducted to verify that ASTM F 609-79 (1984) is an appropriate test method and that the coefficient of friction is high for brick surfaces. This should not only be conducted in the laboratory, but in the field as well. When this is done, designers can feel more confident about specifying brick as a paving material.

CONCLUSION

The Horizontal Pull Slipmeter (HPS) is an appropriate test method to use to determine the slip resistance of brick pavements. The HPS is easily transportable and can be used in the field. The HPS measures static coefficient of friction which has been found to be relevant to slip resistance. ASTM Subcommittee C 15.02 should seriously consider including this test method in ASTM C 902.

REFERENCES

[1]  Wiley, C.C., Principles of Highway Engineering, McGraw-Hill Book Co., New York, 1935, p. 466.

[2]   Mavin, K.C., "The Slip Resistance of Footways and Sporting
      Surfaces," Third National Local Government Engineers Conference,
      Melbourne, 1985.
[3]   Armstrong, P.L. and Lansing, S.G., "Slip-Resistance Testing:
      Deriving Guidance from the National Electronic Injury
      Surveillance System (NEISS), "Walkway Surfaces:  Measurement of
      Slip Resistance", ASTM STP 649, Carl Anderson and John Senne,
      Eds., ASTM, 1978, pp 3-10.
[4]   Brungraber, R.J., "An Overview of Floor Slip-Resistance Research
      with Annotated Bibliography", NBS Technical Note 895, National
      Bureau of Standards,   Jan. 1976.
[5]   Adler, S.C. and Pierman, B.C., "A History of Walkway Slip-
      Resistance Research at the National Bureau of Standards", NBS
      Special Publication 565, National Bureau of Standards, Dec. 1979.

Edwin G. Hedstrom[1] and Mark B. Hogan[1]

THE PROPERTIES OF MASONRY GROUT IN CONCRETE MASONRY

REFERENCE: Hedstrom, E. G., and Hogan, M. B., "The Properties of Masonry Grout in Concrete Masonry", Masonry: Components to Assemblages, ASTM STP 1063, John H. Matthys, Editor, American Society for Testing and Materials, Philadelphia, 1990.

ABSTRACT: This paper presents the results of a research project on the properties of masonry grout and grouted concrete masonry. Seventy-two grout specimens were tested to investigate the effect of varying the proportions of grout (cement to aggregate ratio) on the physical properties of grout and grouted concrete masonry including compressive strength and modulus of elasticity. Both fine and coarse aggregate grouts were investigated. The investigation included a comparison of the compressive strength of grout specimens saw cut from grouted hollow concrete masonry units versus the procedures of ASTM C1019 [1]. Compressive strength tests of grouted and ungrouted concrete masonry prisms were also conducted. This experimental investigation concludes that grout having leaner proportions (cement to aggregate ratio) than permitted by ASTM C476 [2] may be more compatible with typical concrete masonry units. It is recommended that the strength requirements of grout be developed as an alternate to the current proportion requirements of ASTM C476 in order to encourage designers to specify grout with properties similar to the properties of the concrete masonry units thus achieving compatibility of the components and improve the structural performance of grouted concrete masonry structures. It was also found that grout specimens molded in accordance to ASTM C1019 can be used to predict the strength of grout in grouted concrete masonry.

KEY WORDS: grout, concrete masonry, mortar, prisms, compressive strength, test method, concrete masonry units, design, research, specifications.

[1] Director of Research and Development, and Director of Engineering, respectively, National Concrete Masonry Association, P.O. Box 781, Herndon, Virginia 22070.

Specifications for masonry grout evolved from mortar specifications. The 1953 American National Standards Institute code for masonry [3] specified Type M, S or N mortar proportions for grout used with non-reinforced masonry. Proportions for both fine and coarse grout are contained in the 1960 ANSI code for reinforced masonry [4]. In 1961, a tentative ASTM standard for grout was approved. Throughout this evolution, very little basic research on grout properties was carried out. The tentative standard was based on specifying proportions of grout ingredients, which is still the basis of the current ASTM C476 grout standard. However, a significant number of projects specify grout strength in lieu of proportion requirements which implies that refinements to the current methods should be investigated. The advancement of masonry design dictates the need for basic information on grout, its constituents, and its interaction with other masonry components. Grout and grouting have been the subject of several research projects, however sufficient data to demonstrate the correlation between grout proportions and grout strength has not been available.

OBJECTIVES AND SCOPE

The objectives of this research program are:

1. To study the effect of cement to aggregate ratio on grout properties.
2. To develop recommendations for grout strength requirements as an alternative to proportion requirements.
3. To study the effect that different concrete masonry units have on the properties of grout in grouted concrete masonry.
4. To study the relationship between both modulus of elasticity of concrete masonry, $E_m$ and of grout $E_g$ versus grout cement to aggregate ratio.
5. To study the relationship between strength of grout and modulus of elasticity of grout, $E_g$.

The research included compressive strength tests of concrete masonry prisms, molded grout specimens, grout specimens cut from grouted units, and component materials. Details of the test variables for specimens tested are listed in Tables 1 through 3.

TABLE 1 -- Compressive Strength Tests
On Ungrouted Prisms

| Test Variables | | Test Designation: | | |
|---|---|---|---|---|
| | NFS | NFull | HFS | HFull |
| Strength of CMU | Normal | Normal | High | High |
| Mortar Bedding | Face Shell | Full | Face Shell | Full |
| Type of Mortar | N | N | S | S |

TABLE 2 -- Compressive Strength Tests on Grouted Masonry
Prisms, Grout Specimens Molded and Grout Specimens Cut
Using Normal Strength CMU and Type N Mortar

| Test Variables | Test Designation: | | | | | |
|---|---|---|---|---|---|---|
| | NF3 | NF4 | NF6 | NC4 | NC5 | NC8 |
| Type of Grout | Fine | Fine | Fine | Coarse | Coarse | Coarse |
| Cement to Aggregate Ratio | 1:3 | 1:4 | 1:6 | 1:2.4:1.6 | 1:3:2 | 1:4.8:3.2 |

TABLE 3 -- Compressive Strength Tests on Grouted Masonry
Prisms, Grout Specimens Molded and Grout Specimens Cut
Using High Strength CMU and Type S Mortar

| Test Variables | Test Designation: | | | | | |
|---|---|---|---|---|---|---|
| | HF3 | HF4 | HF6 | HC4 | HC5 | HC8 |
| Type of Grout | Fine | Fine | Fine | Coarse | Coarse | Coarse |
| Cement to Aggregate Ratio | 1:3 | 1:4 | 1:6 | 1:2.4:1.6 | 1:3:2 | 1:4.8:3.2 |

FABRICATION, TESTING METHODS AND RESULTS

Concrete Masonry Units

Properties of component materials used in the test program were
determined in accordance with standard ASTM test methods.  Figure 1
shows the concrete masonry units (nominal dimensions 8x8x16 inches) used
in this project.  The concrete masonry units were tested according to
ASTM C140 [5] standard for compressive strength, water absorption and
density.  The results are listed in Table 4.

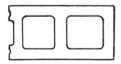

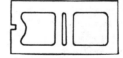

Normal Strength CMU                    High Strength CMU

FIG. 1 -- Hollow Concrete Masonry Units

TABLE 4 -- Summary of CMU Test Data

| Properties | Normal Strength | High Strength |
|---|---|---|
| Compressive Strength (gross area) | 990 psi | 2710 psi |
| Compressive Strength (net area) | 1930 psi | 4740 psi |
| Absorption lb./cu.ft. | 13.6 pcf | 7.4 pcf |
| Unit Weight | 107.3 pcf | 137.6 pcf |
| Minimum Face Shell Thickness | 1.27 in. | 1.33 in. |
| Minimum Web Thickness | 1.00 in. | 0.92 in. |
| Equivalent Web | 2.57 in./ft. | 3.44 in./ft. |
| Gross Area | 118.23 sq.in. | 119.06 sq.in. |
| Average Net Area | 60.37 sq.in. | 68.03 sq.in. |
| Average Net Area, % | 51.1% | 57.2% |
| Variation from Standard Dimension | 0.110 in. | 0.070 in. |

## Mortar

The mortar used in constructing prism specimens was Type N portland cement-lime mortar for normal strength CMU, and Type S portland cement-lime mortar for high strength CMU in accordance to ASTM C270-86b [6]. Two full unit high, stack bond, was selected for prism configuration which is the most commonly used prism configuration for construction quality assurance. Testing of mortar samples was conducted in accordance to ASTM C780 [7] and C270. Results of the mortar tests are shown in Table 5.

TABLE 5 -- Summary of Mortar Test Results

| | Mortar Mix | | | |
| | N | S | N | S |
|---|---|---|---|---|
| Proportions by Volume | | | | |
| Portland Cement | 1 | 1 | 1 | 1 |
| Lime | 1-1/4 | 1/2 | 1-1/4 | 1/2 |
| Mason Sand | 6-3/4 | 4-1/2 | 6-3/4 | 4-1/2 |
| Properties | | | | |
| Unit Weight (pcf) | 129.7 | 131.7 | 130.2 | 134.6 |
| Cone Penetration (mm) | 63 | 66 | 64 | 69 |
| Initial Flow, % | 126 | 129 | 131 | 130 |
| Water Retention, % | 77.8 | 67.1 | 75.2 | 71.2 |
| Air Content, % | 3.1 | 3.7 | 3.0 | 2.5 |
| Strength (psi) | 960 | 2940 | 920 | 2840 |

Masonry Prisms

Masonry prisms used for this investigation were constructed by laying two masonry units in stack bond to form specimens having nominal dimensions of 8-inches-wide by 16-inches-high by 16-inches-long. Mortar joints were tooled with a concave jointer having a 5/8 inch radius. Ungrouted prisms were used as control specimens. Prisms made with normal strength units were laid with Type N, portland cement-lime mortar; prisms made with high strength units were laid with Type S, portland cement-lime mortar. For the ungrouted prism specimens, both face shell mortar bedding and full mortar bedding were used. Prism specimens were cured, capped and tested in accordance with ASTM E447 [8]. Linear Variable Differential Transducers (LVDT) were installed between the plates of the testing machine adjacent to each corner, to measure deformation at selected load levels. Figure 2 illustrates the prisms fabrication method. Prisms with normal strength CMU were tested using Forney Testing Machine with 400,000 pound capacity, while prisms with high strength CMU were tested using the Applied Test System (ATS) testing machine with one million pound capacity. The prisms were tested 28 days after fabrication and the test results are summarized in Table 6. The average compressive strength of ungrouted prisms are expressed as load divided by the average net cross-sectional area of the masonry unit.

TABLE 6 -- Masonry Prism Test Data

| Prism No. | Masonry Prism | | Grout | |
|---|---|---|---|---|
| | Average Compressive Strength psi | Average Modulus of Elasticity $E_m$, psi | Average Strength psi | Average Modulus of Elasticity $E_g$, psi |
| Mortar Type N | | | | |
| NF3 | 3199 | 1500 | 8456 | ---- |
| NF4 | 2879 | 1647 | 6083 | 2585 |
| NF6 | 2177 | 1224 | 2653 | 2506 |
| NC4 | 2741 | 1578 | 8055 | 3248 |
| NC5 | 2938 | 1766 | 6131 | 1501 |
| NC8 | 2377 | 1389 | 3211 | 2432 |
| NFS | 1529 | 1239 | N/A | N/A |
| NFull | 1850 | 1107 | N/A | N/A |
| Mortar Type S | | | | |
| HF3 | 5569 | 2310 | 8448 | ---- |
| HF4 | 5391 | 2104 | 6726 | 3777 |
| HF6 | 4594 | 2136 | 3227 | 2681 |
| HC4 | 5600 | 2540 | 7990 | 3181 |
| HC5 | 4956 | 1840 | 5910 | 2795 |
| HC8 | 4090 | 2227 | 3055 | 2117 |
| HFS | 3131 | 1959 | N/A | N/A |
| HFull | 5066 | 3008 | N/A | N/A |

FACE SHELL          FULL MORTAR          FULLY GROUTED
BEDDING ONLY          BEDDING

SECTION AT MORTAR JOINT

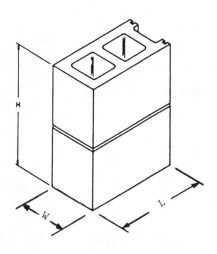

ISOMETRIC OF PRISM

FIG.2 -- Illustration of Prism Fabrication

Three masonry prism compressive strength tests were conducted for each of the sixteen series. Load/deformation data was also recorded for all specimens. Chord modulus of elasticity, $E_m$ was determined based on the slope of the line of best fit of load/deformation data points.

Grout

Grout used in constructing the test specimens consisted of both fine and coarse aggregate grouts. Sufficient water was added to produce a slump of approximately 10 inches and mixing was conducted in accordance to ASTM C476.

a.  Molded Specimens:  Concrete masonry units were used to form grout molds in accordance with the Standard Test Methods for Sampling and Testing Grout, ASTM C1019.  Side dimensions of the prisms measured 3-1/2 x 3-1/2 inches and the height was established at 7 inches.  The units forming the grout mold were lined with absorptive paper to prevent bond of the grout to the concrete masonry units.  Molds were filled to half height with grout and puddled with 1/2" x 1" puddling stick to consolidate.  The remainder of the mold was then filled to the top of the form and puddled.  The tops of the molds were screeded off and covered with a damp cloth to prevent loss of moisture.  After two days the forms were removed and the grout specimens placed in a moist room for curing.

b.  Cut Specimens:  One core of a masonry unit from each form was filled with grout and puddled in two layers as in the method previously described.  The concrete masonry unit containing the grouted core was also placed in a moist room for curing after two days.  Upon completion of curing, the masonry units were removed and prisms measuring 3-1/2 x 3-1/2 x 7-5/8 inches high were saw cut from the grouted core.

Prior to testing at an age of 28 days, the ends of the grouted specimens were capped with high strength gypsum plaster.  The compressive strength tests of the grout specimens were performed on a 400,000 pound capacity testing machine.  Tests to determine the properties of the grout mixes were conducted in accordance with ASTM C1019.

A summary of the average compressive strength test results of molded and cut specimens are shown in Table 7.  Load/deformation data was recorded for selected specimens.  The chord modulus of elasticity of grout, $E_g$, is determined based on the line of best fit of load/deformation data points.  The properties of grout are also shown in Table 7.

TABLE 7 -- Grout Test Data

| Grout Specimen inches | Average Compressive Strength | | | | Initial W/C Ratio | Density pcf | Slump inch |
|---|---|---|---|---|---|---|---|
| | Molded Specimen | | Cut Specimen | | | | |
| | Strength psi | Eg ksi | Strength psi | Eg ksi | | | |
| NF3 | 8456 | ---- | 7062 | ---- | 0.59 | 143.4 | 9.88 |
| NF4 | 6083 | 2585 | 6113 | 3162 | 0.69 | 139.8 | 10 |
| NF6 | 2653 | 2506 | 2267 | 1642 | 1.06 | 135.4 | 10.25 |
| NC4 | 8055 | 3248 | 6639 | 4515 | 0.54 | 146.7 | 10.12 |
| NC5 | 6131 | 1501 | 5888 | 2017 | 0.67 | 148.2 | 10 |
| NC8 | 3211 | 2432 | 2961 | 1509 | 1.04 | 143.5 | 10.12 |
| HF3 | 8448 | ---- | 6930 | ---- | 0.67 | 141.5 | 10.25 |
| HF4 | 6726 | 3777 | 6195 | 2839 | 0.68 | 139.8 | 10.25 |
| HF6 | 3227 | 2681 | 3282 | 3250 | 0.98 | 136.8 | 10 |
| HC4 | 7990 | 3181 | 7480 | 5426 | 0.54 | 146.7 | 10 |
| HC5 | 5910 | 2795 | 5800 | 1618 | 0.68 | 146.8 | 10.12 |
| HC8 | 3055 | 2117 | 3132 | 1553 | 1.02 | 144.1 | 10 |

DISCUSSION OF RESULTS

Molded Specimens Versus Cut Specimens

     The compressive strength of grout is defined by the Standard Test Method for Sampling and Testing Grout, ASTM C1019.  Grout specimens cut or cored from grouted masonry are sometimes considered to be more indicative of the compressive strength of grout in grouted concrete masonry.  The comparison of the strength of cut specimens to the strength of molded specimens indicates how well the standard test method (molded specimens) predicts the compressive  strength of grout in grouted concrete masonry (cut specimens).

     The compressive strength of the molded specimens averaged 9.7% higher than that of the cut specimens.  While this is not a signifi-cant difference, it does appear that the difference is due to more than just typical scatter of data.  The most obvious cause of a strength difference between molded and cut specimens is a difference in  the water/cement ratio (W/C) during the time of set.  All physical characteristics of plastic grout effecting strength including initial W/C ratio were the same for both the molded and cut specimens.  The volume of grout in the molded specimens was significantly less than the volume of the cut specimens and hence the ratio of cross sectional area to perimeter was less for the molded specimen.  These ratios are approximately 0.88 and 1.36 for the C1019 specimens and cast-cut specimens respectively.  The W/C ratio at the time of set would be greater for the cast-cut specimens, especially near the center where the specimen was cut.  Based on the results of this research, it appears that the standard ASTM C1019 method for testing grout reflects strengths representative of the strength of grout in grouted concrete masonry.

## Grout Compressive Strength Versus Aggregate to Cement Ratio

Figure 3 shows the correlation between grout compressive strength and the ratio of total aggregate to cement (A/C) ratio for both fine and coarse grout. The lines plotted in Figure 3 for fine grout and for coarse grout indicate the general relationship between strength and A/C ratio.

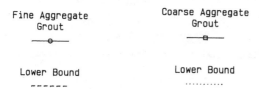

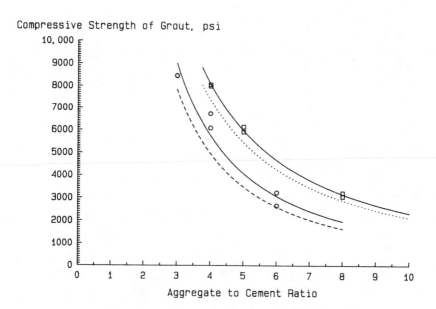

FIG. 3 -- Compressive Strength of Grout Versus
Aggregate To Cement Ratio

A precise relationship between these two parameters would require
additional testing of both fine and coarse grouts; however, using the
available data some general observations can be made:
  o  All of the grouts tested produced compressive strengths
     greater than 2000 psi, which is considered to be a minimum
     grout strength requirement in building codes [9, 10].
  o  The relationship between strength and A/C ratio is different
     for fine and coarse aggregate grout.
  o  Compressive strength reduces as A/C ratio increases. However,
     the reduction is less pronounced as the A/C ratios increase.
  o  Using a lower bound of the data plotted in Figure 3,
     conservative values of grout compressive strength can be
     predicted using Table 8.

TABLE 8 -- Lower Bound Grout Compressive Strength

| Compressive Strength, psi | Total Aggregate To Cement Ratio | |
|---|---|---|
| | Fine Grout | Coarse Grout |
| 6000 | 3-1/2 | 4-5/8 |
| 5000 | 4 | 5-3/8 |
| 4000 | 4-5/8 | 6-1/4 |
| 3000 | 5-1/2 | 7-3/4 |
| 2000 | 7 | 10 |

TABLE 9 -- Conversion Factors

| To Convert From | To | Multiply By |
|---|---|---|
| inch | metre | $2.540 \times 10^{-2}$ |
| kip | newton | $4.448222 \times 10^{3}$ |
| ksi | pascal | $6.894757 \times 10^{6}$ |
| pcf | $kg/m^{3}$ | $16.01846$ |

## Grout Compressive Strength Versus CMU Strength/Absorption

Two different concrete masonry units were used to mold grout specimens: A high strength (4740 psi) unit, with an absorption of 7.4 pcf; and a normal strength (1930 psi) unit, with an absorption ·13.6 pcf. The shape of the units also varied as shown in Figure 1. The properties of the concrete masonry units had no apparent effect on the compressive strength of grout specimens. The bar chart plot in Figure 4 illustrates the comparative strengths of grout specimens which were molded or cut from normal and high strength units for each series.

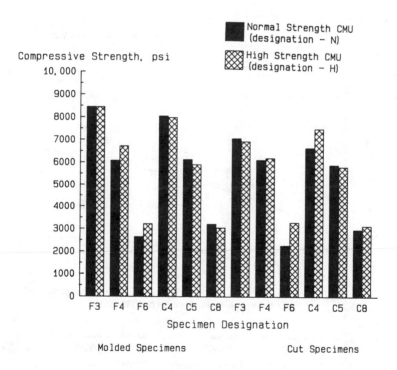

FIG. 4 -- Compressive Strength of Grout Versus
Concrete Masonry Unit Strength

## Compressive Strength Compatibility of Grout and CMU

Compatibility of materials is a key consideration in the design of masonry elements. Figure 5 compares the masonry prism strength to the compressive strength of grout and of masonry units. The intercept of masonry prism strength for both high and normal strength series is the gross area compressive strength of ungrouted prisms. Grout strength is considered to be zero. The vertical dashed lines are the net area compressive strengths of the normal and high strength concrete masonry units. Increasing grout strength above the net area compressive strength of the units has little effect on the prism strengths.

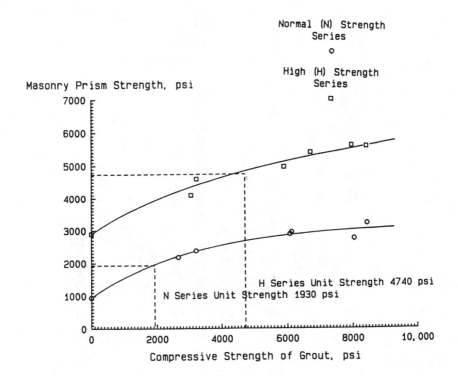

FIG. 5 -- Compressive Strength of Grout Versus
Masonry Prism Strength

## Elastic Moduli and Stress/Strain Properties of Grout

The stress/strain measurements recorded for one grout specimen in each series was found to be linear over the loading range of the specimens tested. Elastic moduli is determined on data points between 5% and 33% of the grout strength, which is the range specified in building codes [9,10]. A plot of this data is illustrated in Figure 6. It has been proposed [9] that the modulus of elasticity of grout ($E_g$) is related to grout strength using the formula:

$$E_g = 57000 \times \sqrt{\text{Grout Strength}} \quad \text{psi}$$

This equation is also plotted in Figure 6. Based on the data presented, the above equation over-estimates grout modulus of elasticity ($E_g$). The line of the best fit passing through the origin for grout modulus of elasticity versus compressive strength of grout is:

$$E_g = 500 \times \text{Grout Strength, psi}$$

The scatter of data plotted in Figure 6 is significant suggesting a poor correlation between grout strength and grout modulus of elasticity. The proposed correlation in building codes which relates grout modulus of elasticity to grout strength (dashed line in Figure 6) does not appear to be substantiated by this data. Given the flatness of the curve over the range of typical grout strengths (above 2000 psi) and considering the scatter of data, a simplified linear relationship between grout strength and modulus of elasticity appears to be more appropriate.

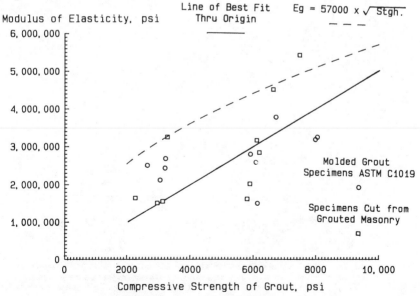

FIG. 6 -- Grout Modulus of Elasticity Versus
Compressive Strength of Grout

## Grout Modulus of Elasticity Versus Aggregate to Cement Ratio

It is common to specify grout by the proportions of ingredients when strength requirements are not specified. Therefore, the designer needs to know the relationship between modulus of elasticity ($E_g$) and grout proportions. Grout modulus of elasticity, $E_g$ data versus grout aggregate to cement ratio is plotted in Figure 7 for both fine and coarse aggregate grout. Unfortunately, the correlation between these two parameters is not good. Additional testing may better define the relationship between modulus of elasticity and grout proportions. Until additional information is available the data in Figure 7 may be used to estimate grout modulus of elasticity from the aggregate to cement ratio of grout.

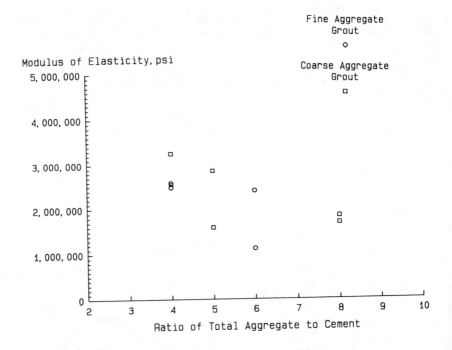

FIG. 7 -- Grout Modulus of Elasticity Versus
Aggregate/Cement Ratio

CONCLUSIONS AND RECOMMENDATIONS

The following conclusions and recommendations are made based on the results of this research [11].

o The Standard Method for Sampling and Testing Grout, ASTM C1019, is a good method for determining the compressive strength of grout in grouted concrete masonry. The compressive strength of grout as defined in building codes should continue to be based on ASTM C1019.

o It is recommended that the ASTM C12.03 committee consider revising the Standard Specifications for Grout for Masonry, ASTM C476. The current specification requires cement to aggregate ratios (proportions by volume) that produce excessive grout strengths which should be changed to achieve better strength compatibility with other masonry components such as concrete masonry units. Furthermore, grout strength requirements should be added to the ASTM C476 standard as an alternate to the proportion specification.

o The ACI/ASCE 530 Committee should consider revising the equation for determining grout modulus of elasticity based on grout strength to be more representative of the correlation between grout strength and grout modulus of elasticity.

o More research is needed to establish a relationship between grout modulus of elasticity and the aggregate to cement ratio.

REFERENCES

[1] ASTM C1019-84, "Standard Method of Sampling and Testing Grout".

[2] ASTM C476-86b, "Standard Specification for Grout for Masonry".

[3] ANSI 41.1, "American Standard Building Code Requirements for Masonry", 1953.

[4] ANSI 41.2, "Building Code Requirements for Reinforced Masonry", 1960.

[5] ASTM C140-75, "Standard Methods of Sampling and Testing Concrete Masonry Units".

[6] ASTM C270-86b, "Standard Specification for Mortar for Unit Masonry".

[7] ASTM C780-85, "Standard Method for Preconstruction and Construction Evaluation of Mortars for Plain and Reinforced Unit Masonry".

[8] ASTM E447-84, "Standard Test Methods for Compressive Strength of Masonry Prisms".

[9]  Building Code Requirements for Masonry Structures
     (ACI/ASCE 530-1989), American Concrete Institute.

[10] Uniform Building Code, Chapter 24 - Masonry, International
     Conference of Building Officials, Whittier, California.

[11] NCMA, Engineered Concrete Masonry Design Committee, "Research
     Investigation of the Properties of Masonry Grout in Concrete
     Masonry", Final Report, August 1988.

Charles E. Buchanan Jr. and B. M. Call

EFFECTS OF AGGREGATE GRADATION ON PROPERTIES   OF MASONRY
CEMENT AND PORTLAND CEMENT-LIME MASONRY MORTARS

---

REFERENCE: Buchanan, C. E. Jr. and Call, B. M.,
"Effects of Aggregate Gradation on Properties
of Masonry Cement and Portland Cement-Lime
Masonry Mortars," Masonry: Components for
Assemblages, ASTM STP 1063, J. H. Matthys, Ed.,
American Society for Testing and Materials,
Philadelphia, 1990.

ABSTRACT: Eight laboratories participated in a
round robin to evaluate variations in aggregate
gradation when tested using Masonry cement and
Portland cement-lime masonry mortars. A control
gradation was used which was near the center of
C144 gradation limits and five other gradations
were used which tested the limits of  C144.   It
appears from the data  that the  limits on C144
can  be broadened  without appreciably altering
the  safety  built  in   for  user  protection.
Although not part of the original test program,
a study was made  of  the yield of  the various
mortars produced,  and  it  was determined that
the  masonry  cements  had  on  average  a 5.3%
increase   in   volume   over   the   Portland
cement-lime mixes.

KEYWORDS:  Masonry  cement,  Portland  cement,
lime, aggregate, gradation

For many years ASTM C144 served the masonry industry
well with stringent limitations on  aggregate gradation.
In the mid sixties  however it became apparent  that the
majority  of  masonry sands  sold  in  the  United States
would not  comply with these gradation  requirements. In
order to  have a  specification which at  least would be
honored  the  majority  of  the  time,  the  limits were
broadened  so  that  most  sands then  being  sold would
comply.  However,  in a few years  some  producers were
taking  advantage  of  the  broader  specification,  and

Charles  E.  Buchanan Jr.,  President, ROAN Laboratories
Inc.,  409  Santee  Drive,  Santee, SC 29142  B. M. Call,
Vice  President Technical  Services,  Pyrament/Lone Star
Industries Inc., P. O. Box 90765, Houston, TX 77290

aggregates were being marketed which were producing inferior masonry mortars. In the mid seventies the original limits were returned to ASTM C144 but with the stipulation that if an aggregate did not meet the gradation requirements, it still could be used if laboratory mixes were made to show compliance with ASTM C270, Specification for Mortar for Unit Masonry.

In 1985, Sub-Committee C12.04 undertook to have a round robin test to evaluate the limits of C144 to ascertain if the limits could be broadened without materially affecting the properties of the masonry mortar, whether made with masonry cement or with Portland cement-lime.

Genstar, of Hunt Valley, MD agreed to prepare the aggregate. A large batch of sand was secured, and it was then screened through the #4, #8, #16, #30, #50, #100, and the #200 sieves. These separated fractions were then forwarded to the eight participating laboratories, which were:

Peerless Cement Co., Detroit, MI - Mr. Don Hill
Lone Star Industries, Houston, TX - Mr. BM Call
Arkansas Cement Co., Foreman, AK - Mr. John Hinkel
Santee Cement Co., Holly Hill, SC - Mr. Tim Conway
US Gypsum Co., Chicago, IL - Mr. Byron Powell
Portland Cement Assoc., Skokie IL - Mr. Jake Ribar
Twin City Testing, St Paul MN - Mr. Charles Britzius
Riverton Lime, Riverton, VA- Mr. John Melander

The gradations used in the six trials are given in Table 1. Also listed are the batch weights used for each aggregate size, the masonry cement, the Portland cement and the lime. These results also are shown in Figure 1. It can be seen that the first gradation was near the average of the C144 limits. The second gradation was similar to the first except no material was retained on the #16 sieve. The third gradation was slightly finer, and was gap graded between the #8 and #16 sieves. The fourth series was coarse. The fifth series was fine. Both the fourth and fifth series were close to the specification limits. The sixth gradation was similar to the first except that it contained ten percent (10%) passing the #200 mesh.

The proportions used were for a type N masonry mortar as defined by ASTM C270 using the proportion method. This was one part masonry cement to three parts aggregate for the masonry cement mixes and one part Portland cement, one part lime, and six parts aggregate for the Portland cement-lime masonry mixes. The laboratory batch weights used were calculated from the unit weights given in note 2 of C270 for one cubic foot of the materials used.

The results obtained from the four laboratories using masonry cement are shown in Table 2, with Table 3 reflecting the same data except that it is expressed as a percentage of the first gradation, the one that was near the center of C144. Tables 4 and 5 are similar to 2 and 3 except that they reflect the data obtained from the Portland cement-lime mixes.

Figure 2 graphically shows the percent ratios for the water demand for both series of tests. You can note that the second, third and sixth series are fairly similar to the control mix. However test four, the most coarse gradation, required much less water, while test five, the finest gradation took considerably more water. The consistency for all these mixes was determined by the flow table, as required by C270.

Figure 3 shows the cone penetrometer results. Very little can be concluded from this except that the second series showed a slightly lower penetration while series five showed a slightly higher one.

Figure 4 shows the percent air, and it is apparent that the second, fourth, and sixth series all showed lower air content than the control, whereas the third series was much higher in air content.

Figure 5 shows the water retention, and very little can be concluded from these data.

Figures 6 and 7 show the compressive strengths at seven and twenty-eight days, and it is apparent that the third and fifth series reduced compressive strength development while the fourth series increased it.

Although not an original part of this study, Table 6 shows data concerning yield and cement content of the various materials. You can see that the masonry cements on average yielded 5.3% more masonry mortar than did the Portland cement-lime mortars, with some mixes approaching 10% increase in yield. Also if one assumes that the masonry cements contained 50% Portland cement, a comparison can then be made concerning cement content per Ml, which shows that the Portland cement-lime mixes are considerably richer than the masonry cement ones. This probably accounts for the increase in compressive strength development shown by the Portland cement-lime mixes.

The average compressive strength of the Portland cement-lime mixes also was increased by laboratory number 3, as they were approximately 75% higher than the other three laboratories.

Based on the data presented here, it appears that acceptable masonry mortars can be prepared from aggregates approaching or slightly exceeding C144 limits. This is especially true for the coarser gradations. It also appears that the proportions for masonry mortars as defined by C270 need to be evaluated as the masonry cement mortars show lower compressive strength, but at the same time have a lower Portland cement content when compared on a volume of mortar produced.

Consequently I would recommend that further studies to be undertaken by C12.04 to see just how far the limits can be moved without bringing harm to the consumer, and that either C12.02 or C12.03 look into the matter of parity with the two types of masonry mortars.

Table 1- ASTM C12.04 Round Robin Test Program to Evalute
Various Sand Gradations

| Sand Number | 1 | 2 | 3 | 4 | 5 | 6 |
|---|---|---|---|---|---|---|
| Fineness Modulus | 2.29 | 2.05 | 2.00 | 3.10 | 1.45 | 2.13 |

| Sieve Number | Percent Passing | | | | | |
|---|---|---|---|---|---|---|
| #4 | 100.0 | 100.0 | 100.0 | 100.0 | 100.0 | 100.0 |
| #8 | 97.5 | 100.0 | 95.0 | 90.0 | 100.0 | 97.5 |
| #16 | 85.0 | 100.0 | 95.0 | 60.0 | 100.0 | 85.0 |
| #30 | 57.5 | 50.0 | 75.0 | 30.0 | 85.0 | 57.5 |
| #50 | 22.5 | 35.0 | 25.0 | 10.0 | 45.0 | 30.0 |
| #100 | 8.5 | 10.0 | 10.0 | 0.0 | 25.0 | 17.5 |
| #200 | 0.0 | 0.0 | 0.0 | 0.0 | 0.0 | 10.0 |

| | Percent Between | | | | | |
|---|---|---|---|---|---|---|
| #4 to #8 | 2.5 | 0.0 | 5.0 | 10.0 | 0.0 | 2.5 |
| #8 to 16 | 12.5 | 0.0 | 0.0 | 30.0 | 0.0 | 12.5 |
| #16 to 30 | 27.5 | 50.0 | 20.0 | 30.0 | 15.0 | 27.5 |
| #30 to 50 | 35.0 | 15.0 | 50.0 | 20.0 | 40.0 | 27.5 |
| #50 to 100 | 14.0 | 25.0 | 15.0 | 10.0 | 20.0 | 12.5 |
| #100 to 200 | 8.5 | 10.0 | 10.0 | 0.0 | 25.0 | 7.5 |
| #200 to pan | 0.0 | 0.0 | 0.0 | 0.0 | 0.0 | 10.0 |

| | Percent Above | | | | | |
|---|---|---|---|---|---|---|
| #4 | 0.0 | 0.0 | 0.0 | 0.0 | 0.0 | 0.0 |
| #8 | 2.5 | 0.0 | 5.0 | 10.0 | 0.0 | 2.5 |
| #16 | 15.0 | 0.0 | 5.0 | 40.0 | 0.0 | 15.0 |
| #30 | 42.5 | 50.0 | 25.0 | 70.0 | 15.0 | 42.5 |
| #50 | 77.5 | 65.0 | 75.0 | 90.0 | 55.0 | 70.0 |
| #100 | 91.5 | 90.0 | 90.0 | 100.0 | 75.0 | 82.5 |
| #200 | 100.0 | 100.0 | 100.0 | 100.0 | 100.0 | 90.0 |

| | Batch Weights, in grams | | | | | |
|---|---|---|---|---|---|---|
| #4 to #8 | 36 | 0 | 72 | 144 | 0 | 36 |
| #8 to 16 | 180 | 0 | 0 | 432 | 0 | 180 |
| #16 to 30 | 396 | 720 | 288 | 432 | 216 | 396 |
| #30 to 50 | 504 | 216 | 720 | 288 | 576 | 396 |
| #50 to 100 | 202 | 360 | 216 | 144 | 288 | 180 |
| #100 to 200 | 122 | 144 | 144 | 0 | 360 | 108 |
| #200 to pan | 0 | 0 | 0 | 0 | 0 | 144 |

Grams Masonry Cement                    420

Grams of Portland Cement                282
Grams of Lime                           120

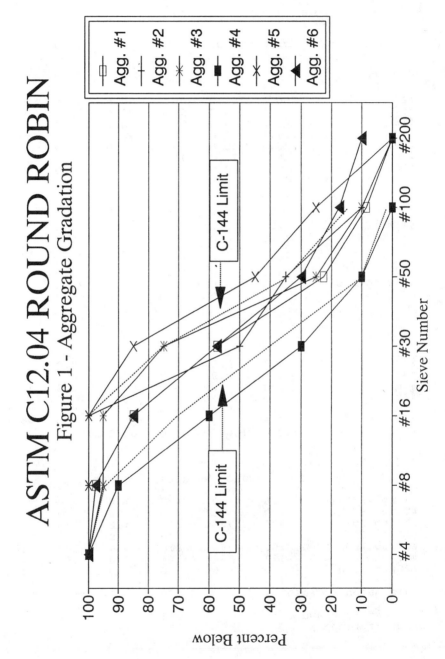

ASTM C12.04 ROUND ROBIN
Figure 1 - Aggregate Gradation

Table 2 Results Obtained by Four Laboratories Using Four Masonry Cements and Six Aggregate Gradations.

| Aggregate | | 1 | 2 | 3 | 4 | 5 | 6 | Range |
|---|---|---|---|---|---|---|---|---|
| Laboratory | | | | | | | | |
| | | | | Milliliters of Water | | | | |
| | 1 | 221 | 222 | 230 | 200 | 270 | 240 | 70 |
| | 2 | 310 | 290 | 295 | 259 | 340 | 295 | 81 |
| | 3 | 245 | 245 | 250 | 210 | 280 | 255 | 70 |
| | 4 | 255 | 250 | 265 | 220 | 290 | 255 | 70 |
| Mean | | 257.8 | 251.8 | 260.0 | 222.3 | 295.0 | 261.3 | 72.75 |
| | | | | Percent Flow | | | | |
| | 1 | 108 | 108 | 108 | 112 | 114 | 115 | 7 |
| | 2 | 105 | 109 | 105 | 114 | 109 | 106 | 9 |
| | 3 | 111 | 107 | 107 | 113 | 110 | 115 | 8 |
| | 4 | 114 | 106 | 111 | 108 | 112 | 110 | 8 |
| Mean | | 109.5 | 107.5 | 107.8 | 111.8 | 111.3 | 111.5 | 4.25 |
| | | | Cone Penetrometer Penetration | | | | | |
| | 1 | 31 | 34 | 37 | 27 | 45 | 35 | 18 |
| | 2 | 63 | 55 | 58 | 71 | 62 | 55 | 16 |
| | 3 | 40 | 34 | 35 | 40 | 44 | 37 | 10 |
| | 4 | 48 | 38 | 44 | 40 | 38 | 34 | 14 |
| Mean | | 45.5 | 40.3 | 43.5 | 44.5 | 47.3 | 40.3 | 7 |
| | | | | Percent Air | | | | |
| | 1 | 20.4 | 21.0 | 22.4 | 17.3 | 22.2 | 15.5 | 6.9 |
| | 2 | 20.2 | 17.0 | 18.6 | 19.5 | 17.3 | 12.3 | 7.9 |
| | 3 | 16.8 | 16.0 | 19.1 | 15.5 | 17.3 | 9.9 | 9.2 |
| | 4 | 15.6 | 15.0 | 16.4 | 14.0 | 13.9 | 10.6 | 5.8 |
| Mean | | 18.25 | 17.25 | 19.13 | 16.57 | 17.68 | 12.08 | 7.05 |
| | | | Percent Water Retention | | | | | |
| | 1 | 84.3 | 84.0 | 85.6 | 83.9 | 90.6 | 88.7 | 6.7 |
| | 2 | 81.9 | 79.0 | 79.0 | 86.0 | 73.0 | 78.3 | 13.0 |
| | 3 | 78.9 | 71.0 | 76.3 | 69.6 | 75.0 | 78.0 | 9.3 |
| | 4 | 81.1 | 77.0 | 79.3 | 78.7 | 72.0 | 80.5 | 9.1 |
| Mean | | 81.55 | 77.75 | 80.05 | 79.55 | 77.65 | 81.38 | 3.90 |
| | | | Seven Day Compressive Strength, psi | | | | | |
| | 1 | 1247 | 1175 | 744 | 1556 | 756 | 1175 | 812 |
| | 2 | 856 | 1021 | 946 | 1158 | 763 | 1225 | 462 |
| | 3 | 1115 | 1212 | 1041 | 1541 | 881 | 1246 | 660 |
| | 4 | 945 | 1018 | 836 | 1243 | 798 | 1150 | 445 |
| Mean | | 1040.8 | 1106.5 | 891.8 | 1374.5 | 799.5 | 1199.0 | 575.0 |
| | | Twenty Eight Day Compressive Strength, psi | | | | | | |
| | 1 | 1333 | 1287 | 1113 | 1808 | 840 | 1410 | 968 |
| | 2 | 1046 | 1304 | 1225 | 1344 | 950 | 1579 | 629 |
| | 3 | 1450 | 1578 | 1312 | 1824 | 1143 | 1638 | 681 |
| | 4 | 1187 | 1230 | 1037 | 1463 | 985 | 1445 | 478 |
| Mean | | 1254.0 | 1349.8 | 1171.8 | 1609.8 | 979.5 | 1518.0 | 630.3 |

Table 3 Results obtained by four laboratories using four
Masonry Cements and Six Aggregate Gradations
Expressed as a Percentage of a Standard Gradation

| Aggregate | | 1 | 2 | 3 | 4 | 5 | 6 | Range |
|---|---|---|---|---|---|---|---|---|
| Laboratory | | | | | | | | |
| | | | | Milliliters of Water | | | | |
| | 1 | 100.0 | 100.5 | 104.1 | 90.5 | 122.2 | 108.6 | 31.7 |
| | 2 | 100.0 | 93.5 | 95.2 | 83.5 | 109.7 | 95.2 | 26.1 |
| | 3 | 100.0 | 100.0 | 102.0 | 85.7 | 114.3 | 104.1 | 28.6 |
| | 4 | 100.0 | 98.0 | 103.9 | 86.3 | 113.7 | 100.0 | 27.5 |
| Mean | | 100.0 | 98.0 | 101.3 | 86.5 | 115.0 | 102.0 | 28.5 |
| | | | | Percent Flow | | | | |
| | 1 | 108.0 | 108.0 | 108.0 | 112.0 | 114.0 | 115.0 | 7.0 |
| | 2 | 105.0 | 109.0 | 105.0 | 114.0 | 109.0 | 106.0 | 9.0 |
| | 3 | 111.0 | 107.0 | 107.0 | 113.0 | 110.0 | 115.0 | 8.0 |
| | 4 | 114.0 | 106.0 | 111.0 | 108.0 | 112.0 | 110.0 | 8.0 |
| Mean | | 109.5 | 107.5 | 107.8 | 111.8 | 111.3 | 111.5 | 4.3 |
| | | | Cone Penetrometer Penetration | | | | | |
| | 1 | 100.0 | 109.7 | 119.4 | 87.1 | 145.2 | 112.9 | 58.1 |
| | 2 | 100.0 | 87.3 | 92.1 | 112.7 | 98.4 | 87.3 | 25.4 |
| | 3 | 100.0 | 85.0 | 87.5 | 100.0 | 110.0 | 92.5 | 25.0 |
| | 4 | 100.0 | 79.2 | 91.7 | 83.3 | 79.2 | 70.8 | 29.2 |
| Mean | | 100.0 | 90.3 | 97.6 | 95.8 | 108.2 | 90.9 | 17.9 |
| | | | | Air | | | | |
| | 1 | 100.0 | 102.9 | 109.8 | 84.8 | 108.8 | 76.0 | 33.8 |
| | 2 | 100.0 | 84.2 | 92.1 | 96.5 | 85.6 | 60.9 | 39.1 |
| | 3 | 100.0 | 95.2 | 113.7 | 92.3 | 103.0 | 58.9 | 54.8 |
| | 4 | 100.0 | 96.2 | 105.1 | 89.7 | 89.1 | 67.9 | 37.2 |
| Mean | | 100.0 | 94.6 | 105.2 | 90.8 | 96.6 | 65.9 | 39.2 |
| | | | | Water Retention | | | | |
| | 1 | 100.0 | 99.6 | 101.5 | 99.5 | 107.5 | 105.2 | 7.9 |
| | 2 | 100.0 | 96.5 | 96.5 | 105.0 | 89.1 | 95.6 | 15.9 |
| | 3 | 100.0 | 90.0 | 96.7 | 88.2 | 95.1 | 98.9 | 11.8 |
| | 4 | 100.0 | 94.9 | 97.8 | 97.0 | 88.8 | 99.3 | 11.2 |
| Mean | | 100.0 | 95.3 | 98.1 | 97.4 | 95.1 | 99.7 | 4.9 |
| | | Seven Day Compressive Strength, psi | | | | | | |
| | 1 | 100.0 | 94.2 | 59.7 | 124.8 | 60.6 | 94.2 | 65.1 |
| | 2 | 100.0 | 119.3 | 110.5 | 135.3 | 89.1 | 143.1 | 54.0 |
| | 3 | 100.0 | 108.7 | 93.4 | 138.2 | 79.0 | 111.7 | 59.2 |
| | 4 | 100.0 | 107.7 | 88.5 | 131.5 | 84.4 | 121.7 | 47.1 |
| Mean | | 100.0 | 107.5 | 88.0 | 132.5 | 78.3 | 117.7 | 54.1 |
| | | Twenty Eight Day Compressive Strength, psi | | | | | | |
| | 1 | 100.0 | 96.5 | 83.5 | 135.6 | 63.0 | 105.8 | 72.6 |
| | 2 | 100.0 | 124.7 | 117.1 | 128.5 | 90.8 | 151.0 | 60.1 |
| | 3 | 100.0 | 108.8 | 90.5 | 125.8 | 78.8 | 113.0 | 47.0 |
| | 4 | 100.0 | 103.6 | 87.4 | 123.3 | 83.0 | 121.7 | 40.3 |
| Mean | | 100.0 | 108.4 | 94.6 | 128.3 | 78.9 | 122.9 | 49.4 |

Table 4 Results Obtained by Four Laboratories Using Four
Portland Cement Lime Combinations and Six
Aggregate Gradations

| Aggregate | | 1 | 2 | 3 | 4 | 5 | 6 | Range |
|---|---|---|---|---|---|---|---|---|
| Laboratory | | | | | | | | |
| | | | Milliliters of Water | | | | | |
| | 1 | 345 | 350 | 355 | 335 | 388 | 340 | 53 |
| | 2 | 333 | 333 | 351 | 300 | 365 | 375 | 75 |
| | 3 | 330 | 360 | 350 | 306 | 375 | 338 | 69 |
| | 4 | 300 | 306 | 312 | 266 | 402 | 320 | 136 |
| Mean | | 327.0 | 337.3 | 342.0 | 301.8 | 382.5 | 343.3 | 80.75 |
| | | | Percent Flow | | | | | |
| | 1 | 112 | 108 | 104 | 122 | 117 | 110 | 18 |
| | 2 | 109 | 108 | 109 | 112 | 109 | 109 | 4 |
| | 3 | 106 | 115 | 106 | 105 | 114 | 106 | 10 |
| | 4 | 106 | 107 | 107 | 106 | 106 | 107 | 1 |
| Mean | | 108.3 | 109.5 | 106.5 | 111.3 | 111.5 | 108.0 | 5 |
| | | | Cone Penetrometer Penetration | | | | | |
| | 1 | 52 | 47 | 48 | 60 | 50 | 50 | 13 |
| | 2 | 34 | 32 | 34 | 31 | 31 | NA | 3 |
| | 3 | NA | NA | NA | NA | NA | NA | NA |
| | 4 | 38 | 37 | 39 | 45 | 45 | 45 | 8 |
| Mean | | 41.3 | 38.7 | 40.3 | 45.3 | 42.0 | 47.5 | 8.83 |
| | | | Percent Air | | | | | |
| | 1 | 3.4 | 3.7 | 5.0 | 2.0 | 4.2 | 3.3 | 3.0 |
| | 2 | 5.1 | 4.9 | 5.0 | 3.6 | 4.1 | NA | 1.5 |
| | 3 | 3.9 | 2.1 | 5.0 | 2.7 | 5.0 | 3.3 | 2.9 |
| | 4 | 11.7 | 11.2 | 12.7 | 10.9 | 8.8 | 8.6 | 4.1 |
| Mean | | 6.03 | 5.47 | 6.92 | 4.80 | 5.53 | 5.07 | 2.12 |
| | | | Percent Water Retention | | | | | |
| | 1 | 85.0 | 82.0 | 82.0 | 82.0 | 85.0 | 89.0 | 7.0 |
| | 2 | 87.0 | 82.0 | 84.0 | 87.0 | 83.0 | NA | 5.0 |
| | 3 | 75.0 | 88.0 | 69.0 | 77.0 | 79.0 | 85.0 | 19.0 |
| | 4 | 87.0 | 86.0 | 84.0 | 87.0 | 90.0 | 88.0 | 6.0 |
| Mean | | 83.50 | 85.00 | 79.75 | 83.25 | 84.25 | 87.33 | 7.58 |
| | | | Seven Day Compressive Strength, psi | | | | | |
| | 1 | 1375 | 1292 | 1248 | 1459 | 993 | 1442 | 466 |
| | 2 | 1416 | 1383 | 1208 | 1850 | 1075 | 1383 | 775 |
| | 3 | 2760 | 2546 | 2420 | 3433 | 1833 | 2933 | 1600 |
| | 4 | 1522 | 1458 | 1282 | 1823 | 1068 | 1465 | 755 |
| Mean | | 1768.3 | 1669.8 | 1539.5 | 2141.3 | 1242.3 | 1805.8 | 899.0 |
| | | | Twenty Eight Day Compressive Strength. psi | | | | | |
| | 1 | 1992 | 1809 | 1759 | 2032 | 1412 | 2023 | 620 |
| | 2 | 1733 | 1717 | 1558 | 2358 | 1437 | 2008 | 921 |
| | 3 | 3378 | 1993 | 2726 | 3776 | 2310 | 3076 | 1783 |
| | 4 | 1875 | 1794 | 1633 | 2202 | 1375 | 1909 | 827 |
| Mean | | 2244.5 | 1828.3 | 1919.0 | 2592.0 | 1633.5 | 2254.0 | 958.5 |

Table 5 Results Obtained by Four Laboratories Using Four
Portland Cement Lime Combinations and Six
Aggregate Gradations Expressed as a Percentage of
a Standard Gradation

| Aggregate | 1 | 2 | 3 | 4 | 5 | 6 | Range |
|---|---|---|---|---|---|---|---|
| Laboratory | | | | | | | |
| | | | Milliliters of Water | | | | |
| 1 | 100.0 | 101.4 | 102.9 | 97.1 | 112.5 | 98.6 | 15.4 |
| 2 | 100.0 | 100.0 | 105.4 | 90.1 | 109.6 | 112.6 | 22.5 |
| 3 | 100.0 | 109.1 | 106.1 | 92.7 | 113.6 | 102.4 | 20.9 |
| 4 | 100.0 | 102.0 | 104.0 | 88.7 | 134.0 | 106.7 | 45.3 |
| Mean | 100.0 | 103.1 | 104.6 | 92.1 | 117.4 | 105.1 | 25.3 |
| | | | Percent Flow | | | | |
| 1 | 100.0 | 96.4 | 92.9 | 108.9 | 104.5 | 98.2 | 16.1 |
| 2 | 100.0 | 99.1 | 100.0 | 102.8 | 100.0 | 100.0 | 3.7 |
| 3 | 100.0 | 108.5 | 100.0 | 99.1 | 107.5 | 100.0 | 9.4 |
| 4 | 100.0 | 100.9 | 100.9 | 100.0 | 100.0 | 100.9 | 0.9 |
| Mean | 100.0 | 101.2 | 98.5 | 102.7 | 103.0 | 99.8 | 4.6 |
| | | Cone PeCone Penetrometer | | | | | |
| 1 | 100.0 | 90.4 | 92.3 | 115.4 | 96.2 | 96.2 | 25.0 |
| 2 | 100.0 | 94.1 | 100.0 | 91.2 | 91.2 | NA | 8.8 |
| 3 | NA | NA | NA | NA | NA | NA | NA |
| 4 | 100.0 | 97.4 | 102.6 | 118.4 | 118.4 | 118.4 | 21.1 |
| Mean | 100.0 | 94.0 | 98.3 | 108.3 | 101.9 | 107.3 | 14.4 |
| | | | Percent Air | | | | |
| 1 | 100.0 | 108.8 | 147.1 | 58.8 | 123.5 | 97.1 | 88.2 |
| 2 | 100.0 | 96.1 | 98.0 | 70.6 | 80.4 | ERR | 29.4 |
| 3 | 100.0 | 53.8 | 128.2 | 69.2 | 128.2 | 84.6 | 74.4 |
| 4 | 100.0 | 95.7 | 108.5 | 93.2 | 75.2 | 73.5 | 35.0 |
| Mean | 100.0 | 88.6 | 120.5 | 73.0 | 101.8 | 85.1 | 47.5 |
| | | Percent Water Retention | | | | | |
| 1 | 100.0 | 96.5 | 96.5 | 96.5 | 100.0 | 104.7 | 8.2 |
| 2 | 100.0 | 94.3 | 96.6 | 100.0 | 95.4 | NA | 5.7 |
| 3 | 100.0 | 117.3 | 92.0 | 102.7 | 105.3 | 113.3 | 25.3 |
| 4 | 100.0 | 98.9 | 96.6 | 100.0 | 103.4 | 101.1 | 6.9 |
| Mean | 100.0 | 101.7 | 95.4 | 99.8 | 101.0 | 106.4 | 11.0 |
| | Seven Day Compressive Strength, psi | | | | | | |
| 1 | 100.0 | 94.0 | 90.8 | 106.1 | 72.2 | 104.9 | 33.9 |
| 2 | 100.0 | 97.7 | 85.3 | 130.6 | 75.9 | 97.7 | 54.7 |
| 3 | 100.0 | 92.2 | 87.7 | 124.4 | 66.4 | 106.3 | 58.0 |
| 4 | 100.0 | 95.8 | 84.2 | 119.8 | 70.2 | 96.3 | 49.6 |
| Mean | 100.0 | 94.9 | 87.0 | 120.2 | 71.2 | 101.3 | 49.0 |
| | Twenty Twenty Eight Day Compressive Strength | | | | | | |
| 1 | 100.0 | 90.8 | 88.3 | 102.0 | 70.9 | 101.6 | 31.1 |
| 2 | 100.0 | 99.1 | 89.9 | 136.1 | 82.9 | 115.9 | 53.1 |
| 3 | 100.0 | 59.0 | 80.7 | 111.8 | 68.4 | 91.1 | 52.8 |
| 4 | 100.0 | 95.7 | 87.1 | 117.4 | 73.3 | 101.8 | 44.1 |
| Mean | 100.0 | 86.1 | 86.5 | 116.8 | 73.9 | 102.6 | 42.9 |

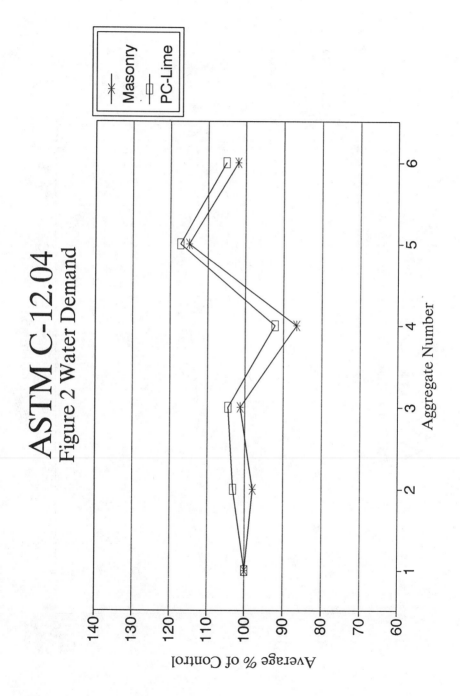

ASTM C-12.04
Figure 2 Water Demand

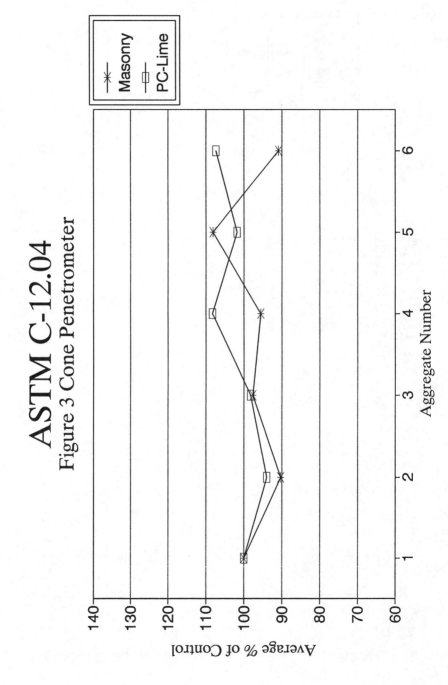

ASTM C-12.04
Figure 3 Cone Penetrometer

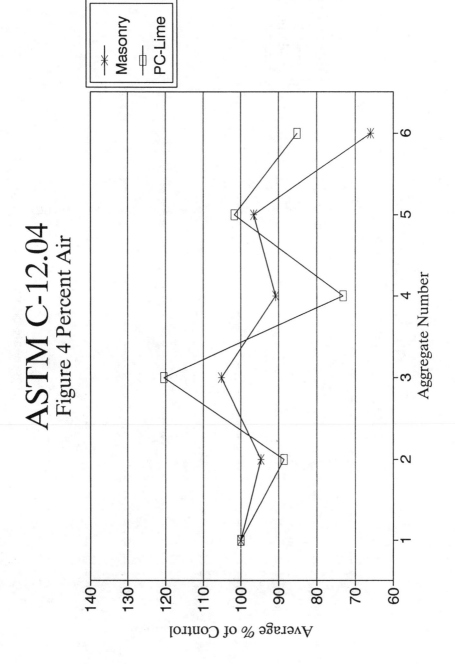

ASTM C-12.04
Figure 4 Percent Air

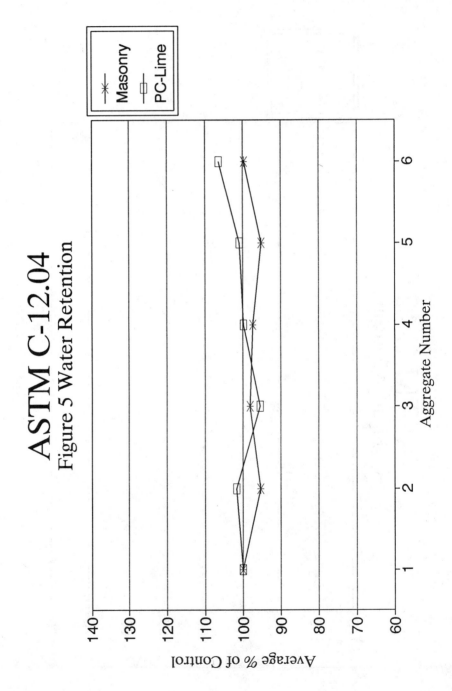

ASTM C-12.04
Figure 5 Water Retention

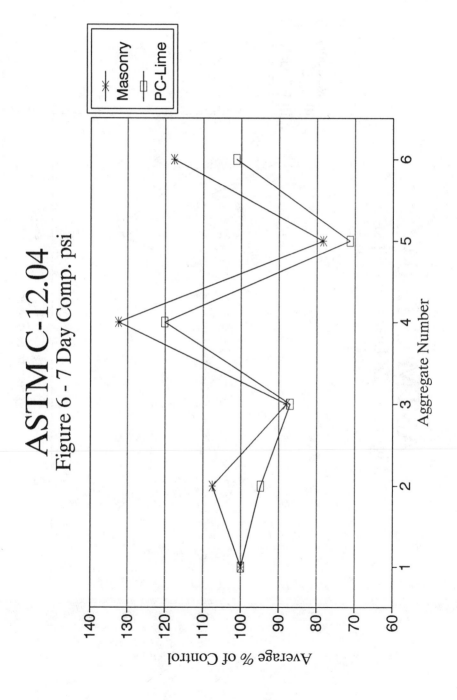

ASTM C-12.04
Figure 6 - 7 Day Comp. psi

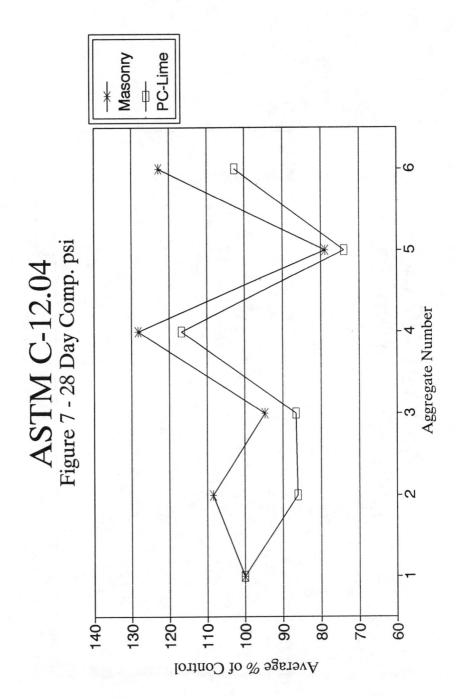

ASTM C-12.04
Figure 7 - 28 Day Comp. psi

Table 6 -  Comparision of Yield and Portland Cement
           Content of Masonry Cement and Portland
           Cement Mortars

| Test # | Ml H20 | % Air | Yield, Ml | Grams PC per Ml | Ml H20 | % Air | Yield, Ml | Grams PC per Ml |
|---|---|---|---|---|---|---|---|---|
| | | Masonry | Cement | | | Lime | | |
| 1 | 221 | 20.4 | 1142 | 0.184 | 345 | 3.4 | 1064 | 0.265 |
| 2 | 310 | 20.2 | 1251 | 0.168 | 333 | 5.1 | 1071 | 0.263 |
| 3 | 245 | 16.8 | 1122 | 0.187 | 330 | 3.9 | 1054 | 0.268 |
| 4 | 255 | 15.6 | 1118 | 0.188 | 300 | 11.7 | 1113 | 0.253 |
| 5 | 222 | 21.0 | 1152 | 0.182 | 350 | 3.7 | 1073 | 0.263 |
| 6 | 290 | 17.0 | 1179 | 0.178 | 333 | 4.9 | 1068 | 0.264 |
| 7 | 245 | 16.0 | 1111 | 0.189 | 360 | 2.1 | 1065 | 0.265 |
| 8 | 250 | 15.0 | 1104 | 0.190 | 306 | 11.2 | 1114 | 0.253 |
| 9 | 230 | 22.4 | 1183 | 0.177 | 355 | 5.0 | 1093 | 0.258 |
| 10 | 295 | 18.6 | 1208 | 0.174 | 351 | 5.0 | 1088 | 0.259 |
| 11 | 250 | 19.1 | 1160 | 0.181 | 350 | 5.0 | 1087 | 0.259 |
| 12 | 265 | 16.4 | 1140 | 0.184 | 312 | 12.7 | 1140 | 0.247 |
| 13 | 200 | 17.3 | 1074 | 0.196 | 335 | 2.0 | 1039 | 0.271 |
| 14 | 259 | 19.5 | 1177 | 0.178 | 300 | 3.6 | 1020 | 0.277 |
| 15 | 210 | 15.5 | 1063 | 0.198 | 306 | 2.7 | 1016 | 0.277 |
| 16 | 220 | 14.0 | 1056 | 0.199 | 266 | 10.9 | 1065 | 0.265 |
| 17 | 270 | 22.2 | 1232 | 0.171 | 388 | 4.2 | 1118 | 0.252 |
| 18 | 340 | 17.3 | 1243 | 0.169 | 365 | 4.1 | 1093 | 0.258 |
| 19 | 280 | 17.3 | 1171 | 0.179 | 375 | 5.0 | 1114 | 0.253 |
| 20 | 290 | 13.9 | 1136 | 0.185 | 402 | 8.8 | 1190 | 0.237 |
| 21 | 240 | 15.5 | 1098 | 0.191 | 340 | 3.3 | 1058 | 0.267 |
| 22 | 295 | 12.3 | 1121 | 0.187 | 375 | | | |
| 23 | 255 | 9.9 | 1047 | 0.201 | 338 | 3.3 | 1056 | 0.267 |
| 24 | 255 | 10.6 | 1055 | 0.199 | 320 | 8.6 | 1097 | 0.257 |
| Mean | 258.0 | 16.82 | 1139.3 | 0.185 | 339.0 | 5.66 | 1082.3 | 0.261 |
| STD | 33.1 | 3.23 | 57.9 | 0.009 | 30.7 | 3.18 | 37.9 | 0.009 |
| Max | 340.0 | 22.40 | 1250.9 | 0.201 | 402.0 | 12.70 | 1189.6 | 0.277 |
| Min | 200.0 | 9.90 | 1046.9 | 0.168 | 266.0 | 2.00 | 1016.4 | 0.237 |

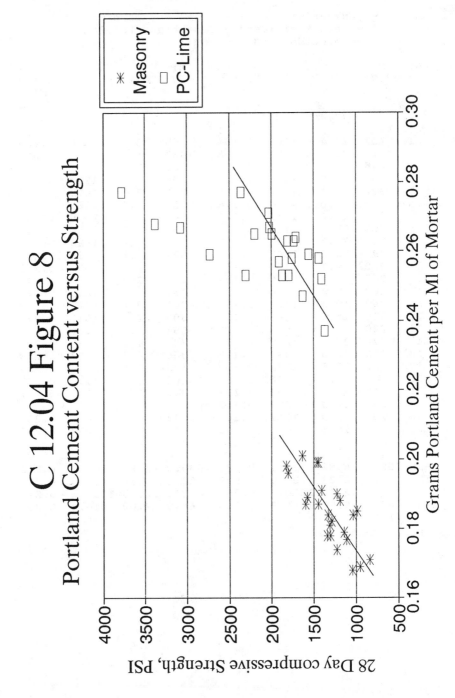

C 12.04 Figure 8
Portland Cement Content versus Strength

DISCUSSION

"Effects of Aggregate Gradation on Properties of Masonry Cement
and Portland Cement - Lime Masonry Mortars"
C. E. Buchanan, Jr., and B. M. Call

Question (Lawrence R. Roberts, W. R. Grace & Co.-Conn.):
The authors are to be commended for a very thorough analysis of
these round robin results. While agreeing that the results tend to
show that the C 144 limits may be more restrictive than absolutely
necessary on the coarse side, at least as measured by these tests,
the data for aggregate blend number 5 would indicate to me that the
limits on the finer side in C 144 are appropriate.

The data presented in Table 6 and in Figure 8 are very interesting,
and seem to indicate much greater cementing efficiency of the
portland cement when present in a masonry cement-based mortar as
compared to a portland-lime mortar. Since the treatment should at
least partially normalize the data with respect to the effect of air
content on strength, do the authors have any other explanation for
this improved efficiency?


Closure from the authors was not submitted by the
publication deadline.

DISCUSSION of

"Effects of Aggregate Gradation on Properties of Masonry Cement and Portland Cement-Lime Masonry Mortars"-- Charles E. Buchanan, Jr. and B. M. Call.

Discussion (Richard C. Meininger, National Aggregates Association): The National Aggregates Association (formerly the National Sand and Gravel Association) on two occasions -- 1954 and 1982 -- surveyed the aggregates industry to gather information on the gradation of sands supplied for Masonry Mortar around the country.

Table D1 summarizes the results of both the 1982 and the older 1954 survey and lists the present requirements of ASTM C 144-87. There are a large number of mortar sands in use which do not meet all of the grading limitations of C 144. Although C 144 does allow for the use of sands with other gradings if, when tested in the laboratory, they are found to comply with the aggregate ratio, water retention, and compressive strength requirements of the property specifications of ASTM Specification C 270 for Mortar for Unit Masonry, many producers are not in a position to have these tests done routinely.

It can be seen in Table D1 that, in general, the averages and standard deviations for percentages passing, fineness modulus, and individual percentages between two sieves from the 1954 and 1982 surveys are comparable. The most significant change is probably the fact that the Pass No. 30 percentage in the more recent survey averages about 72 percent, against 65 percent in the older survey. Similarly, the Pass No. 50 has increased from about 24 to 32 percent. The amount passing the No. 100 remains relatively low, on the average, at about 6 to 8 percent. Minus No. 200 data was not collected in 1954 and in the recent survey only 47 of the 62 reported that value. For this reason it is not included in the Table. All but one of the reported minus No. 200 values are less than 6 percent and the average of the reported values is only about 2 percent.

Also given in Table D1 is the "Mid 80 Percent Range" that would include 80 percent of the sands in each survey. That is, the limits shown include all of the sands, with respect to the indicated property, except the highest 10 percent and the lowest 10 percent.

The "Mid 80 Percent Range" also shows the tendency for sands in the recent survey to be a little higher in Pass No. 30 and Pass No. 50 percentages than those in the older survey. However, it is interesting to note that for both surveys, there are a significant number of sands which have amounts passing the No. 30 and No. 50 sieves which are significantly above the C 144 upper limits of 75 percent and 35 percent, respectively. In the 1982 survey, about one-third of the sands exceed these specification limits. Twenty percent of the sands in the older survey also exceed these limits.

In contrast, the amount passing the No. 100 sieve in most all cases, in both surveys, is below the 15 percent limit in C 144-76. It certainly appears that sand acceptable to the masons, can have a large amount of minus No. 30 and No. 50 fines, but that in most cases, a cut-off at the No. 100 size is necessary. Only about 10 percent of the sands in both surveys show Pass No. 100 percentages more than 15 percent.

Obviously because of the tendency of these masonry sands to be high on passing No. 30 and No. 50 amounts, while at the same time, to have only moderate amounts of material passing the No. 100 sieve, there is a trend toward violation of the limits of 50 percent between the No. 30 and No. 50 sieves and 25 percent between the No. 50 and No. 100 sieves. The "Mid 80 Percent Range" for these two sizes indicates that a limit as high as 60 percent for No. 30 to No. 50 and 45 percent for No. 50 to No. 100 would be necessary to include all but the highest 10 percent of the sands.

Copies of the complete data from these surveys is available from the National Aggregates Association, 900 Spring Street, Silver Spring, MD 20910. Many comments received with the survey indicate that local masons prefer the finer sands and that even though sand meeting the C 144 grading table is available in most areas the masons purchase finer sands. Respondents to the survey in many cases felt that the fine sands are a superior product in terms of the quality of the masonry construction produced. Several other members of ASTM Subcommittee C12.04 on aggregate for masonry mortar have also surveyed sand gradations used in their area and have found that fine gradations are used with no apparent problem with quality. When the C270 tests are run with typical gradations in the mid-80 percent range they show acceptable results. In my opinion, the gradation in ASTM C144 should be moved toward the finer sands which are preferred.

Of the six sand gradings used in the research reported in the paper, No. 3 is closest to being typical for some of the finer sands; however, it did not have enough passing the No. 50 size. Sands No. 5 and 6 have more passing No. 50, but the minus No. 100 in those sands is much too high compared to the survey results. Looking at the results of sand No. 3 expressed as a percent of No. 1 in Tables 3 and 5 with the masonry cement and portland cement-lime, respectively, it appears that the various properties are within plus or minus 15 percent of the standard in most all of the categories. The one exception was percent air for the portland-lime combination where the percent air was a little higher.

Table D1.  Summary of Masonry Mortar Sand Survey Results

| Item | Avg. | 1982 Survey (b) Std. Dev. | 1982 Survey (b) Mid 80%(a) Range | Avg. | 1954 Survey (c) Std. Dev. | 1954 Survey (c) Mid 80%(a) Range | ASTM C144 Nat. Sand (d) |
|------|------|------|------|------|------|------|------|
| Percent Passing |||||||| 
| Pass No. 4 | 99.9 | 0.4 | -- | 100.0 | 0.1 | -- | 100 |
| Pass No. 8 | 98.8 | 1.8 | -- | 98.4 | 2.1 | -- | 95-100 |
| Pass No. 16 | 90.6 | 9.3 | 76-100 | 90.5 | 7.2 | 80-98 | 70-100 |
| Pass No. 30 | 71.5 | 18.0 | 44-98 | 64.7 | 13.6 | 47-84 | 40-75 |
| Pass No. 50 | 32.0 | 15.8 | 16-64 | 23.8 | 13.4 | 8-43 | 10-35 |
| Pass No. 100 | 7.5 | 7.2 | 2-17 | 5.9 | 6.9 | 2-12 | 2-15 |
| | | | | | | | |
| Fineness Modules | 2.00 | 0.43 | 1.3-2.5 | 2.16 | 0.34 | 1.6-2.5 | -- |
| Size Fractions,% |||||||| 
| No. 4 - 8 | 1.1 | 1.6 | -- | 1.6 | 2.1 | -- | -- |
| No. 8 - 16 | 8.2 | 8.0 | 2-22 | 7.9 | 5.7 | 1-16 | 50 max. |
| No. 16 - 30 | 19.1 | 10.9 | 2-32 | 25.8 | 9.8 | 13-40 | 50 max. |
| No. 30 - 50 | 39.5 | 14.5 | 24-60 | 40.9 | 10.8 | 25-53 | 50 max. |
| No. 50 - 100 | 24.5 | 12.1 | 12-45 | 17.9 | 9.6 | 8-33 | 25 max. |

(a)  This range excludes the lowest and highest 10 percent of the sands for the particular property indicated.  That is, ten percent of the sands have values higher than the top of the range and another ten percent have values lower than the bottom of the range.
(b)  Includes 61 natural sands and one manufactured sand.
(c)  Includes 41 natural sands and 4 manufactured sands.
(d)  The grading in C 144 for manufactured sand is the same as natural sand, except 20-40% passing No. 50, 10-25% passing No. 100, and a limit on percent passing the No. 200 of 10% maximum.


Closure from the authors was not submitted by the publication deadline.

Jacob W. Ribar and Val S. Dubovoy

MASONRY CEMENTS - A LABORATORY INVESTIGATION

REFERENCE: Ribar, J. W. and Dubovoy, V. S., "Masonry Cements--
A Laboratory Investigation," Masonry: Components to Assemblages,
ASTM STP 1063, John H. Matthys, Editor, American Society for
Testing and Materials, Philadelphia, 1990.

ABSTRACT: Effects of masonry cement mortars and clay brick units
on flexural bond (ASTM C 1072) and on water penetration of masonry
wall assemblies (ASTM C 514) were investigated. The test variables
included 20 masonry cements and 11 types of brick, each representing materials produced in different areas of the United States.

Regardless of the composition of individual masonry cements,
flexural bond in wall assemblies was generally in excess of
100 psi. Also, single-wythe wall assemblies exhibited
satisfactory watertightness.

Flexural bond tests of prisms built with the above materials
combinations showed that brick properties such as Initial Rate of
Absorption (IRA) and surface texture together with mortar
properties are primary factors in the development of bond strength.

KEY WORDS: air content, bond (mortar to units), bond wrench,
bricks, brick surface texture, compressive strength, initial rate
of absorption, masonry cements, masonry mortars, particle size
distribution, water penetration, water retention.

INTRODUCTION

During the past five decades the masonry community has witnessed
an increasing volume of research with growing emphasis on properties
such as mortar air content, water retention, water content, sand
gradation, admixtures, and curing that influence bond to masonry units.

Mr. Ribar and Mr. Dubovoy are Principal Masonry Evaluation Engineer
and Senior Masonry Engineer, respectively, Construction Technology
Laboratories, Inc., Skokie, Illinois 60077.

Most researchers also agree that properties of brick, such as surface texture and IRA, have a significant effect on bond and water penetration of masonry walls. Yorkdale[1] examined the effect of IRA clay masonry units on flexural bond strength. He concluded that brick IRA does affect the bonding ability of masonry mortar, and noted that the mechanism of adhesion between clay masonry units and mortar is not well understood and should be studied in more depth.

Construction practices and workmanship have consistently been cited as crucial to the success or failure of masonry performance. Fishburn, Walstein and Parsons[2] found that workmanship affected leakage of masonry walls more than any other factor. Youl and Coates[3] concluded that the workmanship factor is more critical in water penetration tests than it is in load tests of masonry panels. Present day investigators often claim that 75 to 80% of masonry failures are caused by design or construction/workmanship errors and only 10 to 15% are materials related.

The research presented in this report provides performance data for a variety of masonry cement mortars to determine their capability to produce bond to masonry units and to provide watertight walls.

SCOPE

The objective of this research was to generate laboratory data on bond of mortar to clay units and water penetration of assemblies using a variety of masonry cement mortars.

Past research on masonry cements has been questioned because of the selection of a specific masonry cement or a small group of masonry cements for testing. The concern is that because of the variety of masonry cements available, test results can be biased by the non-random selection of one particular type of cement. To respond to this issue masonry cement producers were surveyed on a confidential basis to obtain a data base for selection of masonry cements for testing.

Twenty cement companies responded providing data on formulations of masonry cements. Based on these responses, ten Type S cements and ten Type N masonry cements were selected for testing. Selection was based on formulation and geographical availability, and represents masonry cements with a range of compositions. Eleven different bricks representing 11 U.S. manufacturers in diverse areas were combined with these 20 masonry cements. Tests concentrated on two properties, bond to masonry units and water penetration of masonry assemblies.

Primary variables in the study of bond strength were masonry cement formulation, test methods, type of specimens, and test age. Type S and N masonry mortars were prepared in accordance with the proportion requirements of ASTM C 270-88a "Standard Specification for Mortars for Unit Masonry." Bond strength test specimens were fabricated with these mortars and tested using two methods. Brick prisms and half brick couplets were fabricated specifically for flexural and tensile bond strength testing by bond wrench and direct tension techniques. Specimens were also extracted from water penetration assemblies for flexural bond tests.

Initial flow of mortar was established by the bricklayers (field flow). Water was measured by weight and maintained for subsequent batches. Mortar properties are shown in Table A.

RESULTS

Complete results of this study are summarized in Reference 4. The following main findings and conclusions are based on data from the study.

1.  Bond wrench test results indicated that flexural bond strength was a function of several physical parameters of masonry cement or mortar, none of which were applicable to all masonry cements. These included cement fineness, mortar water and air content, and mortar water retention. Surface texture and initial rate of absorption of brick were also found to be major factors affecting bond.

2.  Flexural bond strength values of stack-bond prisms constructed with a solid clay unit ranged from 65 psi to 211 psi (Table 1). Results for 15 of 20 mortars showed flexural bond strengths in excess of 100 psi, a value generally considered to represent a good flexural bond strength. There was essentially no difference between flexural bond strengths at 14 and 28 days.

3.  For Type S masonry mortars, flexural bond strengths of specimens removed from wall assemblies in many cases were significantly greater than bond strengths of companion stack-bond prisms. Flexural bond strengths of specimens removed from the midspan of wall assemblies were usually higher than stack bond prisms and than specimens removed from the edge of wall assembles (Table 1). This effect was not observed for type N masonry mortars.

4.  In most cases, lower bond strengths were obtained for specimens tested in direct tension than for specimens tested using the bond wrench (Table 1).

5.  Flexural bond strength varied significantly with a singular mortar. Mortar 1 in Table 2 shows a range of flexural bond values from 38 to 191 and mortar 54 ranges from 22 to 185. Initial rate of absorption and surface texture of the brick had significant effects on bond (Table 3).

6.  For most specimens, the average water leakage rate during the entire 72-hour water penetration test was lower than that measured during the first 4 hours. This was attributed to continued hydration or carbonation of cementitious components or particle obstruction in passages (Table 4).

7.  Relatively low correlation was found between flexural bond strengths and water leakage rates of wall assemblies (Fig. 1).

8.  During water penetration tests, all walls exhibited a relatively well-defined moisture migration pattern. Generally, dampness was first observed within 8-in. from each vertical edge of the wall. Damp spots were generally observed where initial bond was

disrupted during fabrication of the wall assembly for necessary alignment and adjustment of units. This condition reported in previous research[5] is viewed as workmanship related. Other damp spots were about equally divided between the lower portion and center of the wall.

9. The percentage of wetted area observed in water penetration tests did not necessarily increase with an increase in water leakage rate. This indicates that leaks tended to converge from specific migration paths.

## DISCUSSION

Flexural bond strengths of specimens removed from wall assemblies were greater than bond strength from companion stack-bond prisms. Masonry cements mortars showed satisfactory performance when wall specimens were tested for flexural bond and water penetration and suggest that, with good workmanship, values higher than those achieved from laboratory specimens are possible in construction.

Tests demonstrate that flexural bond of clay masonry is the combined function of both mortar and unit. Bond is effected equally by physical properties of masonry cement, mortar, and initial rate of absorption and surface profile of the clay units.

High IRA can significantly reduce the positive effect of a rough bedding surface texture thus reinforcing need to dampen brick to reduce its dewatering effect on mortar.

## SUMMARY

This investigation has included tests of 20 masonry cements in combination with 11 different clay brick, and has generated 2844 bond strength data points from 545 clay masonry prisms.

Results of this research demonstrate that regardless of type or composition, masonry cement mortars tested are capable of producing good bond to masonry units both in test prisms and masonry wall assemblies. Masonry cements also produce reasonably watertight single wythe masonry walls when proper workmanship and placing techniques are used. This research has also demonstrated that properties of masonry units affect bond at least to the same extent as properties of masonry cements.

## ACKNOWLEDGMENTS

The work described in this paper was sponsored by the Portland Cement Association (PCA) under CTL Project No. H10002. The opinions and findings expressed or implied in this paper are those of the authors. They are not necessarily those of the Portland Cement Association.

REFERENCES

[1] Yorkdale, A. H., "Initial Rate of Absorption and Mortar Bond." ASTM Special Technical Publication 778, pp. 91-99, Dec. 1980.

[2] Fishburn, L. C. Walstein, D., and Parsons, D. E., "Water Permeability of Masonry Walls," National Bureau Standards, BMS Rep. 7, 1938.

[3] Youl, V. A., and Coates, E. R., "Some Studies in Brick-Mortar Bond," Part 1, Austrialian Building Research Congress, 1961.

[4] Dubovoy, V. S. and Ribar, J. W., "Masonry Cements - A Laboratory Investigation," Portland Cement Association Research and Development PCA R&D Serial No. 1846.

[5] Ribar, J. W., "Water Permenance of Masonry:  A Laboratory Study," ASTM Special Technical Publication 778, pp. 200-220, Dec. 1980.

[6] Kuenning, W. H., "Improved Method of Testing Tensile Bond Strength of Masonry Mortars," Research Bulletin 195, Research and Development Laboratories, Portland Cement Association, 1966.

TABLE A – PHYSICAL PROPERTIES OF FIELD FLOW MORTAR PROPORTIONED BY ASTM C270

| Cement Designation | Cement Type | Cube Compressive Strength psi | Mortar Air Content % | Cone Penetration, mm | Water-Cement Ratio |
|---|---|---|---|---|---|
| 1  | S | 2990 | 23.9 | 52 | 0.47 |
| 2  | S | 2380 | 23.8 | 56 | 0.46 |
| 6  | S | 2730 | 18.3 | 58 | 0.48 |
| 30 | S | 2340 | 23.9 | 51 | 0.47 |
| 41 | S | 3410 | 16.1 | 53 | 0.54 |
| 43 | S | 3780 | 9.0  | 54 | 0.47 |
| 49 | S | 2740 | 21.2 | 57 | 0.55 |
| 50 | S | 2200 | 16.5 | 68 | 0.50 |
| 54 | S | 4150 | 16.0 | 66 | 0.51 |
| 56 | S | 3280 | 16.0 | 57 | 0.52 |
| 4  | N | 1080 | 19.8 | 65 | 0.55 |
| 7  | N | 1740 | 19.5 | 58 | 0.54 |
| 11 | N | 750  | 22.6 | 65 | 0.52 |
| 12 | N | 2830 | 10.2 | 57 | 0.58 |
| 13 | N | 1610 | 20.2 | 72 | 0.59 |
| 17 | N | 1180 | 20.5 | 57 | 0.55 |
| 25 | N | 1370 | 24.8 | 61 | 0.50 |
| 26 | N | 1390 | 20.8 | 55 | 0.51 |
| 34 | N | 1780 | 26.4 | 62 | 0.45 |
| 47 | N | 1300 |      | 66 |      |

Table 1 - Results of Bond Tests

Brick K

| Cement Designation | Cement Type | 28-day Cube Comp. Strength psi | Flexural Bond Strength of Masonry Prisms,(1) psi | | Tensile Bond Strength,(2) psi | Flexural Bond Strength(3) of Wall Masonry at 28 days, psi | |
|---|---|---|---|---|---|---|---|
| | | | 14 days | 28 days | | Wall Edge | Specimen Midspan |
| 1 | S | 2990 | 99 | 98 | 75 | 105 | 125 |
| 2 | S | 2380 | 77 | 65 | 34 | 112 | 132 |
| 6 | S | 2730 | 107 | 116 | 47 | 187 | 203 |
| 30 | S | 2340 | 123 | 115 | 74 | 124 | 163 |
| 41 | S | 3410 | 108 | 148 | 116 | 138 | 153 |
| 43 | S | 3780 | 156 | 86 | 286 | 200 | N/A |
| 49 | S | 2740 | 77 | 73 | 90 | 176 | 205 |
| 50 | S | 2200 | 79 | 121 | 66 | 139 | 153 |
| 54 | S | 4150 | 105 | 133 | 115 | 137 | 165 |
| 56 | S | 3280 | 95 | | 123 | 162 | 200 |
| 4 | N | 1080 | 83 | 118 | 75 | 136 | 142 |
| 7 | N | 1740 | 95 | 126 | 55 | 85 | 101 |
| 11 | N | 750 | 95 | 127 | 47 | 74 | 79 |
| 12 | N | 2830 | 117 | 211 | 112 | 98 | 113 |
| 13 | N | 1610 | 171 | 175 | 118 | 142 | 158 |
| 17 | N | 1180 | 96 | 126 | 51 | 138 | 167 |
| 25 | N | 1370 | 164 | 182 | 180 | 159 | 149 |
| 26 | N | 1390 | 108 | 100 | 81 | 107 | 128 |
| 34 | N | 1780 | 146 | 165 | 115 | 102 | 152 |
| 47 | N | 1300 | 109 | 67 | 85 | 76 | 88 |

(1) Determined in accordance with ASTM C 1072-86, "Standard Method for Measurement of Masonry Flexural Bond Strength."

(2) Determined using half-brick couplets at 28 days(6).

(3) Specimens cut from wall and tested in accordance with ASTM C 1072-86 upon completion of water penetration test and drying.

*To convert from psi to MPa multiply by 0.00698.

Table 2 – Flexural Bond Strength of Prisms Prepared with Various Types of Brick

| Brick Designation | IRA grams | Type S Masonry Cement | | | | | | | | | | Type N Masonry Cement | | | | | | | | | |
|---|---|---|---|---|---|---|---|---|---|---|---|---|---|---|---|---|---|---|---|---|---|
| | | Bond strength, psi | | | | | | | | | | Bond strength, psi | | | | | | | | | |
| | | 1 | 2 | 6 | 30 | 41 | 43 | 49 | 50 | 54 | 56 | 4 | 7 | 11 | 12 | 13 | 17 | 25 | 26 | 34 | 47 |
| A | 17 | 88 | 67 | 86 | 92 | 33 | 53 | 81 | 108 | 57 | 62 | 18 | 35 | 47 | 58 | 76 | 50 | 140 | 40 | 64 | 47 |
| B | 20 | 105 | 50 | 88 | 80 | 65 | 52 | 95 | 66 | 71 | 80 | 51 | 83 | 95 | 145 | 82 | 97 | 126 | 31 | 76 | 72 |
| C | 14 | 83 | 65 | 50 | 97 | 121 | 140 | 138 | 145 | 185 | 128 | 82 | 127 | 61 | 184 | 115 | 104 | 132 | 67 | 101 | 116 |
| D | 6 | 103 | 60 | 68 | 115 | 149 | 107 | 143 | 114 | 105 | 118 | 72 | 132 | 72 | 145 | 111 | 102 | 102 | 50 | 72 | 112 |
| E | 12 | 191 | 109 | 130 | 115 | 86 | 61 | 134 | 108 | 123 | 144 | 80 | 62 | 65 | 126 | 141 | 94 | 142 | 70 | 58 | 114 |
| F | 37 | 38 | 34 | 75 | 32 | 22 | 35 | 26 | — | 22 | 49 | 27 | 63 | 14 | 104 | 102 | 47 | 61 | 46 | 32 | 20 |
| G | 2 | 130 | 57 | 57 | 70 | 75 | 54 | 81 | 65 | 68 | 42 | 84 | 63 | 53 | 95 | 69 | 77 | 95 | 29 | 45 | 120 |
| H | 17 | 148 | 66 | 44 | 107 | 168 | 162 | 173 | 116 | 167 | 147 | 99 | 55 | 74 | 76 | 96 | 96 | 148 | 59 | 129 | 141 |
| I | 3 | — | 97 | 67 | 61 | 91 | 117 | 123 | 88 | 104 | 101 | 60 | 77 | 31 | 140 | 105 | 135 | 76 | 72 | 72 | 95 |
| J | 4 | — | 71 | 83 | 103 | 80 | 92 | 135 | 64 | 73 | 98 | 67 | 82 | 67 | 130 | 96 | 76 | 109 | 65 | 88 | 120 |

Each data point is the average of 12 joints tested with the bond wrench.

Tests conducted at 28 days of age.

To convert from psi to MPa multiply by 0.00689.

Table 3 - Effect of IRA and Contour Ratio on Bond Strength

| Brick | IRA, grams | Contour Ratio Rp | Average Flexural Bond Strength[1] psi |
|-------|------------|------------------|----------------------------------------|
| A | 17 | 18 | 67 |
| B | 20 | 12 | 81 |
| C | 14 | 53 | 112 |
| D | 6 | 41 | 103 |
| E | 12 | 46 | 106 |
| F | 37 | 34 | 44 |
| G | 2 | 16 | 71 |
| H | 17 | 81 | 114 |
| I | 3 | 10 | 92 |
| J | 4 | 11 | 89 |

(1) Average of results from 20 cements at 28 days.

$R_p$ represents the ratio of actual surface profile length as traveled by the apparatus stylus to a nominal length, expressed to the nearest thousandth.

To convert from psi to MPa multiply by 0.00689.

Table 4 - Water Penetration of Masonry Assemblies

Brick K

| Cement Designation | Cement Type | Penetration rate in first 4 hr, ml/ft² hr | | Average 72-hr Penetration rate [1] ml/ft² hr | Percent wetted area after 4 hrs [1] |
|---|---|---|---|---|---|
| | | 14 days | 28 days | | |
| 1 | S | 28.7 | 52.2 | 45.0 | 5.9 |
| 2 | S | 37.4 | 69.5 | 41.3 | 21.7 |
| 6 | S | 30.3 | 0 | 0 | 28.9 |
| 30 | S | 4.2 | 3.2 | 1.4 | 12.1 |
| 41 | S | 0 | 14.2 | 7.1 | 20.9 |
| 43 | S | 12.3 | 15.1 | 8.7 | 14.4 |
| 49 | S | 0 | 27.2 | 20.2 | 8.1 |
| 50 | S | 0 | 10.4 | 22.9 | 5.1 |
| 54 | S | 0 | 34.1 | 31.4 | 10.1 |
| 56 | S | 39.7 | 51.0 | 45.9 | 25.0 |
| 4 | M | 28.3 | 32.4 | 14.5 | 3.1 |
| 7 | M | 54.9 | 79.6 | 21.2 | 2.7 |
| 11 | M | 2.0 | 22.7 | 52.5 | 4.4 |
| 12 | M | 23.6 | 30.5 | 27.1 | 3.1 |
| 13 | M | 20.8 | 2.8 | 2.0 | 5.8 |
| 17 | M | 2.0 | 18.1 | 15.1 | 100.0 |
| 25 | M | 0 | 35.7 | 28.4 | 100.0 |
| 26 | M | 0 | 5.9 | 7.3 | 100.0 |
| 34 | M | 0 | 0 | 1.9 | 10.0 |
| 47 | M | 0 | 0 | 30.9 | 7.8 |

To convert from ml/ft$^2$ hr to ml/cm$^2$ hr multiply by 0.00108.

(1)  Determined at 28 days.

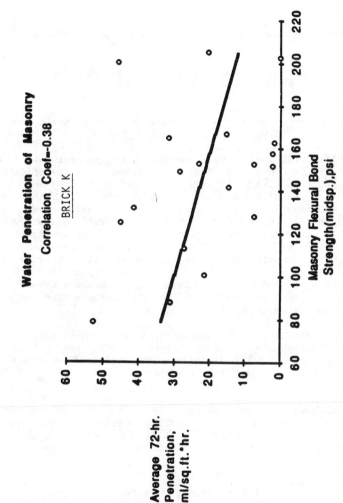

Fig. 1 - Water Penetration as a Function of Flexural Bond Strength of Wall Masonry. (To convert from psi to MPa multiply by 0.00698. To convert from ml/ft² hr to ml/cm² hr multiply by 0.00108.)

DISCUSSION


"Masonry Cements - A Laboratory Investigation -    J. W. Ribar and V. S.
                                                    Dubovoy

Comment (R. E. Klingner, University of Texas at Austin):   In my opinion,
this paper should deliver more specific information.


1)      The following conclusions discussed in the Abstract are not
        sufficiently substantiated in the paper itself:

        a)    "flexural bond in wall assemblies was generally in excess
              of  100  psi."    This  conclusion  is  insufficiently
              substantiated by Table 1, which refers to only one of the
              11 types of brick mentioned in the paper.  Table 2, which
              contains data for different types of brick, shows many
              instances of average strengths considerably below 100 psi.
              For example, Brick A has an average (of the averages) bond
              strength, across all Type S masonry cements, of 72.7 psi,
              with a coefficient of variation of 31%.   Corresponding
              values for Brick F are 37.0 psi and 45%; and for Brick G,
              69.9 psi and 34%.  On this basis, the numbers presented in
              the  paper  do  not  seem  to  substantiate  the  authors'
              conclusion.

        b)    "single-wythe  wall  assemblies  exhibited  satisfactory
              watertightness."  Since neither the water penetration test
              nor the criteria used are identified, this statement by
              itself is not sufficiently convincing.


2)      Flexural bond is usually increased by moist-curing prisms, rather
        than air-curing them.  How were the prisms in this study cured?
        If they were moist-cured, that fact is significant, and should be
        specifically pointed out by the authors.

Answer (Jacob W. Ribar)
We are sorry that you find the paper lacking in specifics.  As an
afterthought we probably would present the data in a different
format.  However, this paper is itself an abbreviated version of
Reference 4.  In answer to the question in paragraph (2) the prisms
were enclosed in plastic wrap for 7 days and air cured for 21 days.

DISCUSSION


"Masonry Cements - A Laboratory Investigation" - J. W.
Ribar and V. S. Dubovoy

Question  (J.  H.  Matthys,  University  of  Texas  at
Arlington):

In  the  abstract  it  is  stated  that  "regardless  of
composition  of  individual  masonry  cements,  flexural
bond  in  wall  assemblages  was  generally  in  excess  of  100
psi."  By  wall  assemblies,  do  you  mean  prisms  cut  from
the  water  permeance  specimens  or  prisms  built  in  the
laboratory?  In  Table  1,  which  applies  to  only  one  of
the  10  bricks  evaluated,  for  the  flexural  bond  strength
of  masonry  prisms  at  28  day  irrespective  of  mortar
type,  25%  of  the  test  gave  bond  strength  less  than  100
psi.  The  same  information  in  Table  2  irrespective  of
masonry  cement  type  indicates  the  following  percentage
of  bond  test  values  less  than  100  psi:  Brick  A  -  90%
of  tests;  Brick  B  -  85%  of  tests;  Brick  C  -  35%  of
tests;  Brick  D  -  30%  of  tests;  Brick  E  -  35%  of  tests;
Brick  F  -  89%  of  tests;  Brick  G  -  90%  of  tests;  Brick
H  -  45%  of  tests;  Brick  I  -  63%  of  tests;  Brick  J  -  74%
of  tests;  and  Brick  K  -  25%  of  tests.  Also  in  Table  3
60%  of  the  brick  tested  gave  average  flexural  bond
strengths  less  than  100  psi.  Obviously  this  data  shows
a  significant  portion  if  not  majority  of  the  28  day
prism  tests  to  give  bond  less  than  100  psi.  Your
statement  seems  to  apply  only  to  prisms  cut  from  water
permeance  walls  as  given  in  Table  1  for  Brick  K  only.
If  so,  does  your  data  for  cut  specimens  from  walls  for
the  other  nine  bricks  evaluated,  which  is  not  given  in
paper,  show  the  same  results?

In  the  abstract  it  is  stated  that  "single-wythe  wall
assemblies  exhibited  satisfactory  watertightness."  The
authors  do  not  state  whether  the  tests  were  conducted
according  to  ASTM  E514-86  or  ASTM  E514-74.  If
conducted  according  to  ASTM  E514-86,  what  is  your
reference  point  to  indicate  from  your  data  the
"satisfactory  statement"  since  E514-86  does  not  give
any  such  reference  point?

Were  the  IRA  of  the  bricks  listed  in  Table  2  based  on
the  "As  Laid"  condition  or  the  "oven  dried"  condition?
What  is  the  IRA  of  Brick  "K"  which  was  the  brick
associated  with  all  the  data  in  Table  1,  Table  4,  and
Figure  1?  What  is  the  contour  ratio  $R_p$  of  Brick  K?

For  the  flexural  bond  strength  of  wall  masonry  at  28
day  in  Table  1  I  assume  this  data  is  from  specimens  cut
from  the  water  permeance  wall  specimens.  If  so,  how
can  this  data  be  28  day  if  the  walls  were  subjected  to

water permeance testing at 28 days for a 72 hour
period?   Were the prisms immediately cut from the 31+
day wet walls and immediately tested or allowed to dry
first?   Were the bond specimens cut from the top or
bottom of the wall?   What joints if any were tooled?
If tooled, what was the position of tooled joints in
bond wrench test?   Was whitewash applied to the water
permeance specimens?   If so for bond specimens cut from
wall, what was the position of whitewashed face in bond
wrench test?

When you say solid clay unit, do you mean a unit with
no holes whatsoever or a unit cored $\leq$ 25% of the gross
area?   Were all of the 10 brick types in your study
cored?

The paper states "Type S and Type N masonry mortars
were prepared in accordance with the proportion
requirements of ASTM C270-88a.   Bond strength test
specimens were fabricated with these mortars.   Initial
flow of mortar was established by bricklayers (field
flow).   Mortar properties are shown in Table 4."   Thus
I assume Table A represents physical properties of
field flow mortars proportioned by ASTM C270.   For
these field flow mortars, how were they cured--in
laboratory air or moist room?   Typical masonry cement
mortar test data that I have seen show ASTM C270
laboratory mortar (Flow $\approx$ 110, moist cured) 28 day
compressive strengths of 1900 to 5100 psi for Type S
masonry cement mortars and 1200-1450 psi for Type N
masonry cement mortars.   On the other hand for field
mortars (Flow $\approx$ 135, lab air cured), 28 day compressive
strengths of 1400-3000 for Type S masonry cement
mortars and 600-1200 for Type N masonry cement mortar.
Was the same mortar sand used for all prisms and wall
assemblages?   What type of sand was used?

What was the construction and curing conditions for the
flexural bond prisms?   Stack bonded?   Prisms wrapped?
Prisms wetted?   Prisms cured in lab air?

In your results Item 2 you indicate three
characteristics:   (a) bond strength from 65 to 211 psi,
(b) 75% of mortar have bond strengths greater than 100
psi, and (c) no essential difference between bond
strength at 14 days and 28 days.   Although these
characteristics may be appropriate for the masonry
cements with Type K brick, they do not seem to be
necessarily appropriate statements as a whole.   For
example the range of bond strength for the Brick Type A
is from 18 psi to 140 psi while for the Brick Type F
from 14 to 104 psi.   The percentage of bond test values
100 psi or greater is only 10% for Brick A, 15% for
Brick B, 65% for Brick C, 70% for Brick D, 65% for
Brick E, 11% for Brick F, 10% for Brick G, 55% for

Brick H, 37% for Brick I, and 26% for Brick J. It appears that the significance of difference between 14 and 28 day strengths for Brick K is a function of the mortar. The percentage difference for Brick K ranges from 1% to 80%. Thirty-five percent of the Brick K tests have a difference between 14 and 28 day tests strengths of 30% or higher. What does this data for the other nine bricks show?

In your results Item 6 you point out as a whole that the average water leakage rate for the 72 hour test was lower than that measured during the first four hours. On the other hand Table 4 for Brick K indicates eight of the 10 Type S masonry cement mortars and seven of the 10 Type N masonry cement mortars give an increase in water penetration rate in the first four hours for the 28 day test as compared to the 14 day test. Why? For Table 4 how many walls per mortar were tested? Were the same walls tested at 28 days that were tested at 14 days?

In Figure 1 you plot average 72 hour water penetration versus bond strength of prisms cut form wall and generate a correlation coefficient of 0.38. Is it not true that this figure only plots Brick K data; i.e., the other nine sets of brick data are not plotted? Does the other brick data indicate the same results?

In discussion it is stated that the flexural bond strengths of specimens removed from wall assemblages were greater than bond strength on companion stack bonded prisms. In your Table 1 for Type N masonry cements, 60% of data indicate strengths of specimens from walls less than stack bonded prism? What do you attribute the difference in behavior between Type N and Type S masonry cement?

Answer (Jacob W. Ribar)
    Dr. Matthys As stated in the first sentence of "Results" on page 3 of the paper, "Complete results are summarized in Reference 4." However in answer to your question only Phases I and II of the research delt with water penetration and hence only Brick K wall prisms were tested.

Your paragraph 2. Reference 4 states that E 514-74 was employed in this research.

Your paragraph 3. All references in our paper to evaluation of materials are based on the results of testing in accordance with ASTM standards. In this case ASTM C 67 "Standard Methods of Sampling and Testing Brick and Structural Clay Tile" Section 9, "Initial Rate of Absorption (Suction)" procedures were followed to establish data. See response to Mr. Walker for remaining questions in this paragraph.

<u>Answer</u> (Jacob W. Ribar) (continued)
Your paragraph 4. Dr. Matthys you are correct in your statement that our prisms were tested later than 28 days. We corrected this error in the footnote to Table I.

Specimens were removed from the upper or top portions of the wall. The concaved tooled joints on the face of the wall were tested in tension; therefore, the whitewashed face of the dried specimen was in compression.

Your paragraph 5. In Reference 4 we state that we used "solid (uncored) brick" or as you so aptly stated, "no holes whatsoever." You will also find that the total research effort employed eleven (11) brick, one (1) in Phases 1 and II and ten (10) brick in Phase III.

Your paragraph 6. No assumptions are necessary. Table A is titled "Physical Properties of Field Flow Mortar Proportioned by ASTM C 270." Mortar cubes were cured in a moist cabinet. The sand from a single shipment was used for all phases of the research. I apologize for not knowing how to answer your question regarding the "type" of sand since ASTM C 144 "Standard Specification for Aggregate for Masonry Mortar" does not list aggregate by types. If you are referring to natural vs manufactured sand then our response is "natural".

Your paragraph 7, 8, 9, 10 and 11 are answered in "Reference 4".

DISCUSSION:

"Masonry Cements--A Laboratory Investigation: - Jacob W. Ribar and Val S. Dubovoy

Question (Dan Walker, CHEMSTAR LIME Company):
Your paper does not give details concerning the materials used in the study or the conditions of test. Could you give a response to the following questions? The 20 masonry cements used (of nearly 100 produced in the country) were picked from various geographical areas, could you state the particular States each were manufactured? Was the same sand used in all the tests, and what was the source? Were the masonry units dampened before used in the prisms? You mention several physical parameters, but do not give the data. For example, what was the cement fineness, mortar water, air content, and water retention of each of the masonry cements or mortars?

In the DISCUSSION of your paper you..."suggest that, with good workmanship, values higher than those achieved from laboratory specimens are possible in construction." How can you be sure of this, when all your data was generated from laboratory conditions and workmanship?

Your Tensile Bond and Water Penetration tests were made on "brick K", but no IRA or Contour data is given for this brick. Would you explain this? Also, were the K Brick moistened before made into prisms? What ASTM method was used for the water penetration tests? Was the mortar for the water penetration tests made according to C 270 procedure, and if so, what type of mixer was used?

Answer (Jacob W. Ribar)
Mr. Walker all the details are not contained in the paper because of ASTM restrictions on the size of the document. It should be noted that the first paragraph in "Results" does state "complete results are summarized in Reference 4."

Your paragraph 2. Our research showed that prisms removed from wall assemblies yielded higher bond values than laboratory prepared prisms. With both assemblies constructed with the same materials and workmanship the conclusion that higher bond values can be achieved in-the-wall as opposed to field constructed prisms is valid.

Your paragraph 3. Your questions are again answered in Reference 4. The IRA of Brick K is shown on Table IV of that document as 9 grams. No contour data were

<u>Answer</u> (Jacob W. Ribar) (continued)
produced for Phases I and II where Brick K was used.
Phase III studied the effects of various masonry units
on flexural bond.  Brick K was not "moistened."

Thank you for your interest and comments.

DISCUSSION


"Masonry Cements - A Laboratory Investigation" - Jacob W.
Ribar and Val S. Dubovoy


James L. Noland, Atkinson-Noland & Associates, Inc.


The reviewers report on a well-conceived and well-
executed experimental program to provide data on the bond
of masonry cement mortars to clay masonry units and on
the water penetration of brick masonry panels built with
masonry cement.  The observations in this discussion
pertain to the flexural bond issue only.

Flexural bond strength is an important attribute which is
related to performance limit states of masonry, e.g.,
moisture penetration through cracked joints and
structural behavior, i.e., response to loads in a cracked
or uncracked state.  As masonry design practice becomes
based upon limit state and probabilistic concepts in the
future, properties such as flexural bond will have to be
expressed in terms of expected values with their
variation stated.  Excessive variation will lead to lower
"phi" factors and penalize the design.  Even for today's
working stress design approach, excessive variation of
properties requires setting a lower minimum strength
limit than may be desirable.

The results reported in Table 1 indicate that the
coefficient of variation of 28-day cube strength is 22%
for Type S masonry cement mortar and 37% for type N
mortar.  The coefficient of variation of 28-day flexural
bond strength of the masonry prisms tested built with
Type S masonry cement mortar is 36% and for those built
with Type N mortar is 31%.

The results in Table 2 are themselves averages of 12
joints.  The data was not presented (due to space
limitations) to enable the variation to be evaluated.
However, the coefficients of variation of these average
values for the flexural bond strengths of prisms for
bricks A, B, C, D, and E is as follows:

| Brick | Type S Masonry Cement | Type N Masonry Cement |
|-------|-----------------------|-----------------------|
| A | 31% | 58% |
| B | 23% | 38% |
| C | 35% | 32% |
| D | 26% | 30% |
| E | 29% | 32% |

Based upon this data, it appears that, in general there is greater variability associated with Type N masonry cement mortar.

The reviewer's opinion is that the variations in properties revealed by this data is excessive especially considering that the masonry cement mortar and other specimens were prepared in a laboratory. There is evidence that flexural bond properties and cube strength of the specimens made from portland cement-lime mortars have coefficients of variation on the order of 20%.

Attainable target values of variation should be set perhaps in the 10%-15% range measures taken to obtain them. This would, of course, require a significant effort and degree of cooperation by producers.

Answer (Jacob W. Ribar)
Mr. Noland Thank you for your gracious compliment. Even though our research included 11 brick and 20 cements, as is always the case with research, it never seems to be enough. It would have been interesting to discover how the cement-lime mortars would have performed with the same materials used in our study.

DISCUSSION

"Masonry Cements -- A Laboratory Investigation" - Jacob  W. Ribar
and Val. S. Dubovoy

Results - In result No. 1, the authors state that flexural bond
strength is a function of masonry cement physical parameters but that
no single parameter was applicable to all masonry cements.  This
suggests that characterizing masonry cement performance using physical
measurements alone is not reliable.  Do the authors think a combination of measurement techniques, for example, chemical and physical,
would improve one's ability to predict masonry cement bond behavior?

Concerning result No. 2, it should be noted that the values given are
for brick K only.  In Table No. 2, masonry cements with brick A vary
from 18 psi to 140 psi and with brick G they vary from 58 psi to 130
psi.

In result No. 3, a comparison is made between specimens removed from
wall assemblies and companion stack bond prisms.  Wall assembly
specimens were tested by C1072 and found to have higher strengths.
These walls were first subjected to water permeance testing.  The
procedure calls for a preconditioning with water, and the test involves
wall exposure to virtually the saturation point for 72 hours.  What
effect does this "wet cure" have on bond?  Can your conclusion that
wall bond performance is greater than air cured prism performance be
applied to walls which have been air cured according to the provisions
of C1072?

In Table No. 1, bond performance at 28 days is 30% greater for Type N
masonry cement mortar compared to Type S.  Is this normally seen for
masonry cement mortars?  To what would you attribute this performance
difference?

In result No. 5, the initial rate of absorption and surface texture
of the brick are said to have had significant effects on bond.  The
reference is Table No. 3.  There is no apparent correlation between
flexural bond and the variables cited.  Tables No. 1 and 2 contain
bond values for all masonry cements with all bricks. These show wide
variations in bond (C1072) for individual masonry cement mortars with
different brick.  For example, with brick K (Table No. 1), bond varies
from a low of 45 psi to a high of 211 psi. Since the initial rate of
absorption and surface texture are the same for all combinations in
Table No. 1, what accounts for the variations?  Bonds for masonry
cement mortars with seven of the ten brick in Table No. 2 show the
lowest being only 1/4 as strong as the highest.  If brick IRA and
surface texture are not the primary factors in creating bond variances
with masonry cement mortar, what factors account for them in Table 2.

Discussion
In the second paragraph the statement is made that bond is affected
equally by physical properties of masonry cement, mortar, initial rate
of absorption and surface profile of the clay units.  There is no
evidence given supporting this conclusion.  What are the correlation
coefficients for each parameter?

In the third paragraph, the authors suggest that high IRA brick need to be dampened. This may be valid for masonry cement mortars. There is no data provided to support this statement for Portland cement-lime mortars. Where can supporting data be found which establishes the need for unit wetting with Portland cement-lime mortars?

Summary - The second paragraph contains a broad conclusion which is not supported by data. In the Results section (Result No. 2) the authors state that "good" bond was 100 psi or better. The chart below lists average bond strengths for each masonry cement mortar in combination with all brick.

TYPE N MORTAR

| Masonry Cement Designation | 4 | 7 | 11 | 12 | 13 | 17 | 25 | 26 | 34 | 47 |
|---|---|---|---|---|---|---|---|---|---|---|
| Average Bond (psi) | 64 | 85 | 59 | 120 | 99 | 88 | 113 | 55 | 74 | 96 |

TYPE S MORTAR

| Masonry Cement Designation | 1 | 2 | 6 | 30 | 41 | 43 | 49 | 50 | 54 | 56 |
|---|---|---|---|---|---|---|---|---|---|---|
| Average Bond (psi) | 110 | 67 | 74 | 86 | 82 | 87 | 113 | 87 | 98 | 96 |

Six Type S and six Type N masonry cements provide a mortar whose performance is at least 10% below the value chosen to represent good bond (100 psi). Is your conclusion based on the fact that each masonry cement produced at least one combination out of 10 that equalled or exceeded the 100 psi? If not, what are the criteria used to arrive at this conclusion?

Answer (Val S. Dubovoy)
Result No. 1. As stated in reference 4, masonry cement bond can be predicted more accurately on a multiple regression basis using physical properties of masonry cements and mortars. It can also be predicted on a multiple regression basis using brick physical characteristics, such as IRA and surface texture.

Result No. 2. The authors agree.

Result No. 3. We cannot conclude quantitatively as to what effect "wet cure" would have on flexural bond. It should be noted, however, that only the areas where moisture migrated through the wall during the entire test period were fully saturated, thus receiving the "wet cure" being referred to. Our results show that out of 20 sets of wall assemblages tested for water penetration, 9 sets exhibited wetted areas ranging from 2.7% to 8.1% of wall area,

Answer (Val S. Dubovoy) (continued)
8 sets - ranging from 10% to 29%, and only 3 sets of walls were fully saturated (see Table 4). These parameters were recorded just after 4 hours of testing - the time period generally associated with highest percentage wetted area before any "self-healing" and subsequent drying of the walls start taking place. Thus, in the majority of cases only limited areas of the wall assemblages received extra "wet cure." Furthermore, our results show wetted areas are generally located within first 8 inches near the edges of the wall assemblages. Had the "wet cure" been an influencing factor, samples removed from these areas would have exhibited bond strength higher or equal to that of the samples removed from the center of the wall assemblages which generally showed much less penetration of moisture. In reality just the opposite happened. The wall samples from center of the assemblies exhibited higher flexural bond. Therefore, it appears that "wet cure" received by the walls during the test does not affect bond.

Table 1 We do not have data to answer your question. Additional research into the properties of these cements would be required.

Result No. 5 Your statement is incorrect. The lowest bond value was 65 psi. The effects of initial rate of absorption and surface texture are discussed in great detail in Reference 4 where very good correlations were established between these two parameters and flexural bond on a multiple regression basis. As far as bond values in Table 1 and 2 are concerned, Table 1 gives the bond values for various masonry cements and the same brick. Table 2 gives the bond values for various brick and masonry cements. If you read Table 2 in only one direction, vertically, the influencing factor is brick. If you read Table 2 horizontally, the influencing factor is masonry cement. Hence, the conclusion stated in "summary" is that both brick and mortar have equal effect on bond.

Discussion
First paragraph - see Result No. 5, second paragraph - Brick Institute of America in its Technical Note 7B recommends wetting brick with IRA in excess of 30 g/min regardless of the type of masonry mortar.

Summary
This conclusion was based on the data in Table 1 where flexural bond is presented for various masonry cement mortars and one brick, thus showing effects of various masonry cements on bond.

Bruce K. Dickelman

A BENTONITE CLAY PLASTICIZER IN MASONRY MORTARS

REFERENCE: Dickelman, B. K., "A Bentonite Clay Plasticizer in Masonry Mortars," Masonry: Components To Assemblages , ASTM STP 1063, John H. Matthys, Editor, American Society for Testing and Materials, Philadelphia, 1990.

ABSTRACT: Select clays modified in the production of proprietary masonry mortar plasticizers yield masonry mortars with improved workability while retaining other physical characteristics desired for masonry construction. The success of the combination has been demonstrated by increased usage during the past thirty years. Extensive product evaluation indicates the compound is economically and technically competitive with or superior to other masonry mortar plasticizers. This report traces the development of the product, general production of the plasticizer, quality control of product by the manufacturer, performance characteristics of mortars when tested in accordance with ASTM C 270, and other special performance characteristics of the mortar, singly or in masonry prism and wall assemblages.

KEYWORDS: bentonite, montmorillonite

Workability of a masonry mortar in hot and or windy weather is a major concern of the masonry contractor. Although ASTM C 270-86 [2] allows for the retempering of mortars and testing has verified that retempering is not detrimental to the integrity of the mortar [2],[3], this process still requires time and labor and passes cost onto the mason contractor.

Florida has one of the most severe climates for masonry construction. To address the need for constant tempering, a Florida mason contractor began experimenting with alternate nonconventional materials in an effort to develop a plasticizer which would increase the board life of masonry mortars. A number of clays were tried in hopes of utilizing their water holding characteristics. Ultimately, he chose bentonite as the base ingredient for a new plasticizer. This plasticizer, when used in conjunction with Type I Portland Cement, displayed superior workability and increased water retention qualities in the hot, windy climate of southwest Florida.

This report will address a single type of clay based plasticizer, which is predominantly comprised of bentonite. The data compiled is from testing performed on a particular bentonite based plasticizer falling within this category. It is, at present and to the knowledge of the author, the only bentonite based plasticizer on the market. Additionally, although predominantly comprised of bentonite, there are other proprietary ingredients added which enhance the basic properties of the bentonite. It is therefore unreasonable to attempt to apply the results and successes of this particular plasticizer to another clay based plasticizer in total. The purpose of this report is to convey the efficacy of a particular modified clay as a plasticizer for masonry mortars and to dispel the current venue that all modified clays are detrimental to the integrity of masonry mortars.

## Material

The subject plasticizer consists predominantly of bentonite. Bentonite is composed of the clay mineral montmorillonite whose initial origin can be traced to volcanic eruptions that occurred millions of years ago. Fine volcanic ash particles were hurled into the atmosphere and carried by the wind many hundreds of miles eastward. The ash was ultimately deposited in discrete layers in a shallow salt sea that existed in the area of the present states of Wyoming, South Dakota and Montana. The ash altered and changed from a fragile, glassy state into the claystone called bentonite, primarily composed of montmorillonite.

The bentonite used is a sodium montmorillonite which is the only type that hydrates to any appreciable extent, thereby developing lubricating properties. It is the expansion and lubricating characteristics that are of primary importance to the effectiveness of this product as a mortar plasticizer. The expansion ensures volume compatibility of the smaller clay plasticizer requirement with that of a much greater lime requirement. The lubricating property increases the workability.

The basic chemistry of a montmorillonite molecular unit consists of two silicon-oxygen sheets with an aluminum hydroxyl sheet sandwiched between them. Certain highly significant effects result from these structural properties.

The central aluminum hydroxyl sheet in montmorillonite (bentonite) is never fully aluminum-hydroxyl. Part of the aluminum is substituted by magnesium and/or iron. When aluminum is substituted by magnesium, an imbalance is created that must be satisfied. This imbalance is referred to as a net negative charge which develops on the surface of each molecular platelet. This net negative charge is satisfied by the loose attachment of positive ions, which in this case is sodium.

In its unhydrated form, millions of montmorillonite platelets are stacked in a deck of cards sort of arrangement. When water is introduced, the sodium ion becomes soluble and is partially removed from the surface of the platelet, thus increasing the negative charge on the platelet surface. This increase in the surface charge causes the stacked platelets to repel one another, thus resulting in expansion and a tremendous increase in the platelet surface area.

Because of the negative charge on the platelets, water molecules adsorb to the platelets forming a "hydration shell" around each platelet. Due to the dipole character of water molecules, the positive end aligns itself to the negative platelet. This extends the negative end outward and provides another surface of negative charges on which can be built another layer of oriented water molecules. At the actual clay mineral surface, the molecules will be highly oriented and the degree of orientation will decrease going outward as the relative effect of thermal movement becomes greater.

With this product, the existence of billions of hydrated clay platelets in the mortar is responsible for the plasticity of the mortar. The hydration of the bentonite also accounts for a significant volume increase. A good quality sodium bentonite will expand up to fifteen times its original volume in potable water. The high calcium environment of a portland cement mortar will limit the volume increase to about eight to nine times the original volume. It is this swelling capability that allows seven pounds of this product to replace fifty pounds of hydrated lime with no loss of overall mortar volume.

## Manufacture

The occurrence of the bentonite used in the plasticizer in Wyoming, South Dakota and Montana predicates its manufacture and processing in those states.

Bentonite is normally mined in shallow deposits which occur over a broad area of land. Exploration crews locate and evaluate these deposits and determine the most likely areas for exploration. A bentonite deposit usually contains several layers and types of bentonite clays. Each has valuable, but sometimes different properties. In order to mix and match these properties to suit a particular product application, each layer must be evaluated through test drilling, then precision mined and stored in carefully segregated stockpiles. Quality control inspectors direct the building and use of the clay stockpiles so that each remains separate and uncontaminated by other clays.

Clay is moved from selected stockpiles, where it has been segregated by composition and blended to fit the exacting specifications of specific products.

Continuous testing of the clay blend as it moves through processing steps maintains the desired balance of clay properties. Each plant has a fully-staffed laboratory with the latest equipment to perform these tests accurately and efficiently.

In its raw form, as much as 30% of bentonite clay by weight is water. This is ultimately reduced to around 12% using rotary kilns, fluid bed dryers and other drying equipment. The bentonite is then processed through large roller mills and ground to a minimum of 65% passing a U.S.A. Standard Sieve Number 200 mesh fineness. Additives are precision metered into the mill to enhance the bentonite's natural properties. Exacting manufacturing controls at each step ensure that the end product performs uniformly and predictably.

TEST PROGRAM

This paper shows the properties of bentonite clay plasticized masonry mortar in comparison to conventional mortar. The properties investigated are: Compressive strength, water permeability, freeze-thaw durability, bond strength to masonry and shrinkage. The test data presented is an excellent representation of the characteristics of the product.

TEST METHODS

Portland Type I Cement and lime mortar mixes were used, where applicable, for all control batches and were proportioned in accordance with ASTM C270-80 [1] using the proportion specification requirements -- volumetric proportions of cementitious materials and sand according to the different types of mortars. The bentonite clay plasticized mortars were proportioned in accordance with the manufacturer's recommendation (see Table 1).

Proportion specification was chosen to limit the variables in the testing as much as possible. The bentonite clay plasticizer tested is designed to be a one 7-pound bag to one 50-pound bag of hydrated lime direct replacement. Maintaining like proportions of all other materials and varying only the plasticizer assured a more direct evaluation and comparison of the plasticizer variable.

TABLE 1 -- Mix Parameters
(Volumetric Parts)

| Mortar Type | Portland Cement | Hydrated Lime | Bentonite Plasticizer | Masons Sand |
|---|---|---|---|---|
| Lime - O | 1 bag | 2 bags | n/a | 9 cu. ft. |
| Lime - N | 1 bag | 1 bag | n/a | 6 cu. ft. |
| Lime - S | 2 bags | 1 bag | n/a | 9 cu. ft. |
| Lime - M | 2 bags | 1/2 bag | n/a | 6 cu. ft. |
| Bent - O | 1 bag | n/a | 2 bags | 9 cu. ft. |
| Bent - N | 1 bag | n/a | 1 bag | 6 cu. ft. |
| Bent - S | 2 bags | n/a | 1 bag | 9 cu. ft. |
| Bent - M | 2 bags | n/a | 1/2 bag | 6 cu. ft. |

Compressive Strength

The compressive strengths reported (see Table 2) are 28 day results of Type O, N, S, and M bentonite clay plasticized mortars. The results indicate that each of the mortar types meet and exceed the minimum compressive strength requirements of the property specification of ASTM C 270-80 [1].

Table 3 reports the effects of time in relationship to compressive strengths of bentonite clay plasticized mortars. A Type S mortar was chosen for compressive tests of; 7 days, 28 days, 90 days, 6 months, 1 year and 2 years. The results indicate a progressive increase in compressive strengths over the life of the test. All test samples exceeded the minimum compressive strength requirements of the property specification of ASTM C 270-80 [1].

TABLE 2   - Compressive strength at 28 days
of Bentonite Clay Plasticized mortars (ASTM C 109-80) [9]

| Mortar Type | Compressive Strength * psi (kPA) | | Average Compressive Strength psi (kPA) | |
|---|---|---|---|---|
| B-O-1 | 560 | (81.22) | | |
| B-O-2 | 575 | (83.39) | | |
| B-O-3 | 585 | (84.84) | 579 | (84.00) |
| B-O-4 | 540 | (78.32) | | |
| B-O-5 | 590 | (85.57) | | |
| B-O-6 | 625 | (90.65) | | |
| | | | | |
| B-N-1 | 1425 | (206.67) | | |
| B-N-2 | 1450 | (210.30) | | |
| B-N-3 | 1490 | (216.10) | 1443 | (209.21) |
| B-N-4 | 1375 | (199.42) | | |
| B-N-5 | 1450 | (210.30) | | |
| B-N-6 | 1465 | (212.47) | | |
| | | | | |
| B-S-1 | 2625 | (380.71) | | |
| B-S-2 | 2610 | (378.54) | | |
| B-S-3 | 2700 | (391.59) | 2318 | (336.23) |
| B-S-4 | 1975 | (286.44) | | |
| B-S-5 | 2090 | (303.12) | | |
| B-S-6 | 1910 | (277.01) | | |
| | | | | |
| B-M-1 | 3215 | (466.28) | | |
| B-M-2 | 3310 | (480.06) | 3288 | (476.92) |
| B-M-3 | 3340 | (484.41) | | |

*   2" x 2" cubes

TABLE 3 -- Long term study of compressive strengths
of Bentonite Clay Plasticized mortar (ASTM C 109-80) [9]

|  | Sample 1 psi (kPA) | | Sample 2 psi (kPA) | | Sample 3 psi (kPA) | |
|---|---|---|---|---|---|---|
| 7 days | 1575 | (228.43) | 1525 | (221.17) | 1605 | (232.78) |
| 28 days | 2145 | (311.09) | 2080 | (301.67) | 2110 | (306.02) |
| 90 days | 2305 | (334.30) | 2270 | (329.22) | 2315 | (335.75) |
| 6 months | 2500 | (362.58) | 2445 | (354.60) | 2415 | (350.25) |
| 1 year | 2705 | (392.31) | 2765 | (401.02) | 2690 | (390.14) |
| 2 years | 2810 | (407.54) | 2795 | (405.37) | 2775 | (402.47) |

Water Permeability

Four test walls approximately 50 inches (127 cm) in length,
56 inches (142 cm) in height, and 8 1/2 inches (21.6 cm) wide were
constructed in one day by two union masons on inverted steel
channels with water trapping flashing built into the samples. The
construction was done in accordance with applicable procedures as
outlined in the Brick Institute of America technical notes [2].
Workmanship on the walls was deemed "average". Each test wall was
constructed from mortar batches mixed in a commercial mortar
mixer. Each of the test walls utilized concrete block and fired
clay brick selected from the same manufacturing lot. Test walls
were allowed to cure together for a minimum of 28 days prior to
testing.

Three test walls were constructed, one each of a Type "O", a
Type "N", and a Type "S" mortar, consisting of Portland Type I
Cement, bentonite clay plasticizer and mason sand. The fourth
wall was constructed using a Type "N" mortar consisting of
Portland Type I Cement, Type S hydrated lime and mason sand.

Water penetration tests performed in accordance with ASTM E
514-74 [10], "Test Method for Water Permeance of Masonry", through
full size masonry walls (Table 4) show that bentonite clay
plasticized mortars produce a wall with water permeance ratings of
"G", as registered by the rating table of ASTM E 514-74 [10].

TABLE 4 — Comparative study of water permeance
of Bentonite Clay Plasticized mortar
vs.
Cement - Lime mortar (ASTM E 514-74) [10]

| Mix Number | Mortar Type | Appearance of Moisture | Visible $H_2O$ | 1st .05 Liter/Hr. Lower | 1st .05 Liter/Hr. Upper | (ml) $H_2O$ Lower Flashing | (ml) $H_2O$ Upper Flashing | (ml) Ext. of Damp Area | (ml) Rate of $H_2O$ flow at end of One Day | Original Permeance Rating |
|---|---|---|---|---|---|---|---|---|---|---|
| B-O | O | 3 hours | 24 hours | 72 hours | — | 24– 850<br>48– 910<br>72–1541 | 24– 52<br>48– 980<br>72– 965 | PC–18.2<br>24–20.1<br>48–28.2<br>72–29.6 | 43.5 | G |
| B-N | N | 2 hours | 25 hours | 48 hours | — | 24– 998<br>48–1460<br>72–1125 | 24– 27<br>48– 105<br>72– 995 | PC–19.4<br>24–25.8<br>48–26.0<br>72–29.9 | 42.7 | G |
| B-S | S | 2 hours | 25 hours | 48 hours | — | 24– 765<br>48–1220<br>72–1090 | 24– 50<br>48– 114<br>72–1065 | PC–16.5<br>24–18.2<br>48–25.9<br>72–24.6 | 40.6 | G |
| R-N | N | 2 hours | 46 hours | — | — | 24–<br>48– 34<br>72– 410 | 24–<br>48– 14<br>72– 39 | PC– 6.2<br>24– 7.1<br>48– 8.9<br>72–11.4 |  | E |

Freeze-Thaw Durability

A series of test specimens were subjected to freeze-thaw durability testing in accordance with ASTM C 666-80 Test Method for Resistance of Concrete to Rapid Freezing and Thawing [5]. Three, 3-inch (7.62 cm), by 4-inch (10.16 cm), by 16-inch (40.64 cm) mortar specimens were constructed for each mortar type. The mortar types tested were: Type "O", Type "N" and Type "S", consisting of Portland Type I Cement, bentonite clay plasticizer and mason sand; and a Type "N" mortar consisting of Portland Type I Cement, Type S hydrated lime and mason sand. The test results (see Table 5) indicate comparable resistance to freeze-thaw cycling between the clay plasticized mortars and the conventional mortar after 108 cycles.

TABLE 5 -- Comparative study of freeze-thaw resistance
of Bentonite Clay Plasticized mortar
vs.
Cement - Lime mortar (ASTM C 666-80) [3]

| Test Specimen | Cycles | Weight of Specimens lbs (kg) | | Relative Dynamic Modulus of Elasticity | Remarks |
|---|---|---|---|---|---|
| R-N | 0 | 12.57 | (5.71) | Original | – |
| B-O | 0 | 12.50 | (5.68) | Original | – |
| B-N | 0 | 12.28 | (5.57) | Original | – |
| B-S | 0 | 12.66 | (5.75) | Original | – |
| R-N | 36 | 12.51 | (5.68) | 96.0 | Scaling |
| B-O | 36 | 12.25 | (5.56) | 93.9 | Scaling |
| B-N | 36 | 12.23 | (5.55) | 94.0 | Scaling |
| B-S | 36 | 12.46 | (5.66) | 96.7 | Scaling |
| R-N | 72 | 12.42 | (5.64) | 87.4 | Increased Scaling |
| B-O | 72 | 11.89 | (5.40) | 80.7 | Severe Scaling |
| B-N | 72 | 12.15 | (5.52) | 86.0 | Increased Scaling |
| B-S | 72 | 12.15 | (5.52) | 90.3 | Increased Scaling |
| R-N | 108 | 12.30 | (5.59) | 72.1 | Severe Scaling |
| B-O | 108 | 11.56 | (5.25) | 71.8 | Severe Scaling |
| B-N | 108 | 12.05 | (5.47) | 73.8 | Crumbling of edges & Severe Scaling |
| B-S | 108 | 12.02 | (5.46) | 72.4 | Severe Scaling |

NOTE:  Tests terminated due to severe deterioration of specimens.

Samples:  Average of three beams per mortar mixture.

R-N -- Type "N" Portland Cement - Lime Control Mix
B-O -- Type "O" Bentonite clay plasticized mix
B-N -- Type "N" Bentonite clay plasticized mix
B-S -- Type "S" Bentonite clay plasticized mix

Bond Strength to Masonry

The bond strength of bentonite clay plasticized mortar was tested using two separate ASTM test procedures.

Type "O", Type "N" and Type "S" bentonite clay plasticized and Type "O", Type "N" and Type "S" conventional mortars were tested in accordance with ASTM C 321-77 [6] Bond Strength of Chemical-Resistant Mortars and ASTM C 1072-86 [7] Measurement of Masonry Flexural Bond Strength. Comparable bond strengths were achieved for both mortar types (see Table 6 and Table 7, respectively).

TABLE 6 -- Comparative study of bond strength
of Bentonite Clay Plasticized mortar
vs.
Cement - Lime mortar - (ASTM C 321-77) [6]

| Mix Number | Mortar Type | 28 Day Strengths psi | (kPA) |
|---|---|---|---|
| B-O | O | 85 | (12.33) |
| B-N | N | 118 | (17.11) |
| B-S | S | 97 | (14.07) |
| R-O | O | 76 | (11.02) |
| R-N | N | 104 | (15.08) |
| R-S | S | 110 | (15.95) |

TABLE 7 -- Comparative Study of Flexural bond strength
of Bentonite Clay Plasticized mortar
vs.
Cement - Lime mortar - (ASTM C 1072-86) [7]

| Mix Number | Mortar Type | 28 Day Strengths psi | (kPA) |
|---|---|---|---|
| B-O | O | 58.5 | (8.48) |
| B-N | N | 82.1 | (11.91) |
| B-S | S | 84.3 | (12.23) |
| R-O | O | 69.6 | (10.09) |
| R-N | N | 85.2 | (12.36) |
| R-S | S | 78.5 | (11.39) |

Shrinkage

Testing was performed to determine the shrinkage characteristics of a bentonite plasticized mortar. ASTM Standard Test Method C 157-80 [8] Length Change of Hardened Hydraulic-Cement Mortar and Concrete was used to test Type "O", Type "N" and Type "S" bentonite clay plasticized and Type "N" conventional mortars. The results (see Table 8) indicate very comparable values for all types of bentonite plasticized mortars to the control "N" mortar.

Additionally, testing following the same ASTM Test Method was performed to determine the long term shrinkage characteristics of bentonite plasticized mortars. The test reported shrinkage results for 7 days, 28 days, 90 days, 6 months, 1 year and 2 years for an "N" mortar. The results of this testing indicate a very reduced shrinkage tendency for the bentonite clay plasticized mortar (see Table 9).

TABLE 8 -- Shrinkage of mortar - Percentage of length change
Comparative study of Bentonite Clay Plasticized
vs.
Cement - Lime mortars (ASTM C 157-80) [8]

| Age of Drying | Bentonite clay plasticized | | | Cement - Lime |
| | Type O | Type N | Type S | Type N |
|---|---|---|---|---|
| 4 day | 0.017 | 0.025 | 0.020 | 0.016 |
| 7 day | 0.019 | 0.028 | 0.025 | 0.025 |
| 14 day | 0.025 | 0.029 | 0.026 | 0.025 |
| 28 day | 0.027 | 0.029 | 0.032 | 0.029 |

TABLE 9 -- Long term study of shrinkage
of Type "N" Bentonite Clay Plasticized mortar -
Percentage of length change   (ASTM C 157-80) [8]

| Age of Drying | Set 1 | Set 2 | Set 3 |
|---|---|---|---|
| 7 day | 0.086 | 0.006 | 0.011 |
| 28 day | 0.095 | 0.006 | 0.011 |
| 90 day | 0.099 | 0.006 | 0.012 |
| 6 months | 0.112 | 0.008 | 0.016 |
| 1 year | 0.114 | 0.008 | 0.018 |
| 2 year | 0.114 | 0.008 | 0.020 |

Conclusion

The results of the test program and actual field experience with this bentonite clay plasticizer prove that it is now and has been for quite some time, a viable product for the masonry industry from a performance standpoint. Due to its smaller proportioning, it offers a certain convenience to the mason. Seven bags of this plasticizer weigh less than one bag of hydrated lime. The smaller packaging also allows competitive pricing capability to any location in North America.

This test program and the successful marketing of the product over the last thirty years serve to point out a deficiency with the current ASTM Specifications in regards to masonry mortar. At present, there is no means by which this product can be evaluated for acceptance by ASTM C 270-86 [2]. ASTM C 270-86 [2] is titled "Specification for Mortar for Unit Masonry", but it is inadequate in regards to the evaluation and its virtual exclusion of this clay plasticizer and other non-conventionally proportioned mortar plasticizers due to its specification of particular mortar components to be used in mortar. It is therefore somewhat lacking as a generic specification for mortar for unit masonry.

ASTM Subcommittee C12.09, "Modified Mortars", has been formed to address this very point, not simply for plasticizers, but for a vast assortment of modifiers and admixtures. The development and approval by ASTM of this specification is important. There are a number of products being used in the masonry industry for which there is no current ASTM accepted method of evaluation. Some of these products are being used on the strength of testing performed in accordance with specific ASTM test procedures and some are being used without this same testing. A specification such as the proposed Modified Mortars specification would provide a valuable and accepted evaluation medium for the engineering sector. Such a medium would help assure the performance acceptability of not only a particular bentonite clay based plasticizer, but of any product presented to the masonry industry for use in mortar.

It is imperative that this vehicle be provided for the assessment of alternate plasticizers. This test program indicates that the existing test procedures provided by ASTM for the evaluation of performance characteristics of masonry mortars are applicable to this bentonite clay plasticizer. Comparable results of alternate plasticizers tested in tandem with accepted cement-lime mortars in all major performance criteria should warrant ASTM recognition. This test program proves that a particular bentonite clay plasticizer does perform comparably to the accepted hydrated lime.

REFERENCES

[1]  ASTM C 270-80  "Specification for Mortar for Unit Masonry".

[2]  ASTM C 270-86 X1.7.7.1   "Specification for Mortar for Unit Masonry".

[3]  Copeland, R. E. and Saxer, E. L., "Tests of Structural Bond of Masonry Mortars to Concrete Block".

[4]  Brick Institute of America Technical Note.

[5]  ASTM C 666-80,  "Test Method for Resistance of Concrete to Rapid Freezing and Thawing".

[6]  ASTM C 321-77,  "Standard Test Method for Bond Strength of Chemical-Resistant Mortars".

[7]  ASTM C 1072-86,  "Standard Method for Measurement of Masonry Flexural Bond Strength".

[8]  ASTM C 157-80,  "Test Method for Length Change of Hardened Hydraulic Cement Mortar and Concrete".

[9]  ASTM C 190-77,  "Test Method for Compressive Strength of Hydraulic Cement Mortars".

[10] ASTM E 514-74,  "Test Method for Water Permeance of Masonry".

DISCUSSION

"A Bentonite Clay Plasticizer in Masonry Mortars" -
Bruce K. Dickelman

Question (Dan Walker, CHEMSTAR LIME Company):
You report that the plasticizer consists of a sodium,
montmorillonite clay. These type clays will contain
from 1.5 to 2..5 percent sodium and potassium, which in
a seven (7) pound bag will contain the equivalent of
about .25 pounds of sodium carbonate. It is common
knowledge that portland cement hydrates release calcium
hydroxide. Therefore a lot of the sodium in the mortar
will be exchanged for the calcium in the portland
cement. What happens to the sodium when it is freed
into the ionic state? Sodium salts in mortars are
notoriously known to cause efflorescing. Have you
investigated mortars made with the clay for this
problem?

It is known that aluminum minerals, such as
montmorillonite clay, will react under moist conditions
with sulfates and calcium to form ettringite. This
formation causes a great deal of swelling to occur.
Mortars made with clays will certainly have all these
elements present. Have you investigated the long term
expansion effects of clay mortars in high moisture
conditions?

You indicate there are other additives in your clay
plasticizer. Would one such additive be an air
entraining material? If so, is this distinction so
noted on the selling package, such as is required in
C 207, Lime for Masonry Mortar? How much air content
will be found in a mortar made with clay plasticizers?

**Response to Question 1:** (Dickleman)
It is a well known fact that montmorillonite clays contain sodium. The main objective of this ion is to increase the hydration and swelling of the clay. The degree of hydration is dependent on size, the species and the charge of the exchangeable ions, as well as the magnitude and location of the layer charge within the adjacent silicate sheets and finally, the amount of water used to hydrate the clay. If the water content is approximately 90% to 95%, the sodium tends to promote the development of many oriented water layers on the interlamellar surfaces. This hydration may produce swelling to the extent of complete dissociation of the individual crystals, thus increasing surface area and decreasing particle size. This high degree of dispersion results in a high viscosity and the sodium is in a free ionic state. In the case of sodium montmorillonite in a typical mortar formulation the water content is not sufficient to dissociate the sodium into its ionic state. However, there is enough water to achieve hydration keeping the sodium closely held to the surface of the clay and allowing it to function as a mortar plasticizer. The amount of water present in a typical mortar would have to be five to six times greater for the release of the sodium in its ionic state to occur.

Another factor effecting the sodium ion and hydration is the degree of mixing. In the practice of mixing a mortar, the agitation that is employed is not sufficient for releasing the sodium in its ionic state.

Finally, with respect to the actual amount of sodium in montmorillonite clay, you must concentrate on the total weight of the finished product. Most sodium montmorillonites contain 2.5% $Na_2O$. This is equal to a sodium content of 1.833%. When you consider the amount of clay being added to the total mortar, this amount is quite small. Given a typical Type S mortar formulation of: 94 lbs (1505 kg) portland cement, 3.5 lbs (56 kg) bentonite clay based plasticizer, 360 lbs (5,764 kg) mason sand, and 83 lbs (1,328 kg) water, the actual percent sodium added to the system by the bentonite is only 0.064 lbs (1.02 kg), or 0.012% of the total system weight. In short, even if all of the sodium in the plasticizer were released into the system, which cannot occur, the amount would be too small to contribute significantly to efflorescence.

**Response to Question 2:** (Dickleman)
Two restrictions apply to the lack of formation of ettringite with the combination of montmorillonite clay, sulfates and calcium. The first again is moisture. As I explained above, the actual moisture content used in relation to the clay is actually quite small. Secondly, ettringite, which is $6CaO.Al_2O_3.3SO_3.33H_2O$, would have to react with the aluminum in the octahedral layer of the montmorillonite. In order for the montmorillonite to react the aluminum would have to be released from its octahedral layer. This cannot occur with the lack of water, and mixing, etc.. The aluminum is held very strongly in the octahedral layer.

**Response to Question 3:** (Dickleman)
Although predominantly comprised of bentonite, there are indeed other ingredients included in the production of this product. These ingredients, including the addition of air entraining materials, if any, are proprietary and will remain so.

Hydrated lime is a cementitious material and therefore must conform with the requirements of ASTM C226-86, "Standard Specification for Air-Entraining Additions for Use in the Manufacture of Air-Entraining Portland Cement". It is this specification that grandfathered the referenced, imposed requirement in ASTM C207-79 "Standard Specification for Hydrated Lime for Masonry Purposes". This bentonite clay based plasticizer is not considered cementitious and therefore, does not fall under the requirements of this specification. On a more fundamental basis, there is no vehicle by which this plasticizer can be recognized by ASTM C270-88 "Standard Specification for Mortar for Unit Masonry", so mandatory compliance to any ASTM specification is moot.

The laboratory air content of mortars made with this bentonite clay based plasticizer generally range between 12% and 18%, depending upon the mortar Type.

Robert E. Gates, Robert L. Nelson, and Michael F. Pistilli

THE DEVELOPMENT OF READY MIXED MORTAR IN THE UNITED STATES

---

REFERENCE:  Gates, R. E., Nelson, R. L., and Pistilli, M. F.,
"The Development of Ready Mixed Mortar in the United States,"
Masonry:  Components to Assemblages, ASTM STP #1063, John H.
Matthys, Editor, American Society for Testing and Materials,
Philadelphia, 1990.

ABSTRACT:  The introduction of Ready-Mixed or Extended-Life
Mortar to the masonry industry in the United States in the
early 1980's prompted the American Society for Testing and
Materials to initiate development of a specification for
Ready-Mixed Mortar.  The development of this specification
was under ASTM Committee C12, on Mortars, and Sub- Committee
C12.03, on Specifications.  The Sub-Committee commissioned a
task group, designated C12.03.7, to develop the
specification.  The authors of this paper are some of the
members of that task group.  The data presented is a
compilation of information that was submitted by members of
the task group in evaluation of ready-mixed mortar.  Both
laboratory- and field-developed data have been submitted.
    Ready-Mixed Mortar is best defined as "Mortar consisting
of cementitious materials, aggregate water, and set control
admixtures which are measured and mixed at a central location
using weight or volume control equipment.  This mortar is
delivered to a construction site and shall have a workability
period in excess of 2-1/2 hours.".
    The data that is presented in this paper represents
testing of Ready-Mixed Mortar for basic physical properties
such as compressive strength, air content, water retention,
and workability.  Additional properties such as bond strength
of ready-mix mortars to masonry units, bond strength of ready
mixed mortar to steel, water permeance, and compressive
strength of masonry prisms, employing ready-mixed mortar have
also been examined.

    Robert E. Gates, Product Manager, Concrete and Masonry
Admixtures, Construction Products Division, W. R. Grace & Co. -
Conn., 62 Whittemore Avenue, Cambridge, Massachusetts  02140-1692
    Robert L. Nelson, President, Robert L. Nelson & Associates,
Inc., 856 Cortbridge Road, Inverness, Illinois  60067.
    Michael F. Pistilli, Technical Director, Gifford-Hill &
Company, Inc., 2200 East Devon Avenue, Suite 111, Des Plaines,
Illinois  60018.

KEYWORDS:  Mortar, ready mix, extended life, retarded, set
controlled

## INTRODUCTION

Ready-Mixed Mortar has been in production in the United States for
the past 8 years.  Techniques of production by the suppliers and
refinement of admixtures by the manufacturers have created a product
system which has resulted in a controlled mortar.  During this time
period, contractors and producers have accumulated test data verifying
the technical excellence of this mortar.

The formation of the Extended Life Mortar Association, which at
the time had 30+ members, is just another step in the growth of the
industry.

The system Ready-Mixed Mortar is currently described as "mortar
consisting of cementitious materials, aggregate, water and
set-controlled admixtures which are measured and mixed at a central
location using weight or volume control equipment.  This mortar is
delivered to a construction site and shall have a workability for a
period in excess of 2 1/2 hours."

## TEST PROCEDURES

The evaluation of ready mixed mortar is not considered by the
authors to be difficult.  However without guidelines that are
typically presented in an ASTM Specification or Standard Method, a
standard set of rules for ready mixed mortar had to be established.
It was decided by the Task Group that the general procedure that would
be followed would be to employ ASTM standard methods where ever
applicable.  It was felt that mortar is mortar,  whether it is ready
mixed or site mixed.  If one makes this assumption and you look at the
mortar as just mortar then the evaluation becomes very simple. ASTM
C-780-87(1),Standard Method for Preconstruction and Construction
Evaluation of Mortars for Plain and Reinforced Unit Masonry provided
the general test criteria that could be used by all in the evaluation.
In addition ASTM C-270-87(2),  Standard Specification for Mortar for
Unit Masonry,  was also included as a guide to the types of mortar as
well as the test methods to be employed.

It is important to stress that the Task Group members involved in
the evaluation program believe that the only difference between the
ready mixed mortar and the conventional mixed mortar is that the board
life of the ready mixed mortar could be much greater than that of the
conventional mixed mortar.  One additional guideline was adopted for
use.  A publication by a task group of the Ready Mixed Mortar and
Stucco Association titled "Quality Control & Testing Procedures for
Ready Mixed Mortars"(3) provided the needed standard of the techniques
used to fabricate, cure, store, and test the specimens reported in
this paper.  Further discussion of these methods will be presented in
the specific section.

## PROPERTIES OF THE PLASTIC MORTAR

The consistency of the plastic mortar is normally the first physical property to be determined. In most of the testing, consistency of the mortar was determined by use of the cone penetrometer in conformity with ASTM C-780-87(1). There were a number of instances where the consistency was also determined by use of the flow table in accordance with the procedure outlined in ASTM C-109-86(4). Much of the test data presented in this paper is a result of testing programs by various testing laboratories as well as admixture suppliers and was done in the field where practical utilization of a flow table is impossible. The cone penetrometer results generally fall into the 45mm to 65mm range.

The air content of the mortar was also determined in conformity with the procedures in ASTM C-780-87(1) which refers to ASTM C-231-82, Standard Test Method for Air Content of Freshly Mixed Concrete by the Pressure Method.(5) Average values for the air content of the mortars have ranged from approximately 12% to 20%. A very important characteristic of the ready mixed mortar is its ability to retain the entrained air over the expected life.(6)

Unit weight of the plastic mortar was also determined in general conformity with ASTM C-138-81(7), Standard Test Method for Unit Weight, Yield and Air Content (Gravimetric) of Concrete. In most cases the measuring vessel was the base of the air content apparatus, a 0.25 cu. ft. calibrated container. The unit weight results are used as a check on the work being done. The normally expected parameters continue to apply to the mortar in that when air content increases the unit weight is expected to decrease. Also the unit weight was used to adjust the mix proportions because of either over or under yield of the cubic yard volume.

Typically the final test made to the plastic mortar was the fabrication of the compressive strength specimens. ASTM C-780-87(1) allows for the use of a multiple of test specimen shapes and sizes, from 2 inch cubes to 2"x4" and 3"x6" cylinders. As mentioned earlier, most of this work generated from field production mortar, mortar that was commercially produced in volumes greater than two cubic yards and subsequently used at a construction site, rather than laboratory batched mortar. The test specimen normally used was the 3"x6" cylinder. After a preliminary look it was decided to use plastic molds exclusively. In cases where cubes are reported, either brass or steel molds were employed. This decision was prompted by the excessive water absorption by cardboard molds. When ready mixed mortar is placed in a non absorptive mold, the mortar will remain in a plastic state for the maximum possible time. This is because the mortar contains a set controlling admixture (broadly classified as a retarder) that's effectivness and performance is controlled by the loss of water. Even though the cardboard molds are produced to satisfy the ASTM requirement for single use molds, the extended period of curing for the extended life mortar could cause wetting of the cardboard with a resulting deformation of the mold shape or breakdown of the release

compound.  This would create a faulty specimen unsuitable for
testing.  In 1988, the Extended Life Mortar Association(3) published
recommendations that once fabricated the cylinders remain undisturbed
for eight(8) days or until the mortar has completely set.  Some of the
data presented in this paper was developed prior to the publication
but in all cases the mortar in these tests remained in the molds for a
minimum of four(4) days.  In many cases the eight day storage was not
practicable.  The use of cube molds is not prohibited.  However, when
one is working with multiple mortar mixes each day, and test specimens
are made and then left for 4 to 8 days in the mold, it becomes cost
prohibitive to own the required number of molds.

BOND TO STEEL TEST METHOD

     The bond strength of the mortar to steel was determined by means
of modifying existing test methods to accommodate the need of this
specific test. The overall requirement was for a test specimen that
could be examined in tension.  To accomplish this the briquet specimen
described in ASTM C-190-85(8), Standard Test Method for Tensile
Strength of Hydraulic Cement Mortars, was selected to be adapted to
satisfy the need for a test method.  Metal coupons were machined to an
approximate size of 1" x 2" so as to fit within the necked down area
of the briquet.  A dam was installed in each of the molds so as there
would be no contact between the mortar cast in each half of the mold.
This would allow the metal coupon to be the only force binding the two
halves of the briquet. After proper curing the briquets were tested in
conformity with the procedure outlined in the ASTM C-190-85(8) test
method.  The data obtained was not the tensile strength of the mortar
but the force required to dislodge the metal coupon from the mortar
under pure tensile force.


MATERIALS

     Much of the data presented in this paper is an accumulation from
many different sources.  It would be very time and space consuming to
present all of the individual test and qualification data for each and
every component used in all of the tests. This data was obtained from
the following sources:
     Addiment,Inc., 6555 Button Gwinnett Dr., Atlanta, GA. 30340
     American Admixtures, 5909 N. Rogers Ave. Chicago, IL. 60646
     W. R. Grace & Co., 62 Whittemore Ave., Cambridge, MA. 02140

     For informational purposes a summary of the properties of the
materials used in the evaluations will be presented.

     Cements used in all tests were either Portland Type 1 conforming
to ASTM C-150-85a(9) or Masonry Cement conforming to ASTM
C-91-83a(10).  When masonry cement was used the appropriate mortar was
formulated from the specific Type of masonry cement that was used.
Type N mortars conforming to ASTM C-270-87(2) were made with Type N
masonry cement, however for continuity of testing Type S mortars were
created from Type S masonry cement but not from Type N masonry cement
supplemented with Portland Cement.

Mix proportions for the various mortars produced adhered for the most part to the ASTM C-270-87(2) specifications. There were a few instances where a mortar was proportioned to provide for economics and in these cases the mixes were over-sanded. The compressive strength requirement was satisfied and no other deleterious properties were noticed.

The sands used for the testing programs were locally available sand conforming to ASTM C-144-87(11) as to gradation. The range of Fineness Modulus varied from 1.9 to 2.3, depending on in which part of the country the testing was performed.

Brick selected to construct the masonry assemblages was also selected from locally available materials. In all cases the brick were clay units and were selected because of specific absorption properties and were classified by the initial rate of absorption (I.R.A.). Within the test series the brick are identified by an I.R.A.range, either low, medium or high. ASTM C-67-87(12) test methods for I.R.A. based on oved dried specimens were employed at either the authors' laboratories or by an independent laboratory. The masonry prisms that were constructed were comprised of six bricks and five joints. The results presented are the average of the five joints tested.

RESULTS

SERIES A

The compressive strength of the mortar in this Series of testing show that the ready mixed mortar developed equal or greater compressive strength than the reference  or control mixes at the 28 day age (Series A, Table 1). Compressive strengths reported were from samples taken immediately after mixing in each case. Strengths were slightly lower at the very early ages and equal to the control at the mid age range. It should be noted in this series of tests the mortars used were proportioned to the ASTM C-270-87(2) specifications and the compressive strengths far exceeded the alternate strength requirement.

The stack bonded masonry prisms examined in this series of tests provided somewhat erratic compressive strength results. All of the prisms made for this series of tests were made immediately after mixing of the mortars. The specimens were made on plastic bags and upon completion of the fabrication the bags were pulled up over the specimens and tightly closed until the time of test. The results (Tables 2A-C, Series A) show that the ready mixed mortar produced generally lower strength than the reference mixes at 24 hours age. From a point 3 days and on toward 28 days the extended life mortar designed for 36 hours life is generally similar to the reference mix. As the 28 day age is approached the ready mixed mortar tends to develop a higher compressive strength than the reference. The mortar that was admixed to produce a 2 1/2 day life also showed similar results to the reference in that the compressive strengths were similar in the 7 to 28 day range.

Table 3, Series A, provides the results of shrinkage testing done in conformity with ASTM Method C-157-86(13). The results show that the ready mixed mortar made with masonry cement developed 13.8% less shrinkage than the masonry cement control mixes at 28 days age and more than 5% less shrinkage at the 64 week test age. When the portland cement-lime test mix is compared to a portland cement-lime control mortar, the ready mixed mortar developed 26% less shrinkage at 28 days and 23% less at 64 weeks.

Freeze- thaw testing was performed in general conformity with ASTM Method C-666-84(14). Specimens for this test were made immediately after mixing and then stored in saturated lime water from the time of their removal from the molds until they were 14 days old when the freeze thaw testing was begun. This test is designed for examination of concrete and the freezing and thawing is done rapidly in water, opposed to the ASTM C-67-87(12) method which describes a 50 cycle test that is conducted at one cycle per day. In the C-666-84(14) test method cycling is conducted at approximately 7 cycles per day to 300 cycles total. The freeze thaw durability of the Type N ready mixed mortar produced from Masonry Cement was similar to the control mix. Table 4, Series A presents the results which show the portland cement - lime control to have failed in about 1/3 of the time as the Masonry Cement control or test mix. The ready mixed mortar provided equivalent results to the Masonry Cement control.

Bond strength of mortar to masonry (in this case brick) units was measured by what has become ASTM Method C-1072-86(15), Standard Method for Measurement of Masonry Flexural Bond Strength. The results were obtained from stack bond prisms fabricated immediately after mixing. The results reported in Table 5, Series A were developed prior to ASTM establishing the C-1072-86(15) test method but the procedures were the same. The range of results was very large, but within a single test specimen the range was within reason. Table 5, Series A presents the results which show the ready mixed mortar to yield equal or greater strengths than the control mixes.

Results of the special bond to steel (Series A, Table 6) conducted as previously described indicate that comparable or better strength was achieved in the ready mixed mortar made with portland cement - lime when compared to the control mix. Here again the specimens were made immediately after mixing. Even with substantially greater amounts of entrained air the ready mixed mortar provided at least 50% better strength. A ready mixed mortar made with masonry cement developed strength slightly lower than the control mix made with the portland cement lime mix. It was expected that this resulted from the lower compressive strength of the masonry cement mortar that was shown in Table 1, Series A.

The final set of data to be discussed in Series A is presented in Table 7. This test program was conducted to determine the differences in water permeance of standard wall sections constructed from the test as well as control mortars. A test wall was constructed from each mortar in conformity with ASTM E-514-74(16). Results show that the walls constructed from ready mixed mortar provided the same or better resistance to water penetration than did the wall sections built from the conventionally mixed mortar. The time that any leakage was noted was 4 times longer with the ready mixed mortar and it took five times as long to reach the point of maximum leakage. When the time of maximum leakage was determined, the rate determined for the ready mixed mortar was only 1/3 as great as that for the conventionally mixed mortar.

SERIES B

This series of tests was performed on mortars made with a blast furnace slag cement. All of this data was generated in the laboratory under standard laboratory conditions. In the blast furnace slag cement evaluation, three batches of mortar were made to examine the potential for using blast furnace slag cement in the commercial production of mortar. The use of cements conforming to the requirements of ASTM C-595-86(17), Specification for Blended Hydraulic Cements is permitted in ASTM C-270-87(2), Standard Specification for Mortar for Unit Masonry. However, during the development of the specification for ready mixed mortar there was concern from the Sub-Committee that the use of other than portland or masonry cement should not be allowed until sufficient data documenting its use was published. This data is part of what was developed for that purpose.

Based on historical production information provided by ready mixed mortar producers in Germany, mixes were proportioned to the typical values that are presented in Table 1, Series B. At first the cement content appears to be insufficient however it was reported that a workable Type N mortar was produced. Results indicate that the 72 hour mortar that was produced did indeed maintain it's expected workability over the 72 hours. It was also noted that there was little loss of air entrainment during the entire workability period. Compressive strengths of the three batches were similar and averaged 1065 psi (7.3 MPa). Tests examining flexural shear strength in conformity with the German Standard DIN 18 555(18) were uniform at 350 psi (2.4 MPa) which is acceptable for a Type N mortar according to these standards.

SERIES C

This series evaluates the use of fly ash as a substitute for portland cement in a mortar mix. Table 1, Series C presents the laboratory generated data comparing a reference or control mix to a test mix. In this evaluation it was decided to make the mortars from only portland cement to avoid any other parameters that might affect the outcome of the test.The results show that the test data developed over a 41 hour test period. The mix that contained the fly ash substitution maintained better workability during the test period. A slightly greater loss of entrained air was noted with the fly ash but the workability was basically unchanged. Compressive strengths were much greater than the mortar mix design called for with an average compressive strength of 2125 psi (14.6 MPa). Flexural bond strength was also determined and results of 103 psi (0.7 MPa) were much greater than expected and considered to be excellent.

Table 2, Series C presents the results of field testing of mixes containing different amounts of fly ash and a fixed portland cement content. Sample preparation, storage and subsequent laboratory curing was done in accordance with standard methods. Type S mortar proportions were used and the portland cement content was held constant. The volume of fly ash added was compensated for in the sand. Mortars were mixed to equal consistency as measured by the cone penetrometer. Air contents were measured and the mix with the greater amount of fly ash developed slightly more air than the other mix. This is reflected in the lower, but acceptable compressive strength. The flexural bond strength was determined for the mix containing the lower amount of fly ash and it was also found to be in the range of what was considered to be acceptable.

SERIES D

This series of portland cement-lime mixes contains both field prepared mortars as well as those that were made in the laboratory. In each case, the ready mix mortar was commercially produced and then delivered to the laboratory. The reference mortar was made in the laboratory mixer. Table 1, Series D presents the physical properties of mortars as well as the 28 day compressive strengths of the mortar as sampled initially after lab mixing or delivery as applicable. The Type N mixes exhibited similar workability however the consistency data determined by cone penetrometer showed a greater than expected variance. The workability was adjudged by the mason in attendance to build prisms and wall sections.

The air content of the ready mixed mortars was greater than that of the conventionally mixed mortar. It was expected that the compressive strength of the mix with the higher air would exhibit a lower compressive strength as was the case with the Type N mixes, however the opposite was observed in the Type S mixes This is different from the results presented in Table 1, Series A, but different mixes were used.

The Type S mortar mixes also exhibited the same trend in workability verses consistency measurement. At equal workability again determined by the mason the ready mixed mortar showed a much lower cone penetration result. In this test, the ready mixed mortar had a higher air content as is the case with most ready mixed mortar. In each case the water retention of the mortar was increased with the use of the set controlling admixture.The data presented in Table 1, Series A shows the opposite result.

Table 2, Series D presents the flexural bond strength data for the two test mixes. In all of the tests in this series a medium-to-high or high Initial Rate of Absorption clay brick was used. The Type N mixes followed the trend of the compressive strength, that is the flexural strength of the ready mixed mortar was higher than that of the conventionally produced material. The Type S mixes also followed the previously established trend for this set and the ready mixed mortar produced a lower flexural strength.

The final table, Table 3, Series D presents the results of water permeance testing performed in conformity with ASTM E-514-74(16) on a two wythe wall utilizing hollow load bearing masonry units and fired clay brick. Ome wall was constructed for each case. The Type N mortar results show a wide variance in the data with the ready mixed mortar providing a much greater resistance to the water penetration. The test showed that the wall built with ready mixed mortar took almost 5 times as long to begin leaking as did the wall built with conventional mortar. Leakage rates were essentially existant in the ready mixed mortar wall.

The walls built with the Type S mortars showed that the ready mixed mortar provided a 100% increase in the ability of the wall to resist water penetration and leakage. First dampness and leakage times for the wall containing the ready mixed mortar were noted to be twice that of the conventionally produced mortar wall. After the second day of testing for permeance each wall was generally thoroughly wet as the reference wall showed to be 80% damp ant the test wall was 95% damp. The Type S mortar walls did achieve an equal rating as to their permeance characteristic.

CONCLUSION

The results of the test programs demonstrate that for the products tested ready mixed mortars compare favorably to conventional mortar. The data presented shows that in general, ready mixed mortar produced a more workable mortar at a lesser consistency. Equal or better bond strengths and equal or lower water penetration through the masonry were demonstrated.

The results of the prism strength testing showed that the ready mixed mortar developed lower strength at 24 hours than the control mix. Although the strength is lower, it is not low enough that it would restrict the rate at which the wall are constructed. The ready mixed mortar produced lower prism strengths at one day but it was shown that there is a cross-over point where the ready mixed mortar catches up to the conventional mortar at 3 to 7 days when the strengths are similar.

Because the ready mixed mortar is air entrained the freeze thaw durability is markedly improved over that of the non air entrained portland cement - lime, conventionally produced mortar. The amount of air entrainment typically found in the ready mixed mortar is in the 14% to 20% range which has been shown to account for some loss in the compressive strength. Although a loss in strength should be of concern the loss here will be offset by the increase in durability.

The results of the bond to steel testing demonstrate that the use of the ready mixed mortar will provide equal or better bond to the reinforcing as well as other embedded metal in the mortar. Even with the increased air contents of the ready mixed mortars better bond strength to the steel coupons was determined.

In summary, the authors feel that ready mixed mortar prepared in a central batching plant with the use of specially formulated set controlling admixtures will provide a mortar that is at least equal to if not better than the conventionally produced site mixed mortar. It is important that the ready mixed mortar be properly manufactured and delivered to the work site. The most important quality of the ready mixed mortar is the quality control that must be associated with it's production. Guidelines as to the production and quality control have been published by the Extended Life Mortar Association and in general following these will result in a better masonry construction.

REFERENCES

1  American Society for Testing and Materials:  C-780-87,
   "Preconstruction and Construction evaluation of Mortars for Plain
   and Reinforced Unit Masonry", ASTM Standards, Vol 04.05.

2  American Society for Testing and Materials:  C-270-86,
   "Specification for Mortar for Unit Masonry", ASTM Standards,
   Vol 04.05.

3  "Quality Control & Testing Procedures for Ready Mixed Mortars",
   Extend Life Mortar Association, Cincinnati, OH, 1988.

4  American Society for Testing and Materials:  C-109-86, "Standard
   Test Method for Compressive Strength of Hydraulic Cement Mortars",
   ASTM Standards, Vol. 04.01.

5  American Society for Testing and Materials:  C-231-82, "Standard
   Test Method for Air Content of Freshly Mixed Concrete by the
   Pressure Method", ASTM Standards, Vol. 04.02.

6  Ready Mixed Mortar in the United States, "8th Inter, Brick/Block
   Masonry Conference", 1988, Nelson, R. L., Schmidt, S., Munro, C.,
   Pistilli, M., Gates, R., Seyl, J., and Lauber,  R.

7  American Society for Testing and Materials:  C-138-81, "Standard
   Test Method for Unit Weight, Yield, and Air Content (Gravimetric)
   of Concrete", ASTM Standards, Vol. 04.02.

8  American Society for Testing and Materials:  C-190-85, "Standard
   Test Method for Tensile Strength of Hydraulic Cement Mortars",
   ASTM Standards, Vol. 04.01.

9  American Society of Testing and Materials:  C-150-85a, "Standard
   Specification for Portland Cement", ASTM Standards, Vol. 04.01.

10  American Society for Testing and Materials:  C-91-83a, "Standard
    Specifications for Masonry Cement", ASTM Standards, Vol. 04.01.

11  American Society for Testing and Materials:  C-144-87, "Standard
    Specification for Aggregate for Masonry Mortar", ASTM Standards,
    Vol. 04.05.

12  American Society for Testing and Materials:  C-67-87, "Standard
    Methods of Sampling and Testing Brick and Structural Clay Tale",
    ASTM Standards, Vol. 04.05.

13  American Society for Testing and Materials:  C-157-86, "Standard
    Test Method for Length Change of Hardened Hydraulic Cement Mortar
    and Concrete", ASTM Standards, Vol. 04.01.

REFERENCES (Con't)

14  American Society for Testing and Materials:  C-666-84, "Standard
    Test Method for Resistance of Concrete to Rapid Freezing and
    Thawing", ASTM Standards, Vol. 04.02.

15  American Society for Testing and Materials:  C-1072-86, "Standard
    Method for Measurement of Masonry Flexural Bond Strength", ASTM
    Standards, Vol. 04.05.

16  American Society for Testing and Materials:  E-514-74, "Standard
    Test Method for Water Permeance of Masonry", ASTM Standards,
    Vol. 04.07.

17  American Society for Testing and Materials:  C-595-86, "Standard
    Specification for Blended Hydraulic Cements", ASTM Standards,
    Vol. 04.01.

18  Deutsches Institut for Normung e.V. Berlin, "Testing of Mortars
    Containing Mineral Binders, DIN Standards, March 1986.

# MIX PARAMETERS - SERIES A [6]

## Volumetric Parts

| Control Mixes | Type-N Mixes | | Type-S Mixes | |
|---|---|---|---|---|
| Mix Id. | C1 | C2 | C3 | C4 |
| Portland Cement | 1 | | 1 | |
| Lime | 1 | | 1 | |
| Masonry Cement | | 1 | | 1 |
| Sand | 6 | 3 | 5 | 3 |

## Weighed Batches

| Ready Mixed Mortar Mixes | Type-N Mixes | | Type-S Mixes | |
|---|---|---|---|---|
| Mix Id. | T1 | T2 | T3 | T4 |
| Portland Cement, lbs(kg) | 400 (181) | | 500 (226) | - |
| Lime, lbs(kg) | 100 (45.3) | | 100 (45.3) | - |
| Masonry Cement, lbs(kg) | | 560 (254) | | 650 (294) |
| Sand, lbs(kg) | 2230 (1011) | 2230 (1011) | 2200 (997) | 2200 (997) |
| Water, gal(L) | 33 (125) | 33 (125) | 35 (132) | 35 (132) |
| Admixtures | As recommended by the manufacturer | | | |

# SERIES A - TABLE 1 - ASTM C780 [6]
## Compressive Strength of Mortar, psi (MPa)

| Age | Mix C2 2 Hr. Control Masonry Cement (1:3) Type-N | Mix T2 36 Hr. Mortar Ready Mix Type-N | Mix T3 60 Hr. Mortar Ready Mix Type-S | Mix C3 2 Hr. Control Portland-Lime (1:1:5) Type-S |
|---|---|---|---|---|
| 28 Day | 2108 (14.5) | 2362 (16.3) | 3367 (23.2) | 2585 (17.8) |
| | 2037 (14.0) | 2716 (18.7) | 2900 (20.0) | 2585 (17.8) |
| | 2009 (13.9) | 2376 (16.4) | 3183 (21.9) | 2609 (18.0) |
| | | | 3593 (24.8) | |
| Average | 2051 (14.1) | 2485 (17.1) | 3261 (22.5) | 2593 (17.9) |
| Air Content % | 17.8 | 16.2 | 17.0 | 5.0 |

# SERIES A - TABLE 2A - ASTM C447 [6]
## Average Compressive Strength of Masonry Prisms
### Low IRA Brick
### Gross Strength, psi (MPa)

| Age | Mix C2<br>Type N Control<br>Masonry | Mix T2<br>36 Hr. Mortar<br>Type-N | Mix T3<br>60 Hr. Mortar<br>Type-S | Mix C3<br>Type S Control<br>Portland-Lime |
|---|---|---|---|---|
| 24 hours | 2857 (19.7) | 1143 ( 7.9) | 952 ( 6.56) | 3953 (27.3) |
| 3 day | 2770 (19.1) | 3608 (24.9) | 2520 (17.4 ) | 4571 (31.5) |
| 7 day | 2910 (20.1) | 4583 (31.6) | 4126 (28.4 ) | 5111 (35.2) |
| 14 day | 2629 (18.1) | 5653 (39.0) | 4312 (29.7 ) | 4800 (33.1) |
| 28 day | 3505 (24.2) | 6255 (43.1) | 4339 (29.9 ) | 5166 (35.6) |

# SERIES A - TABLE 2B - ASTM C447 [6]
## Average Compressive Strength of Masonry Prisms
### Medium IRA Brick
### Gross Strength, psi (MPa)

| Age | Mix C2<br>Type N Control<br>Masonry | Mix T2<br>36 Hr. Mortar<br>Type-N | Mix T3<br>60 Hr. Mortar<br>Type-S | Mix C3<br>Type S Control<br>Portland-Lime |
|---|---|---|---|---|
| 24 hours | 2344 (16.2) | 803 ( 5.5) | 459 ( 3.16) | 2699 (18.6) |
| 3 day | 2497 (17.2) | 2160 (14.9) | 826 ( 5.7 ) | 2890 (19.9) |
| 7 day | 3013 (20.8) | 3108 (21.4) | 1621 (11.2 ) | 4154 (28.6) |
| 14 day | 2887 (19.9) | 3000 (20.7) | 2451 (16.9 ) | 4250 (29.3) |
| 28 day | 3558 (24.5) | 4638 (32.0) | 2581 (17.8 ) | 4651 (32.1) |

## SERIES A - TABLE 2C - ASTM C447 [6]
### Average Compressive Strength of Masonry Prisms
### High IRA Brick
### Gross Strength, psi (MPa)

| Age | Mix C2<br>Type N Control<br>Masonry | Mix T2<br>36 Hr. Mortar<br>Type-N | Mix T3<br>60 Hr. Mortar<br>Type-S | Mix C3<br>Type S Control<br>Portland-Lime |
|---|---|---|---|---|
| 24 hours | 1303 ( 9.0) | 879 ( 6.1) | 228 ( 1.57) | 1916 (13.2) |
| 3 day | 1898 (13.1) | 1832 (12.6) | 1039 ( 7.2 ) | 1772 (12.2) |
| 7 day | 2511 (17.3) | 2543 (17.5) | 2088 (14.4 ) | 2795 (19.3) |
| 14 day | 2050 (14.1) | 2050 (14.1) | 2574 (17.7 ) | 3370 (23.2) |
| 28 day | 2095 (14.4) | 3093 (21.3) | 2764 (19.1 ) | 3375 (23.3) |

## SERIES A - TABLE 3 - ASTM C157 [6]
### Shrinkage of Type N Mortar - % Length Change

| Drying<br>Time | Mix C4<br>Control Mix<br>Masonry Cement | Mix T3<br>Ready Mixed<br>60 Hr. Mortar | Mix C3<br>Control Mix<br>Portland-Lime |
|---|---|---|---|
| 4 days | -0.031 | -0.038 | -0.035 |
| 7 days | -0.037 | -0.042 | -0.040 |
| 14 days | -0.054 | -0.052 | -0.058 |
| 28 days | -0.058 | -0.050 | -0.068 |
| 8 weeks | -0.062 | -0.061 | -0.073 |
| 16 weeks | -0.074 | -0.079 | -0.083 |
| 36 weeks | -0.089 | -0.085 | -0.095 |
| 64 weeks | -0.095 | -0.090 | -0.117 |

# SERIES A - TABLE 4 - ASTM C666 [6]
## Freeze-Thaw Resistance of Ready Mix Mortar

| Mortar Mixture | Relative Dynamic Modulus of Elasticity In Percentage of the Dynamic Modulus of Elasticity at Zero Cycles | Air Content | Average Percentage of Weight Loss |
|---|---|---|---|
| Portland-Lime Type-N (Control) | Destroyed at 36 cycles, zero reading | 5.1% | 18.7% after 36 cycles |
| Masonry Cement Type-N (Control) | 92% after 113 cycles | 17.1% | 3.8% after 113 cycles |
| Ready Mix Mortar Type-N (Test) | 89% after 113 cycles | 17.6% | 2.8% after 113 cycles |

Test results after 113 cycles of rapid freezing and thawing, (average of three beams per mortar mixture).

# SERIES A - TABLE 5 - ASTM C1072 [6]
## 28 Day Bond Strength, psi (MPa)

| IRA* | Mix C2 Masonry Control Type-N | Mix T2 36 Hr. Mortar Type-N | Mix T3 60 Hr. Mortar Type-S | Mix C3 Portland-Lime Control Type-S |
|---|---|---|---|---|
| 3 | 83 (0.57) | 145 (1.00) | 108 (0.74) | 102 (0.70) |
| 8 | 50 (0.34) | 72 (0.50) | 82 (0.57) | 56 (0.39) |
| 15 | 37 (0.26) | 132 (0.91) | 165 (1.14) | 67 (0.46) |

(*) Note: Initial rate of absorption (IRA), grams water/one minute.

# SERIES A - TABLE 6[6]

## Bond Strength of Ready Mix Mortar to Steel

| Mix | Mix Type | | Mix Id. | Air Content % | Failure Load* Lbs(kg.) |
|-----|----------|--------|---------|------------------|------------------------|
| 1 | Control Portland-Lime | Type-N | C1 | 4.7 | 236 (107.0) |
| 2 | Control Portland-Lime | Type-S | C3 | 4.3 | 302 (137.0) |
| 3 | Ready Mix Portland-Lime | Type-N | T1 | 20.7 | 428 (194.1) |
| 4 | Ready Mix Portland-Lime | Type-S | T3 | 22.9 | 455 (206.4) |
| 5 | Ready Mix Masonry Cement | Type-N | T2 | 18.0 | 190 ( 86.2) |
| 6 | Ready Mix Mortar Masonry Cement | Type-S | T4 | 22.8 | 285 (129.3) |

*Average of three specimens

# SERIES A - TABLE 7[6]

## Water Permeance - ASTM E514

| Mix Id. | C3 Reference Wall | T3 Ready Mix Mortar Wall |
|---------|-------------------|--------------------------|
| First dampness | 1 minute | 1 minute |
| 25% area damp | 20 minutes | 20 minutes |
| 50% area damp | 30 minutes | 30 minutes |
| 100% area damp | 1 hour | 1 hour |
| Visible water | 1 minute | 1 minute |
| Begin leakage | 5 minutes | 22 minutes |
| Maximum rate of leakage | 3.8 litre/hr. | 1.3 litre/hr. |
| Time of maximum rate | Second hour | Eleventh hour |
| Rate at end of one day | 2.5 litre/hr. | 1.1 litre/hr. |
| Stabilized rate | 1.73 litre/hr. | 1.07 litre/hr. |

# SERIES B - TABLE 1
## Blast Furnace SLAG Cement Evaluation

**Mix Design**

| | | | | | | | | | | | | |
|---|---|---|---|---|---|---|---|---|---|---|---|---|
| Slag Cement, lbs (Kg) | 420 (190) | | | | 420(190) | | | | 420 (190) | | | |
| Sand, lbs (Kg) | 3090 (1403) | | | | 3090(1403) | | | | 3090 (1403) | | | |
| Water, gal (L) | 48.4 (182.8) | | | | 48.4 (182.8) | | | | 48.4 (182.8) | | | |
| Admixture | A12 | | | | A12 | | | | A12 | | | |
| **Properties** | Initial | 24 Hour | 48 Hour | 72 Hour | Initial | 24 Hour | 48 Hour | 72 Hour | Initial | 24 Hour | 48 Hour | 72 Hour |
| Penetration, MM | 50 | 43 | 41 | 39 | 49 | 46 | 43 | 38 | 50 | 45 | 42 | 37 |
| Air, % | 20 | 19 | 18 | 18½ | 20 | 17 | 17 | 16½ | 21 | 18 | 17 | 16 |
| Mortar Temp. °F(°C) | 64 (18) | | | | 63 (17) | | | | 64 (18) | | | |

**Strength, psi @ 28 days**

| | | | |
|---|---|---|---|
| Compress, psi (MPa) | 1090 (7.5) | 1190 (8.2) | 915 6.3 |
| Flexural, psi (MPa) | 350 (2.4) | 375 (2.6) | 320 (2.2) |

# SERIES C - TABLE 1
## Fly Ash Evaluation

| | Control Mix | | | | Test Mix | | | |
|---|---|---|---|---|---|---|---|---|
| Cement, lbs (Kg) | 530 (240) | | | | 500 (226) | | | |
| Fly Ash, lbs (Kg) | – | | | | 50 (22.6) | | | |
| Sand, lbs (Kg) | 2150 (973) | | | | 2150 (973) | | | |
| Water, gals (L) | 35.7 (135) | | | | 34.5 (130) | | | |
| Admixture | A12 | | | | A12 | | | |
| Properties | Initial | 17 Hour | 24 Hour | 41 Hour | Initial | 17 Hour | 24 Hour | 41 Hour |
| Penetration, MM | 69 | 57 | 56 | 37 | 73 | 59 | 58 | 48 |
| Air Content, % | 19½ | 16 | 17 | 14 | 21 | 18½ | 16½ | 14 |
| Unit Weight, pcf | 111.7 | 115.3 | 115.4 | 117.3 | 110.5 | 113.7 | 115.6 | 117.9 |
| Temperature, °F(°C) | | | | | | | | |
| Mortar | 65 (18) | 60 (16) | 61 (16) | – | 63 (17) | 59 (15) | 62 (17) | – |
| Ambient | 69 (21) | 65 (18) | 69 (21) | – | 68 (20) | 65 (18) | 69 (21) | – |
| Strength, psi (MPa) | | | | | | | | |
| Mortar Compress. | | | | | 2125 (14.6) | | | |
| Prism Bond | | | | | 103 (0.7) | | | |

# SERIES C - TABLE 2
## Field Evaluation of Fly Ash

**Mix Parameters**

| | | |
|---|---|---|
| Cement, lbs (Kg) | 620 (281) | 620 (281) |
| Fly Ash, lbs (Kg) | 75 (33.9) | 100 (45.3) |
| Sand, lbs (Kg) | 2130 (964) | 2050 (928) |
| Water, lbs (L) | 37.2 (141) | 36.0 (136) |
| Admixture | A2 | A2 |

**Properties**

| | | |
|---|---|---|
| Penetration, MM | 60 | 60 |
| Air Content, % | 18 | 21 |
| Unit Weight, pcf | 116.0 | 110.3 |
| Strength, psi | | |
| Mortar Compress | 2140 (14.7) | 1925 (13.3) |
| Prism Bond | 80 (0.55) | -- |

# SERIES D - TABLE 1 - ASTM C780
## Compressive Strength of Mortar, psi (MPa)

| | Ref. 1 Type N Mortar | Test 1 Type N Mortar | Ref. 2 Type S Mortar | Test 2 Type S Mortar |
|---|---|---|---|---|
| Initial Penetration, mm | 60 | 48 | 63 | 52 |
| Water Retention % | 89 | 92 | 80 | 91 |
| Air Content % | 3.8 | 17.0 | 4.4 | 19.2 |
| Unit Weight, pcf | N.R. | 117.4 | 126.5 | 106.4 |
| Compress Strength, psi (MPa) | 1740(12.0) | 2710(18.7) | 2160(14.9) | 1500(10.3) |

N.R. = Not reported

# SERIES D - TABLE 2
## Flexural Bond Strength

|                              | Ref. 1    | Test 1    | Ref. 2    | Test 2    |
|------------------------------|-----------|-----------|-----------|-----------|
| Mortar Type                  | Type N    | Type N    | Type S    | Type S    |
| IR.A. gm/min/30 in$_2$       | 17        | 17        | 24        | 24        |
| AGE at Test                  | 28 days   | 28 days   | 32 days   | 32 days   |
| Flexural Strength, psi (MPa) | 47 (0.3)  | 96 (0.6)  | 157 (1.1) | 63 (0.4)  |

# SERIES D - TABLE 3
## Water Permeance - ASTM E514

|                            | Ref. 1      | Test 1     | Ref. 2    | Test 2      |
|----------------------------|-------------|------------|-----------|-------------|
| First dampness             | 1¼ hours    | 8 hours    | 20 min.   | 45 min.     |
| % Damp Area                |             |            |           |             |
| 1st Day                    | 32          | 5          | 20        | 45          |
| 2nd Day                    | 36          | 6          | 80        | 95          |
| 3rd Day                    | 37          | 6          | --        | --          |
| Begin leakage              | 8 hours     | 39 hours   | 3 hours   | 6½ hours    |
| Flow from lower flashing   | 200ml       | 0ml        | --        | --          |
| Rate at end of one day     | 415ml       | --         | --        | --          |
| Stabilized rate            | 35ml/hr     | --         | 145ml/hr  | 407ml/hr    |
| Permeance Rating           | F           | E          | F         | F           |

DISCUSSION


"The Development of Ready Mixed Mortar in the United States" - R. E. Gates, R. L. Nelson, and M. F. Pistilli

Question ( J. H. Matthys, University of Texas at Arlington):

In the Introduction the authors indicate contractors and producers have accumulated volumes of test data verifying the technical excellence of this mortar.    I suggest that the authors provide appropriate references to this data so that evaluation of its significance can be made by one having an interest.

In Test Procedures the authors indicate adoption of a guideline that provided the much needed standard of the techniques used to fabricate, cure, store and test the specimens.    This document as referenced is dated 1988. Is all the data presented in this paper based on these guidelines - or only some of the data.    If only some - how does data development differ?

Under Properties of Plastic Mortar authors state that a very important characteristic of ready mix mortar is its ability to retain the entrained air over the expected life.    Please furnish the readers a reference to substantiate this.

You state most of this work generated from field production mortar rather than laboratory batched mortar. Please clarify the difference between these two categories for ready mix mortar.    Also indicate specifically which data in the paper is field mortar and lab mortar.

Under Materials you indicate data presented comes from many different sources.    References for the individual data needs to be given so if desired, others can obtain the appropriate details of the data needed to make an evaluation.    Why did you accept data on oversanded mixes?    For such mixes which strength requirement was satisfied:    compression or bond?

You refer to low, medium or high IRA range of brick used.    What is the specific range for each of these classifications?    Were the IRA values based on as laid condition or oven dried?

For the reported compressive strength in Series A of ready mix mortar specimens and prisms (36 hour & 60 hour mortars), were the specimens made immediately after initial mixing or after certain suspension times, i.e., six hours, 12 hours, etc.?    This question on suspension times really applies to all the ready mix mortar data mentioned in this paper.  Others that have conducted tests in the ready mix mortar area (Matthys, etc., ASTM STP 992) have shown potentially significant changes in properties depending on suspension times.  What were the specific curing conditions and construction conditions for the prisms?

For the water permeance test of walls in Series A, how many walls were tested for each mortar type?

In Series B you state all data generated in the laboratory under generally standard lab conditions. Please indicate which presented data did not meet standard lab conditions and what deviations existed for each.

For specimens in Series C, how were specimens constructed, cured, and tested?

For reference mortars in Series D, were the mortars proportioned based on C-270 and then mixed to a flow for lab mortar or field mortar?  Is the only difference in quantity of materials between the test mortars and the reference mortar the ready mix admixture used?

For Series D the mortar compression strength for the Ref. Type N mortar was less than that of the corresponding strength for Type N Ready Mix.  For Type S mortar the opposite was true.  Why?

The presented water penetration results on walls of Series D is an average of how many wall tests?  Were the walls single wythe brick/block or multiwythe brick/block?  What are the properties of the units used?

The first sentence of the conclusion should probably read, "The results of the test programs demonstrate that for the products tested, ready mix mortar compares favorably to conventional mortars."  The statement that higher bond strengths and lower water penetration for masonry made with ready mix mortar is not substantiated by Type S mortar in Table 2 and 3 of Series D.

Answer (Robert E. Gates) Your questions have been answered as follows:

1.  There has been much data accumulated by the producers and users
    of ready mixed mortar. Many have indicated that they prefer not
    to publish the data. For those who have indicated a willingness,
    the authors will try to provide assistance to get this
    information published on a timely basis.

2.  The test procedure had not been published formally until 1988.
    Prior to that time similar procedures were followed.

3.  There is no reference to substantiate this. These are the
    author's opinions of general field testing and observations.
    Hopefully as more of the data referenced in Item 1 becomes
    available, this item will be provided.

4.  Field produced mortar refers to mortar which was commercially
    produced in volumes exceeding 1 cu.yd. (in most cases a minimum
    of 2 cubic yards) while lab batched mortar might be about a 1
    cu.ft. volume.

5.  References have been provided within the revised text of the
    paper as to where to obtain individual material data.

6.  The IRA values were based on ovendried condition. Specifics of
    this testing performed by H. H. Holmes Testing Laboratories, Inc.
    in Wheeling, IL for W. R. Grace & Co., 62 Whittemore Ave.,
    Cambridge, MA  02140-1692 are available on request.

7.  Prisms were made within 1 hour after mixing. Prisms were covered
    in plastic (bagged) after fabrication until time of test,
    excepting a short period for capping.

8.  One wall section for each condition was fabricated and tested by
    the National Concrete Masonry Association.

9.  The non-standard lab condition refers to ambient temperatures of
    17°C and 18°C (62.6°F and 64.4°F) which is lower than the
    standard 23°C ± 1.7°C (73.4°F ± 3°F) laboratory control
    temperature.

10. Specimens in Series C were fabricated from commercially produced
    ready mixed mortar sampled at a job site, then taken to the
    mortar producer's laboratory where the specimens were
    fabricated. Curing for compressive strength specimens was done
    in lime saturated water. Prisms were cured in plastic bags.

11. The mason who produced the mortar in a portable mixer visually
    and physically adjudged the mortar to be satisfactory for his
    intended use. The physical properties of the mortar were then
    determined by ASTM standard methods.

12. The Type S data set and the Type N data set were made on different days, from different cements at different laboratories.  The combined data should not be viewed, only the individual data sets.

13. Multiwythe walls were used employing brick and block.

14. It does.

Sylvester Schmidt, Marshall L. Brown, and Roger D. Tate

QUALITY CONTROL OF MORTARS: CUBES VS. CYLINDERS

REFERENCE: Schmidt, S., Brown, M.L., and Tate, R.D., "Quality Control of Mortars: Cubes vs. Cylinders," Symposium on Masonry: Components to Assemblages, ASTM STP 1063, John H. Matthys, Ed., American Society for Testing and Materials, Philadelphia, 1990.

ABSTRACT: Use of 75x150-mm (3x6-in.) and 100x200-mm (4x8-in.) cylinders was found to be more consistent than 50x100-mm (2x4-in.) cylinders in predicting 50-mm (2-in.) cube compressive strengths. The use of 100-mm (4-in.) cubes generally produced the same results as 50-mm (2-in.) cubes but not in every case. The compressive strength results for cylinders can be conservatively assumed to be 85% of 50-mm (2 in.) cube strengths for all mortars tested except for conventionally-mixed masonry cement mortars, which produced cylinder strengths which were 93% as strong as their companion cube strengths. Ready-mixed mortars produced no difference in the relationship between cylinders and cubes from conventionally-mixed mortars.

KEYWORDS: conventional mortars; ready-mixed mortars; compressive strength; size effects; shape effects; quality control

In the development of an ASTM standard for ready-mixed (including retarding admixtures) mortar, questions arose regarding the relationship between cylinder and cube compressive strengths. An existing standard, ASTM Standard Specification for Preconstruction and Construction Evaluation of Mortars for Plain and Reinforced Unit Masonry C 780-87, which governs the process of evaluating conventionally-mixed (without admixtures) mortars, states when cubes and cylinders are to be compared for a particular mixture, cylinder compressive strengths may be considered to be equal to 85% of cube compressive strengths. But during the development of the Standard, the relationship was not investigated for ready-mixed mortars. Does the retarding effect of the admixtures used in ready-mixed mortars affect the

The authors are from Addiment, Inc., a construction chemicals company, P.O. Box 47520, Atlanta, Georgia 30362. S. Schmidt, President; M. L. Brown, Manager of Technical Services; and R. Tate, Manager of Laboratories.

relationship between cubes and cylinders?  Also,  what effect does cylinder and cube size have on the  compressive strengths obtained?  Can strength results from different size cylinders be used interchangeably?  In order to find answers to these questions, this study was undertaken.

The  objectives of this research were to determine whether cylindrical  specimens  could  be  considered  as  satisfactory alternatives to cube specimens for measuring the quality of ready-mixed mortars, and to determine the relationship between cubes and cylinders for both conventionally-mixed and ready-mixed mortars.

Four different mixtures were investigated:

1) conventionally-mixed portland cement/lime mortar;
2) ready-mixed portland cement/lime mortar;
3) conventionally-mixed masonry cement mortar; and
4) ready-mixed masonry cement mortar.

The mix designs used for the four mortars as well as selected plastic properties measured after mixing are shown in Table 1.

For the conventional mixtures, type S mortars were produced using the proportion specification of ASTM Standard Specification for Mortar for Unit Masonry C 270-88a with loose unit weights of 1152 kg/cubic meter (72.0 lb./cft.) for portland cement, 448 kg/cubic meter (28.0 lb./cft.) for hydrated lime, 1053 kg/cubic meter (65.8 lb./cft.) for masonry cement, and 1498 kg/cubic meter (93.6 lb./cft.) for masonry sand.  Loose unit weights were determined using a 0.007 cubic meter (1/4 cubic foot) container. The moisture content of the masonry sand was extremely high, being 18%.

Consistency (workability), air content, and compressive strength were measured for each mortar.  Consistency was measured by flow (ASTM Standard Test Method for Compressive Strength of Hydraulic Cement Mortars C 109-87) as well as by cone penetration (ASTM C 780-87).  Due to their respective or potential ASTM specifications, the percent flow method was used to govern the consistency of the conventionally-mixed mortars (110% +/- 5%), and the cone penetration method was used to govern the consistency of the ready-mixed mortars. Air content was measured for all mortars using the pressure method.  Specimens were molded into three cylinder and two cube sizes [50x100-mm (2x4-in.), 75x150-mm (3x6-in.), and 100x200-mm (4x8-in.) cylinders, and 50-mm (2 in.) and 100-mm (4-in.) cubes].  The cylinder molds were made of plastic and the cube molds were made of steel.

Initial curing was conducted differently for ready-mixed and conventionally-mixed mortars, based on general practices used for them in the field.  The ready-mixed mortar specimens were stored in laboratory air in their molds which were sealed with lids for the first 7 days of curing and then stripped from their molds and submerged in lime-saturated water.  Conventionally-mixed mortars were stripped from their molds at an age of 24 hours and immediately submerged in lime-saturated water.

Compressive strengths were measured at 14, 28, 42, and 56 days
of age  (1,2). Strength  results and sample statistics for both
conventionally-mixed and ready-mixed mortars are presented in
Tables 2 and 3, respectively. Each of the values reported in the
Tables are averages of five specimens, except for the 100-mm (4-
in.) cubes, which were averages of three.

Figures 1 through 4 show compressive strength versus time for
each mortar tested.   Each Figure displays the results for the
different geometric shapes and sizes chosen in this study. After
visually inspecting the Figures, the results seem reasonable,
except possibly the ready-mixed portland cement/lime mortar at 14
days of age (Figure 2). It shows that the 50-mm (2-in.) cube
strengths were lower than two of the three cylinder strength
averages at that age.  This is not consistent with what would be
expected when considering the influence of specimen slenderness on
compressive strength. It is not certain why this phenomenon
occurred. This indicates that the relationship between cubes and
cylinders for ready-mixed portland cement/lime mortars in the
first two weeks may vary from the generally accepted relationship
that cubes should produce higher strengths than 1:2
(diameter:length) cylinders, based on geometric differences.

In order to properly analyze the data, a statistical analysis
was carried out at Clemson University, Clemson, SC, using
Statistical Analysis System (SAS) (1), taking into account any
differences in sample size as well as the variability within each
set of mortar strength results.  The first analysis incorporated
all of the data measured in the study; the second analysis deleted
any data which might be considered anomalous.

INFLUENCE OF CUBE SIZE

For the limited amount of data collected for the 100-mm (4-
in.) cube size, it was determined that 50-mm (2-in.) and 100-mm
(4-in.) cubes produce equivalent strengths for ready-mixed and
conventionally-mixed masonry cement mortars (Table 4). However,
different strengths were determined between the two cube sizes for
conventionally-mixed portland cement/lime mortar.  The difference
may be attributable to the use of a somewhat different compaction
procedure for the two cube sizes.  The 50-mm (2-in.) cubes were
compacted as described in ASTM C 109, and the 100-mm (4-in.) cubes
were compacted using the same size tamper, requiring 128 tampings
instead of the conventional 32 tampings.  Therefore, it cannot be
conclusive from this study that 50-mm (2-in.) and 100-mm (4-in.)
cubes can be used interchangeably. No 100-mm (4-in.) cubes were
molded for ready-mixed portland cement/lime mortar.

INFLUENCE OF CYLINDER SIZE

Comparing the compressive strength results for the three sizes
of cylinders  for each of the four mortars, it was statistically
determined that 50X100-mm (2X4-in.) cylinder strengths were not
always equivalent to 75X150-mm (3x6-in) and 100X200-mm (4X8-in.)
cylinders.  This was probably due to several factors. Since the
50X100-mm (2X4-in.) cylinders are very small, they may be more
sensitive to differences in compaction variations than the larger

cylinders.  Also, small variations in alignment during capping and
testing are more significant to such small cylinders.  Since the
50X100-mm (2X4-in.) cylinder data was not always consistent with
the strengths obtained for the other size cylinders, all 50X100-mm
(2X4-in.) cylinder strengths were deleted from the second analysis
when any data which appeared to be anomalous was eliminated.  For
75X150-mm (3X6-in.) and 100X200-mm (4X8-in.) cylinders, strengths
were found to be statistically equivalent for similar mortars,
regardless of age, with the exception of the ready-mixed
portland/lime mixtures, where the results for the 75X150-mm (3x6-
in.) cylinders were abnormally high in comparison to the 50-mm (2-
in.) cube strengths.  Also within the data for ready-mixed
portland/lime mortars, the 14 day results showed no consistent
correlation between any of the cylinders or cubes.  This may be
due to the early-age retardation of ready-mixed mortars, which
might have had a greater effect on the portland/lime mortars than
it did on the masonry cement mixtures.  Since the 75X150-mm (3X6-
in.) cylinders and all of the 14-day findings for ready-mixed
portland cement/lime mortar were somewhat inconsistent with the
other trends in this study, their results were deleted in the
second analysis.

RELATIONSHIP BETWEEN CUBES AND CYLINDERS

    Table 5 shows the relationships between cylinder and 50-mm (2-
in) cube strength averages.

    Table 6 shows the statistically-generated predictive
relationships between cylinders and 50-mm (2-in.) cubes, initially
for all of the data combined, and then excluding selected results
which appeared to be anomalous. It shows the relationships for all
mortars as a group, separated by mix location (conventionally- or
ready-mixed), and separated by each location-mixture combination.
The coefficient of determination ($R^2$) also was reported, which is
a measure of how well a particular predicted relationship models
the actual data.  Its absolute value can range from 0.0 to 1.00. A
perfect correlation between actual data and a predictive model
will give a value of 1.00, and a very inappropriately chosen
predictive model will give a value near 0.00. The $R^2$ value is also
a function of the number of measurements taken (n) as well as
where the data lies along the predictive relationship.

    Interestingly enough, it appears that the elimination of
selected results in the second analysis did not change the
relationships found between cylinders and cubes significantly.
Therefore, the remaining discussion will center around the first
part of Table 6 which presents findings based on the inclusion of
all data measured.

    When all of the mortar strength results were combined, the
general relationship developed between cylinders and cubes was
determined to be

    Cyl. Str. = 0.83 (Cube Str.)    ($R^2$ = 0.99, n = 48)

with a 95 % probability that the true relationship lies between
0.81 and 0.85.  This would indicate that the relationship

expressed in ASTM C 780 of 0.85 between cylinders and 2-in. cubes
is a reasonable single point value for this study as well.

However, to understand more closely how this relationship may
change for individual cases, it was decided to separate the mortar
results into two categories: conventionally-mixed and ready-mixed
mortars.

The relationship developed between cylinders and 50-mm (2-in.)
cubes for conventionally-mixed mortars was

Cyl. Str. = 0.83 (Cube Str.)   ($R^2$ = 0.99, n = 24)

with a 95 % probability that the true relationship lies between
0.80 and 0.86.

For ready-mixed mortars, the relationship was found to be

Cyl. Str. = 0.83 (Cube Str.)   ($R^2$ = 0.99, n = 24)

with a 95 % probability that the true relationship lies between
0.80 and 0.86 -- exactly the same as for conventionally-mixed
mortars. This again indicates that the relationship expressed in
ASTM C 780 of 0.85 is reasonable.

Another step in data segregation was investigated which
separated the mortars by mixture as well as where they were
produced. For conventionally-mixed masonry cement mortar, the
relationship was determined to be

Cyl. Str. = 0.93 (Cube Str.)   ($R^2$ = 1.00, n = 12)

with a 95 % probability that the true relationship lies between
0.90 and 0.96. This result was surprisingly high.

For conventionally-mixed portland cement/lime mortars, the
relationship was found to be

Cyl. Str. = 0.81 (Cube Str.)   ($R^2$ = 1.00, n = 12)

with a 95 % probability that the true relationship lies between
0.78 and 0.83.

For ready-mixed masonry cement mortar,

Cyl. Str. = 0.84 (Cube Str.)   ($R^2$ = 1.00, n = 12)

with a 95 % probability that the true relationship lies between
0.81 and 0.86.

And finally, for ready-mixed portland cement/lime mortar, the
relationship was

Cyl. Str. = 0.83 (Cube Str.)   ($R^2$ = 1.00, n = 12)

with a 95 % probability that the true relationship lies between 0.78 and 0.87.

All of the relationships except those for conventionally-mixed masonry cement mortar showed that the value of 0.85 presented in ASTM C 780 is a relatively conservative single point value for predicting 2-in. cube strengths from cylinders. The much higher value obtained in this study for conventionally-mixed masonry cement must be studied further before any conclusion can be reached regarding why such a result was obtained.

CONCLUSIONS

The following conclusions can be drawn from this study:

1) Mortar strengths obtained through the use of 50X100-mm (2X4-in.) cylinders are not always equivalent to values obtained for 75X150-mm (3X6-in.) or 100X200-mm (4X8-in.) cylinders. Therefore, it is recommended that only 75X150-mm (3X6-in.) or 100X200-mm (4X8-in.) cylinders be used to measure mortar cylinder strengths.

2) Mortar strengths obtained in this study through the use of 100 mm (4-in.) cubes were not always equivalent to values obtained with 50 mm (2-in.) cubes. A difference in tamping technique for the two sizes may have been the reason for the difference in some of the strengths between the two cube sizes. Since a difference was detected in their strengths, the authors must recommend that only 50-mm (2-in.) cubes be used to measure mortar cube strengths.

3) a. Grouping all of the data together, the relationship between cylinders and cubes was determined to be

Cyl. Str. = 0.83 (Cube Str.).

3) b. Separating the data into two categories, conventionally-mixed and ready-mixed mortars, yielded two identical relationships which were found to be the same as the one shown above.

3) c. Separating the data further into four categories, such that each mixture was analyzed individually, the relationships were found to be different from the one shown above. For conventionally-mixed masonry cement mortar, the relationship   was found to be

Cyl. Str.= 0.93 (Cube Str.);

For conventionally-mixed portland cement/lime mortar it was

Cyl. Str.= 0.81 (Cube Str.);

For ready-mixed masonry cement mortar,

Cyl. Str.= 0.84 (Cube Str.);

and for ready-mixed portland cement/lime mortar,

Cyl. Str.= 0.83 (Cube Str.).

With the exception of conventionally-mixed masonry cement mortar, the cylinder to cube strength relationships determined in this study suggest that the cylinder to cube relationship of 0.85 expressed in ASTM C 780 conservatively and satisfactorily predicts 50-mm (2-in.) cube strengths from cylinder strengths for both conventionally-mixed and ready-mixed mortars.

4) More data should be generated for conventionally-mixed masonry cement mortar before suggesting a different relationship from what is recommended in ASTM C 780.

ACKNOWLEDGEMENT

The authors would like to thank Dr. Lawrence Grimes of Clemson University, Clemson, SC, for his assistance in the statistical analysis of the data developed in the study.

REFERENCES

[1] SAS Institute Inc., SAS/GRAPH Users Guide, Version 5 Edition, Cary, NC: Sas Institute, Inc., 1985, 596 pp.

TABLE 1. Mix Designs

| Material | Conventionally-Mixed | | Ready-Mixed | |
| | Portland/Lime | Masonry | Portland/Lime | Masonry |
|---|---|---|---|---|
| Cement* | 500 lbs. | 530 lbs. | 500 lbs. | 550 lbs. |
| Lime | 96 lbs. | none | 100 lbs. | none |
| Sand (SSD) | 2520 lbs. | 1930 lbs. | 2140 lbs. | 1930 lbs. |
| Water | 49.0 gals. | 49.3 gals. | 43.2 gals. | 49.0 gals. |
| Admixture #1 | none | none | 78.0 oz. | none |
| Admixture #2 | none | none | none | 69.0 oz. |
| Percent Flow | 105 % | 109 % | 98 % | 114 % |
| Cone | | | | |
| Penetration | 52 mm | 67 mm | 58 mm | 68 mm |
| Air Content | 5.7 % | 19.8 % | 18.4 % | 19.5 % |
| Unit Weight | 127.0 pcf | 105.6 pcf | 112.6 pcf | 107.6 pcf |

*Type I Portland Cement or Masonry Cement

Table 2. Compressive Strengths for Conventionally-Mixed Mortars

| Specimen Type | | Portland/Lime Mortar | | | | Masonry Cement Mortar | | | |
|---|---|---|---|---|---|---|---|---|---|
| | | Age in Days | | | | | | | |
| | | 14 | 28 | 42 | 56 | 14 | 28 | 42 | 56 |
| 2-in. Cubes | Average* | 1420 | 1700 | 1920 | 1990 | 750 | 850 | 880 | 900 |
| | Stan. Dev. | 87 | 17 | 65 | 56 | 43 | 22 | 11 | 12 |
| | COV (%) | 6.1 | 1.0 | 3.4 | 2.8 | 5.7 | 2.6 | 1.3 | 1.3 |
| | n | 5 | 5 | 5 | 5 | 5 | 5 | 5 | 5 |
| 4-in. Cubes | Average* | N/A | 1520 | 1560 | N/A | N/A | 930 | 980 | N/A |
| | Stan. Dev. | | 53 | 39 | | | 40 | 6 | |
| | COV (%) | | 3.5 | 2.5 | | | 4.3 | 0.6 | |
| | n | | 3 | 3 | | | 3 | 3 | |
| 2-in. Cylinders | Average* | 1210 | 1490 | 1530 | 1550 | 630 | 760 | 800 | 820 |
| | Stan. Dev. | 49 | 27 | 16 | 15 | 21 | 18 | 26 | 33 |
| | COV (%) | 4.0 | 1.8 | 1.1 | 1.0 | 3.3 | 2.4 | 3.2 | 4.0 |
| | n | 5 | 5 | 5 | 5 | 5 | 5 | 5 | 5 |
| 3-in. Cylinders | Average* | 1170 | 1480 | 1500 | 1510 | 640 | 810 | 870 | 870 |
| | Stan. Dev. | 30 | 29 | 19 | 29 | 18 | 10 | 17 | 15 |
| | COV (%) | 2.5 | 2.0 | 1.3 | 1.9 | 2.8 | 1.2 | 1.9 | 1.8 |
| | n | 5 | 5 | 5 | 5 | 5 | 5 | 5 | 5 |
| 4-in. Cylinders | Average* | 1190 | 1470 | 1470 | 1530 | 660 | 820 | 860 | 910 |
| | Stan. Dev. | 29 | 26 | 25 | 23 | 5 | 10 | 20 | 24 |
| | COV (%) | 2.4 | 1.7 | 1.7 | 1.5 | 0.7 | 1.5 | 2.4 | 2.8 |
| | n | 5 | 5 | 5 | 5 | 5 | 5 | 5 | 5 |

*psi (MPa = psi X .006895)

Table 3. Compressive Strengths for Ready-Mixed Mortars

| Specimen Type | Statistics | Portland/Lime Mortar | | | | Masonry Cement Mortar | | | |
|---|---|---|---|---|---|---|---|---|---|
| | | 14 | 28 | 42 | 56 | 14 | 28 | 42 | 56 |
| 2-in. Cubes | Average* | 2020 | 2950 | 2990 | 3140 | 1210 | 1430 | 1590 | 1740 |
| | Stan. Dev. | 65 | 48 | 110 | 101 | 23 | 55 | 91 | 91 |
| | COV (%) | 3.2 | 1.6 | 3.7 | 3.2 | 1.9 | 3.9 | 5.7 | 5.2 |
| | n | 5 | 5 | 5 | 5 | 5 | 5 | 5 | 5 |
| 4-in. Cubes | Average* | N/A | N/A | N/A | N/A | N/A | 1410 | 1680 | N/A |
| | Stan. Dev. | | | | | | 21 | 66 | |
| | COV (%) | | | | | | 1.5 | 3.9 | |
| | n | | | | | | 3 | 3 | |
| 2-in. Cylind- ers | Average* | 1850 | 2360 | 2390 | 2380 | 900 | 1230 | 1300 | 1360 |
| | Stan. Dev. | 30 | 99 | 71 | 69 | 40 | 36 | 36 | 81 |
| | COV (%) | 1.6 | 4.2 | 3.0 | 2.9 | 4.7 | 2.9 | 2.8 | 6.0 |
| | n | 5 | 5 | 5 | 5 | 5 | 5 | 5 | 5 |
| 3-in. Cylind- ers | Average* | 2170 | 2460 | 2490 | 2540 | 1060 | 1250 | 1380 | 1440 |
| | Stan. Dev. | 43 | 39 | 71 | 48 | 34 | 9 | 50 | 50 |
| | COV (%) | 2.0 | 1.6 | 2.9 | 1.9 | 3.2 | 0.7 | 3.6 | 3.4 |
| | n | 5 | 5 | 5 | 5 | 5 | 5 | 5 | 5 |
| 4-in. Cylind- ers | Average* | 2080 | 2360 | 2390 | 2390 | 1070 | 1240 | 1330 | 1420 |
| | Stan. Dev. | 15 | 27 | 20 | 26 | 34 | 23 | 38 | 25 |
| | COV (%) | 0.7 | 1.1 | 0.8 | 1.1 | 3.2 | 1.9 | 2.9 | 1.8 |
| | n | 5 | 5 | 5 | 5 | 5 | 5 | 5 | 5 |

*psi (MPa = psi X .006895)

Table 4. Average Relationships Between 2-In. and 4-In. Cubes

Conventionally-Mixed Mortars--------------------------------------------

| | Portland/Lime Mortar | | | | Masonry Cement Mortar | | | |
|---|---|---|---|---|---|---|---|---|
| | 14 | 28 | 42 | 56 | 14 | 28 | 42 | 56 |
| (4-in./2-in.) | N/A | .89 | .81 | | N/A | N/A | 1.09 | 1.11 N/A |

Ready-Mixed Mortars-----------------------------------------------------

| | Portland/Lime Mortar | | | | Masonry Cement Mortar | | | |
|---|---|---|---|---|---|---|---|---|
| | 14 | 28 | 42 | 56 | 14 | 28 | 42 | 56 |
| (4-in./2-in.) | N/A | N/A | N/A | | N/A | N/A | .99 | 1.06 N/A |

Table 5. Average Relationships Between Cylinders and 2-In. Cubes

Conventionally-Mixed Mortars------------------------------------------

| | Portland/Lime Mortar | | | | Masonry Cement Mortar | | | |
|---|---|---|---|---|---|---|---|---|
| | 14 | 28 | 42 | 56 | 14 | 28 | 42 | 56 |
| 2-in. Cylinders | .85 | .88 | .80 | .78 | .84 | .89 | .91 | .91 |
| 3-in. Cylinders | .82 | .87 | .78 | .76 | .85 | .95 | .99 | .97 |
| 4-in. Cylinders | .84 | .87 | .77 | .77 | .88 | .97 | .98 | 1.01 |

Ready-Mixed Mortars-------------------------------------------------

| | Portland/Lime Mortar | | | | Masonry Cement Mortar | | | |
|---|---|---|---|---|---|---|---|---|
| | 14 | 28 | 42 | 56 | 14 | 28 | 42 | 56 |
| 2-in. Cylinders | .92 | .80 | .80 | .76 | .74 | .86 | .82 | .78 |
| 3-in. Cylinders | 1.07 | .83 | .83 | .81 | .88 | .87 | .87 | .83 |
| 4-in. Cylinders | 1.03 | .80 | .80 | .76 | .88 | .87 | .84 | .82 |

Table 6. Statistical Relationships Between Cylinders and 2-In. Cubes

| | Cylinder Strength 2-In. Cube Strength | 95% Coverage Probability | R² | n |
|---|---|---|---|---|
| All Data Included: | | | | |
| General Relationship | 0.83 | 0.81-0.85 | 0.99 | 48 |
| Conventionally-Mixed | 0.83 | 0.80-0.86 | 0.99 | 24 |
| Ready-Mixed | 0.83 | 0.83-0.86 | 0.99 | 24 |
| Conventionally-Mixed,Masonry | 0.93 | 0.90-0.96 | 1.00 | 12 |
| Conventionally-Mixed, Portland/Lime | 0.81 | 0.78-0.83 | 1.00 | 12 |
| Ready-Mixed, Masonry | 0.84 | 0.81-0.86 | 1.00 | 12 |
| Ready-Mixed, Portland/Lime | 0.83 | 0.78-0.87 | 1.00 | 12 |
| Some Data Excluded:* | | | | |
| General Relationship | 0.82 | 0.80-0.84 | 1.00 | 27 |
| Conventionally-Mixed | 0.83 | 0.79-0.87 | 0.99 | 18 |
| Ready-Mixed | 0.85 | 0.83-0.87 | 1.00 | 8 |
| Conventionally-Mixed,Masonry | 0.95 | 0.91-1.00 | 1.00 | 27 |
| Conventionally-Mixed, Portland/Lime | 0.80 | 0.78-0.83 | 1.00 | 27 |
| Ready-Mixed, Masonry | 0.85 | 0.82-0.88 | 1.00 | 27 |
| Ready-Mixed, Portland/Lime | 0.76 | 0.76-0.81 | 1.00 | 27 |

*All 50-mm (2-In.) Cubes, and 14 Day and 75X150-mm (3X6-In.) Cyls. for Ready-Mixed Portland Cement/Lime Mortar Deleted (See text).

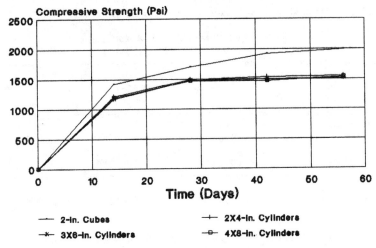

FIG. 1 -- Conventionally-Mixed Portland/Lime Mortar
(MPa = psi X .006895)

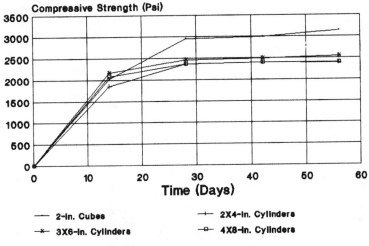

FIG. 2 -- Ready-Mixed Portland/Lime Mortar
(MPa = psi X .006895)

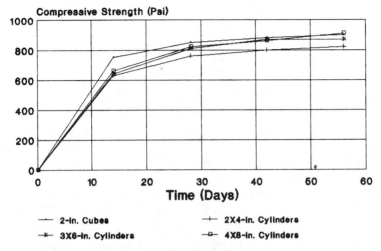

FIG. 3 -- Conventionally-Mixed Masonry Cement Mortar
(MPa = psi X .006895)

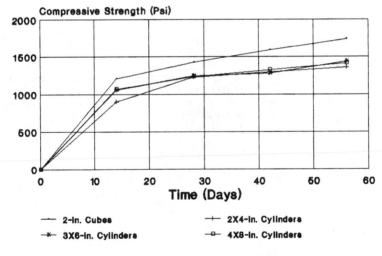

FIG. 4 -- Ready-Mixed Masonry Cement Mortar
(MPa = psi X .006895)

DISCUSSION

"Quality Control of Mortars:  Cubes vs. Cylinders" - S.
Schmidt, M. L. Brown, and R. D. Tate

Question   (J. H. Matthys, University of Texas at
Arlington):

The paper mentions in the curing process submerging all
specimens  (cubes and cylinders) for all mortar types in
lime-saturated water for approximately 21 to 27 days.
ASTM C-780-87 and ASTM C-1142 (Ready Mix Mortar) only
stipulate moist room curing.  In my experience lime
saturated water curing is not typically used in addressing  either C-780 or C-270 test requirements.  Thus I
would suppose that the 0.85 relationship of cylinders to
cubes mentioned in C-780 is based on non-lime cured
specimens.   Do you think that saturated lime water
curing could potentially give a different relationship
for cylinders to cubes as compared to normal moist
curing?

You presented data on two ready mix mortars:  (1) ready
mixed portland cement lime mortar, and (2) ready mixed
masonry cement mortar.  There are several manufacturers
of ready mix mortar additives.  Was the same manufacturer's additive used in both mixes reported?  What was the
design life of these ready mix mortars - i.e., 24 hour
mortar, 36 hour mortar, etc.?

Previous work conducted on ready mix mortars by Matthys,
etc. (ASTM STP 992), indicated potentially significant
changes in ready mix mortar properties depending on when
specimens were constructed with respect to initial mortar
mixing, i.e., suspension time of mortar.  A suspension
time of zero would mean the specimens were made immediately after initial mortar mixing.  A suspension time of
12 hours would mean the specimens were made 12 hours
after initial mixing.  For the data in your paper what
was the suspension time for the ready mix mortar?  Do you
have or know of any data with respect to cubes versus
cylinders for various suspension times of ready mix
mortars?

Closure   (M. L. Brown):

It is interesting that virtually the same relationship
was determined between cylinders and cubes for ASTM C-780
using moist-cured specimens as indicated by Dr. Matthys
and this study using lime-water submerged specimens (0.85
and 0.83, respectively).  This seems to indicate that no
difference in strength development should be expected
between lime-water submerged and moist-cured specimens.

Since it is difficult and expensive for producers and users of products such as mortar and concrete to maintain a moist cure room for everyday quality control, it is important to know how these mortars perform when cured in lime-water baths, which would be easier and more economical.

The ready mix mortar additive used for the masonry cement mortar was different from the additive used for the portland cement/lime mortar. Both additives were produced by the same manufacturer. The additive used for the portland cement/lime mortar mixture contained a retarder and air entraining agent, whereas the masonry cement mortar additive contained only a retarding agent (the masonry cement already contained an air-entraining agent). Both the ready mixed masonry cement and the portland cement/lime mortars were designed to have 36 hour working times.

All of the cube and cylinder specimens were molded immediately after mixing. Only a limited amount of data has been published investigating the influence of delayed molding of ready mix mortar specimens on the relationship between cylinder and cube strengths (Matthys, ASTM STP 992). However, specimens have been molded in the past at various time delays by jobsite quality control inspectors and testing laboratories. The outcome of such samplings has not been made available to the authors of this study for comment or comparison.

John H. Matthys

PERFORMANCE OF MORTARS PRODUCED UNDER
THE PROPORTION SPECIFICATION OF ASTM C-270-86a

REFERENCE: Matthys, John H., "Performance of
Mortars Produced Under the Proportion
Specification of ASTM C-270-86a," Masonry:
Components to Assemblages: ASTM STP 1063, J.
H. Matthys, Ed., American Society for Testing
and Materials, Philadelphia, 1990.

ABSTRACT: ASTM C-270-86a "Mortar For Unit
Masonry" lists different types of mortars
(i.e., M, S, N, O) that can be produced from a
proportion or a property specification. This
is the first ASTM C-270 standard that has
allowed the use of masonry cements as the only
cementitious material for all the mortar types
under the proportion specification. The
Construction Research Center at the University
of Texas at Arlington has conducted a
comparative performance study of Portland
cement lime (Type S lime) mortars to ASTM
equivalent masonry cement (MC) mortars to
determine whether mortars produced by the
proportion specification meet the performance
specification requirements of ASTM C-270.
Eleven mortars consisting of eight masonry
cement mortars and three Portland cement lime
mortars were evaluated.

KEYWORDS: conventional mortars, compressive
strength, water retention, air content,
Portland cement lime, masonry cement

In Spring 1987 the Construction Research Center at
the University of Texas at Arlington began a research
project sponsored by the National Lime Association
entitled, "Conventional Masonry Mortar Investigation".
The objective of that study was to evaluate the

Dr. John H. Matthys is Professor of Civil
Engineering and Director of the Construction Research
Center at The University of Texas at Arlington, Box
19347, Arlington, Texas 76019-0347.

equivalency of ASTM C-270-86a [1] Type S and Type N mortars produced using either Portland cement lime or masonry cement alone as the cementitious material. The reason for this study was due to the current changes in the mortar specifications found in ASTM C-270-86a.

Mortar for masonry work in the U.S.A. is typically specified in the contract documents by reference to ASTM C-270. The current specification covers mortar for use in both nonreinforced and reinforced masonry construction. Four types of mortar (M, S, N, O) are allowed in two alternate specifications: (1) proportion specification, and (2) property specification. The proportion specification basically specifies parts by volume of cementitious and noncementitious material to be used for a particular mortar type, i.e., a recipe specification. The property specification gives criteria on performance in terms of compressive strength, water retention, and air content that a laboratory mortar (Flow $\approx$ 100 to 115) must satisfy. Either the proportion or property specification governs as specified. When neither is specified, the proportion specification governs. Normally, the proportion specification of C-270 is used in contract documents in specifying mortar in the U.S.A. Prior to 1986 in the proportion specification, masonry cement had to be combined with Portland cement for Type M or S mortars; for N and O mortars masonry cement could be used alone.

With the adoption of the new ASTM C-270-86a standard any listed mortar type for either reinforced or unreinforced masonry is allowed to be produced using masonry cement alone as the binder by either the proportion or the property specification. This new provision in the proportion specification is related to the development of the new Type M and Type S masonry cements. Several groups in the building industry were concerned whether the mortars produced using commercially available masonry cements are really equivalent to Portland cement lime mortars particularly with respect to both mortar properties and masonry assemblage performance characteristics such as masonry flexural bond, masonry watertightness, masonry compressive strength, and masonry shear strength. Many people in the building community feel that such masonry cement mortars are not equivalent to Portland cement lime mortars even though the masonry cement mortars satisfy the proportion or property specification of ASTM C-270-86a and the material properties of ASTM C-91 (Specifications for Masonry Cements) [2]. Are like mortars produced under the proportion specification of C-270 equivalent? Material suppliers, the masonry industry, and the consumer should know the answer to this question. This paper deals only with the results of the mortar properties. The mortars were produced under the proportion specification of ASTM C-270-86a and

then examined in light of the property specification of
ASTM C-270-86a with respect to compression strength, air
content, and water retention as shown in Table 1.

TABLE 1 -- Property Specification ASTM C-270-86a

| Mortar Type PCL, MC | Strength 28 Day Min, kPa* | Water Retention Min, % | Air Max, % PCL, MC |
|---|---|---|---|
| M | 17500 | 75 | 12, 18 |
| S | 12600 | 75 | 12, 18 |
| N | 5250 | 75 | 14, 18 |
| O | 2450 | 75 | 14, 18 |

*7 kPa = 1 psi

MATERIAL SELECTION

The program developed to address the concern of
mortar equivalency consisted of a comparative study of
Type S and Type N masonry cement mortars (using masonry
cement alone as the only cementitious material) to Type
S and N Portland cement lime mortars.  Four Type S
masonry cement mortars, four Type N masonry cement
mortars, one Type S Portland cement lime mortar, one
Type N Portland cement lime mortar and one Type O
Portland cement lime mortar were evaluated.

In obtaining the materials for this study, in
general the requirements were that the materials be:

1.   Representative
2.   Commercially available
3.   Geographically dispersed
4.   Statistically evaluated

In the United States there are approximately 90
brands of masonry cements.  The procedure used to obtain
typical masonry cements was as follows:

1.   Divide the United States into four zones:
     West, Midwest, Northeast and Southeast
2.   Consider the masonry cements of greatest sales
     volume
3.   Purchase from dealer outlets

This procedure resulted in examining seven MC from
the West, 11 MC from the Midwest, five MC from the
Northeast, and nine MC from the Southeast (21 Type N
masonry cements, 11 Type S masonry cements).  These MC
were then tested for air content, water retention, and
compressive strength.  These three properties are
criteria for MC  listed in ASTM C-91  (Specification for

Masonry Cement) [2]. These three properties were statistically evaluated for the given masonry cement types. From this evaluation for both the Type S and N masonry cements, one brand was chosen that represented an upper end value, one representing a lower end value, and two representing approximately average values.

The four Type S MC and four Type N MC were purchased from commercial building supply centers across the U.S.A. and shipped to the University of Texas at Arlington, for investigation.

The Portland cement selected for use in the PCL mortars was a local (Dallas, Texas) cement meeting ASTM C-150 (Specification for Portland Cement) [3]. The masonry sand was a local sand meeting ASTM C-144 (Specification for Aggregate for Masonry Mortar) [4]. The lime was a Type S dolomitic hydrate that had average water requirements and average plasticity and met ASTM C-207 (Specification For Hydrated Lime For Masonry Purposes) [5].

MORTAR SPECIMENS AND TEST PROCEDURES

Mortar mix proportions were based on the stipulated mortar type per ASTM C-270-86a using the proportion by volume specification as shown in Table 2. Sand was examined for gradation and moisture content. All mortar mixing was conducted in a six cubic foot paddle type mortar mixer. Mixing time was standardized. Half the water and half the sand were charged into the mixer and mixed for two minutes. Then the cement and the lime (or masonry cement) were added and mixed for two minutes. The remainder of the sand and additional water was added and mixed for two minutes. A trial flow test was run to check the desired flow of $\approx$ 115. Additional water was added if required. A final one minute mixing was conducted. This mortar was then used for conducting mortar tests for compression strength, water retention, and air content.

Standard ASTM C-270-86, 5.1 cm. compression cubes were made in sets of six for 28 day moist cured test specimens for the eleven mortar types (eight masonry cement and three Portland cement lime mortars). Specimen construction, curing, and testing followed ASTM C-270 and ASTM C-109 [6].

For each of the eleven mortar mixes, water retentivity was evaluated using ASTM standard water retention apparatus as described in ASTM C-91. Flow was measured using the flow table described in ASTM C-230 [7].

For each of the eleven mortar mixes, air content was measured using the 400 ml. weight method prescribed in ASTM C-270-86a. Densities of materials were obtained on all materials used.

TABLE 2 -- Mortar Volume Proportions[a]

| Mix | Proportions | | | |
| --- | --- | --- | --- | --- |
| | Masonry Cement | Portland Cement | Lime | Sand |
| A-N | 1.0 | -- | -- | 3.0 |
| B-S | 1.0 | -- | -- | 3.0 |
| C-N | 1.0 | -- | -- | 3.0 |
| D-S | 1.0 | -- | -- | 3.0 |
| E-N | 1.0 | -- | -- | 3.0 |
| F-S | 1.0 | -- | -- | 3.0 |
| G-N | 1.0 | -- | -- | 3.0 |
| H-S | 1.0 | -- | -- | 3.0 |
| I-S | -- | 1.0 | 0.5 | 4.5 |
| J-N | -- | 1.0 | 1.0 | 6.0 |

[a]The proportions given are based on volume. The weight printed on bag of masonry cement is equivalent to one cubic foot. For Portland cement one cubic foot = 42.6 kg.; for lime one cubic foot is taken as 18.1 kg.

TEST RESULTS

The results of the mortar tests are given in Table 3.    The mortar column indicates the mortar mix designation and mortar type, i.e., A-N indicates mix A and Type N mortar. Mixes A-C-E-G are Type N mortars using Type N masonry cement alone as the binder.   Mixes B-D-F-H are Type S mortars using Type S masonry cement alone as the binder.   Mix I-S is a Type S Portland cement lime mortar and mix J-N is a Type N Portland cement lime mortar.

The flow column is the average flow table reading computed value for two flow tests per mortar mix. Typically the mortar flow was in the order of 115.

Computed air contents for Type N and Type S masonry cement mortars averaged 14.3% and 14.2%, respectively. Type S and Type N Portland cement lime mortars' air contents were 6.7% and 5.6%, respectively.   All of the mortar types examined met the maximum air content requirements of ASTM C-270-86a property specification.

TABLE 3 -- Mortar Mix Properties

| Mortar Type | Mortar Designation | Mortar Properties | | | |
|---|---|---|---|---|---|
| | | Flow % | Air % | Water Retention % | Cube Strength* kPa |
| N | A-N | 115.8 | 12.3 | 73.3 | 10059 |
| N | C-N | 115.0 | 15.1 | 88.0 | 8442 |
| N | E-N | 114.0 | 12.8 | 77.9 | 9149 |
| N | G-N | 116.6 | 15.5 | 90.6 | 10087 |
| S | B-S | 112.8 | 15.2 | 88.3 | 16905 |
| S | D-S | 107.8 | 13.0 | 81.4 | 26397 |
| S | F-S | 114.8 | 12.5 | 87.2 | 35700 |
| S | H-S | 115.6 | 16.0 | 87.0 | 13587 |
| S | I-S | 117.8 | 6.7 | 88.8 | 18228 |
| N | J-N | 116.0 | 5.6 | 85.8 | 8351 |

*7 kPa = 1 psi

The water retention on the Type N masonry cement mixes ranged from 73 to 91; the Type S masonry cement had a narrower spread from 81 to 88. The Portland cement lime mixes Type S and N gave retention values of 88 and 85, respectively. Only the masonry cement mix A-N failed to meet the ASTM C-270-86a minimum water retention requirement of the property specification.

The 5.1 cm. mortar compression cubes for the Type N masonry cement mixes varied from 8442 kPa to 10059 kPa with an average of 9436 kPa. The Type N Portland cement lime mortar compression strength was 7973 kPa. Type S masonry cement mixes cubes ranged from 13587 kPa to 35700 kPa with an average of 20216 kPa. Type S Portland cement lime mortar compression strength was 18228 kPa. All mixes easily satisfied the ASTM C-270-86a requirement for compression cube strength of 5250 kPa for Type N mortar and 12600 kPa for Type S mortar. All the Type N masonry cement mortar compression cube values were higher than the Type N Portland cement lime mortar. Two of the Type S masonry cement mortar values were higher than the Type S Portland cement lime mortar compression strength.

CONCLUSIONS

1. The Portland cement lime mortars, proportioned by the ASTM C-270-86a proportion specification met the property specification requirements of ASTM C-270-86a.

2.  The masonry cement mortars, proportioned by the ASTM
    C-270-86a met the property specification
    requirements of ASTM C-270-86a with the lone
    exception being mix A-N with respect to water
    retention.

3.  Based on 1 and 2 above one would have to say that
    mortars produced under the proportion specification
    of C-270-86a using either Portland cement lime or
    masonry cement alone as the binder are equivalent
    with respect to meeting the required ASTM mortar
    properties in the property specification of ASTM C-
    270-86a.

ACKNOWLEDGEMENTS

    This project was financially supported by the
National Lime Association of Arlington, Virginia.
Portland cement was donated by LaFarge Corporation of
Dallas, Texas.

REFERENCES

[1]  "Standard Specification For Mortar For Unit
     Masonry," ASTM C-270-86a, American Society for
     Testing and Materials, Philadelphia, PA   19103.
[2]  "Standard Specification For Masonry Cement," ASTM
     C-91-83a, American Society For Testing and
     Materials, Philadelphia, PA   19103.
[3]  "Standard Specification For Portland Cement," ASTM
     C-150-86, American Society For Testing and
     Materials, Philadelphia, PA   19103.
[4]  "Specification For Aggregate For Masonry Mortar,"
     ASTM C-144-84, American Society For Testing and
     Materials, Philadelphia, PA   19103.
[5]  "Specification For Hydrated Lime For Masonry
     Purposes," ASTM C-207-84, American Society For
     Testing and Materials, Philadelphia, PA   19103.
[6]  "Standard Test Method For Compressive Strength of
     Hydraulic Cement Mortars", ASTM C-109-86, American
     Society For Testing and Materials, Philadelphia, PA
     19103.
[7]  "Standard Specification For Flow Table For Use in
     Tests of Hydraulic Cement", ASTM C-230-83, American
     Society For Testing and Materials, Philadelphia, PA
     19103.

DISCUSSION

"Performance of Mortars Produced under the Proportion
Specification of ASTM C-270-86 a" - J. H. Matthys

Question (S. K. Ghosh, Portland Cement Association):
   The author states that four Type S and four Type N masonry
   cements were chosen, one brand in each case representing an
   upper end value, one representing a lower end value, and two
   representing approximately average values of air content,
   water retention, and compressive strength.  The writer
   wonders if any one product truly represented upper end or
   lower end values of all three properties.  Would the author
   shed more light on this?  Actually, if the author in his
   closure would make available the three properties for the 21
   Type N and the 11 Type S masonry cements considered,
   indicating the ones selected, that would be of much
   interest.

Answer (J. H. Matthys, University of Texas at Arlington):

The properties of the four resulting Type S masonry cements initially chosen in the study
were:

| Cement Brand | Air % | Water Ret. | Compression Strength - psi |
|---|---|---|---|
| 1 | 17.7 | 81 | 2175 |
| 2 | 19.0 | 83 | 2482 |
| 3 | 19.6 | 84 | 2406 |
| 4 | 21.3 | 84 | 3383 |

One represented an upper end, one represented a lower end, and two represented middle
values.

The properties of the four resulting Type N masonry cements initially chosen in the study were:

| Cement Brand | Air % | Water Ret. | Compression Strength - psi |
|---|---|---|---|
| 5 | 16.0 | 82 | 819 |
| 6 | 18.6 | 85 | 1050 |
| 7 | 19.1 | 81 | 1558 |
| 8 | 26.0 | 89 | 951 |

For Type N masonry cements although an individual cement did not represent the upper end in all categories, the upper, middle, and lower ranges were all represented by the combinations used.

*Christopher A. Haver, Diana L. Keeling, Shan Somayaji, Dane Jones and Robert H. Heidersbach*

# CORROSION OF REINFORCING STEEL AND WALL TIES IN MASONRY SYSTEMS

---

REFERENCE: Haver, C.A., Keeling, D.L., Somayaji, S., Jones, D., and Heidersbach, R.H., "Corrosion of Reinforcing Steel and Wall Ties in Masonry Systems, Masonry: Components to Assemblages, ASTM STP 1063, John H. Matthys, Ed., American Society for Testing and Materials, Philadelphia, 1990.

ABSTRACT: The corrosion of metal associated with masonry systems is controversial. Little research has appeared on the subject, even though corrosion has been associated with multimillion dollar failures of masonry facades on buildings and other structures.
     This report reviews the available literature on corrosion in masonry and compares this literature with the more abundant literature on corrosion in concrete. Corrosion patterns and performance are discussed based on analyses of metal removed from a seawall in California and a building facade in Illinois.

KEY WORDS: corrosion, zinc, steel, mortar, chlorides, masonry, concrete, veneer walls, cavity walls, wall ties, metal reinforcement, carbonation.

## INTRODUCTION

The corrosion of metal reinforcement and wall ties in masonry systems has been studied for almost 50 years[1]. Much of this

     C. Haver is a graduate in Metallurgical and Materials Engineering from California Polytechnic State University , San Luis Obispo, CA 93407.    He is presently employed by Corrpro, Inc. in Philadelphia, Pa.    Ms. Keeling is a lecturer in chemistry and engineering, Dr. Jones is a professor of chemistry, Dr. Somayaji is a civil engineering professor and Dr. Heidersbach is head of the Metallurgical and Materials Engineering Department at California Polytechnic State University.

interest stems from the high corrosion susceptibility of these structural supports. Where the performance of masonry structures depends on the strength of metal supports, degradation due to corrosion can be both serious and costly.

Much modern brick masonry construction consists of either cavity or veneer walls. Veneer wall construction involves the use of exterior panels constructed from one to three wythes of brick. The veneer is attached to the steel or reinforced concrete frame of the building using galvanized wall ties and metal studs (Figure 1). Most of the panel weight is supported by the shelf angle. Veneer walls support only their own weight: they are simply decorative facades on small to medium height buildings.

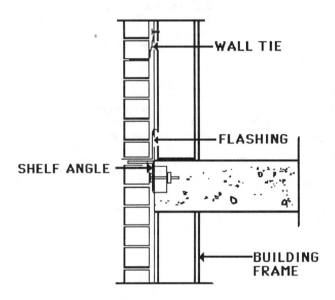

Figure 1: Veneer wall construction.

Cavity wall construction, using a double wythe wall (Figure 2), provides structural strength in low rise buildings. Metal supports used in conjunction with cavity walls include: ties, anchors and lintels. Lintels support structure over windows, doors and overhangs. The wythes are connected across the cavity by metal ties at specified vertical and horizontal intervals. Wall ties transmit lateral loads and stabilize compressively loaded wythes.

Metal Component Corrosion

Problems associated with metal component corrosion in

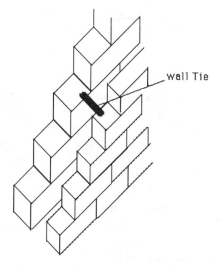

Figure 2: Cavity wall construction.

masonry buildings include Sarabond mortar additive    problems[2-4] and many instances of veneer wall    cracking and spalling[5-9]. Facade problems now represent 30 percent of all construction industry claims[10], and failures such as that shown in Figure 3 have led to increased masonry research.

Figure 3: Masonry walls collapsing due to corroded masonry wall ties. Sheffield, England.

Wall ties and support components can corrode either in the exposed atmosphere of the wall cavity or while embedded in mortar. Corrosion in the wall cavity occurs due to chloride penetration[11], galvanic cell corrosion[12] or moisture penetration. Corrosion of embedded metal is discussed in the next section. Some authorities maintain that chlorides from hydrochloric acid-based cleaners used to remove efflorescence (Figure 4) or from atmospheric sources diffuse through brick and mortar, causing corrosion. Galvanic couples can occur between the dissimilar metals of the wall tie, metal stud attachments, and steel building structure. Small amounts of moisture can cause reactions to occur on the exposed surface of a wall tie. This corrosion can be minimized by ensuring adequate drainage[13]. Moisture is plentiful inside wall cavities because cracks between both bricks and mortar allow rain penetration[14]. Moisture from condensation may also be present[15].

Figure 4: Efflorescence: New York City Police Headquarters.

Building designers expect moisture ingress and protect metal components in the cavity by installing flashing and weepholes to divert moisture from exposed metal.[16]. Such precautions, when installed and maintained correctly, effectively prevent large amounts of water from contacting exposed metal. It is nearly impossible, however, to prevent all moisture from contacting the metal components.

Facades fail for a variety of reasons other than corrosion. Brick facades attached to concrete buildings may expand with time, while concrete may undergo initial shrinkage  Serious stress may build  up with resulting cracks. Wind driven rain may penetrate into wall cavities and  become trapped  in the wythes if drainage is inadequate. Inadequate or mortar-filled expansion joints may create stresses, which result in cracking[7]. These cracks increase water penetration, increasing corrosion of the wall ties and eventual failure. Wall tie failure results in a loss of structural support for panel facades, unevenly distributing the load, creating more cracks.

Embedded  Metal  Corrosion

Published research involving corrosion in masonry structures has been largely concerned with the components previously discussed. Little information is available regarding corrosion of steel embedded in mortar. This lack of information has led consultants and researchers to use the abundant corrosion in concrete literature to explain reinforcement corrosion in masonry.

Many believe that a protective oxide layer protects reinforcement embedded in both concrete and masonry. This theory, introduced by I. Cornet[17], uses potential-pH (Pourbaix) diagrams[18] to explain the formation of a passive oxide film on the steel. Wet mortar is a base (pH=12.5-13.2). According to the Pourbaix diagram of water and iron (figure 5), such a basic solution causes a passive oxide layer ($Fe_2O_3$ or similar compound) to form on the surface of the steel.

The highly alkaline deposits formed by the hydration of the cement in the mortar is neutralized by acidic gases in the atmosphere and by acidic solutions sometimes used to clean exterior masonry structures. These gases, in particular $CO_2$ and $SO_2$, neutralize the hydroxides, decreasing alkalinity[14]. This process is known as carbonation.

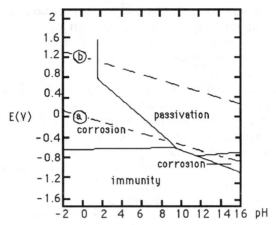

Figure 5: Potential-pH(Pourbaix) diagram of water and iron[18].

Since its inception, this theory has been used in papers and research as proven fact; however, there are problems with the theory. No one has ever isolated the oxide film. Also, the use of Pourbaix diagrams, which are theoretical and thermodynamic in nature, to explain actual phenomena can be misleading. Most basic corrosion textbooks indicate that Pourbaix diagrams are extremely difficult to apply to actual kinetic situations, and that passive films that form may or may not be protective[13,18]. Furthermore, it may be inappropriate to compare the corrosion of iron in water to the corrosion of steel in concrete or mortar.


Corrosion Evaluation

Corrosion and construction consultants use various methods to evaluate the corrosion activity of steel in masonry and concrete[19, 20], including:
- Visual examination, photography, and mapping of cracks and spalls.
- Delamination surveys using a hammer (hollow sound means delamination).
- Half-cell potential mapping in accordance with the ASTM Test for Half Cell Potentials of Uncoated Reinforcing Steel in Concrete(ASTM C876).
- Core sampling for carbonation testing, and determination of chloride depth penetration.
- Chloride content analysis.

Visual examination is effective in locating problem areas. Delamination surveys can locate delamination problems in areas

where visual examination is inconclusive.    Half-cell    potentials
measure    the    potential    between    an    exposed    section    of
reinforcement    and    the    concrete    adjacent    to    the    reinforcement
(Figure 6).    Potentials more negative than -0.35 versus a copper-
copper sulfate electrode supposedly indicate a greater than 90%
probability that reinforcement corrosion exists[21].

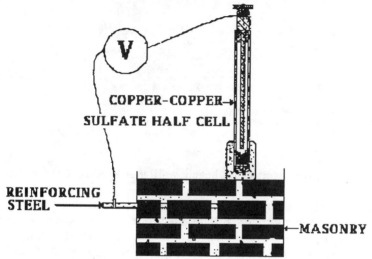

Figure 6: Half cell potential test.

Depth of carbonation can be determined using phenolphthalein to
discover if carbonation has progressed far enough into the mortar
to break down the protective oxide film.    Chloride content
analyses indicate if the chloride content exceeds a level of 0.15-
0.35% by weight of concrete or mortar[19-20,22-23], which is the
supposed threshold at which corrosion of metals embedded in
cement and mortar will occur.

Unfortunately, a scanning electron microscope(SEM) equipped
with Energy Dispersive X-Ray(EDX) analysis capability is seldom
used in conjunction with the corrosion analysis of metal masonry
components.    Such equipment can quickly and accurately indicate
the corrosion product on the surface of the corroding metal, and
more often than not pinpoint the cause of corrosion.

## EXPERIMENTAL WORK

Figure 7 shows a beach wall in Cayucos, California that was tested
for corrosion activity.    This wall was at least seven years

Figure 7: Standing section of beach wall in Cayucos, California.

old and probably much older. Half cell potential readings were taken using a copper-copper sulfate electrode. Core samples were drilled in areas where embedded reinforcement existed. Chloride sampling in accordance with the ASTM Test for Chemical Analysis of Hydraulic Cement (C114-87)[24] was performed on the cores 3 and 8 centimeters(1 and 3 inches) from the surface of the wall and directly adjacent to the grout. Phenolphthalein and pH indicator paper were used to determine the extent of carbonation in the mortar. An SEM equipped with EDX capability was used to determine the surface composition of the reinforcement.

An approximately twelve year-old section of veneer wall from the Howe Developmental Center in Tinley Park, Illinois was also examined for corrosion activity (Figure 8). Mortar samples were drilled for chloride analysis from the exterior surface, the middle, and the rear of the wall section. Phenolphthalein and pH indicator paper were used to determine the extent of carbonation in the mortar. Energy dispersive X-ray(EDX) spectroscopy was used to determine the composition of the surface of a corroding zinc galvanized wall tie.

## RESULTS

The beach wall analyzed in Cayucos was made from masonry blocks (Figure 7).    The wall protected the hillside from wave action, preventing erosion and protecting the houses above.

Both large (Figure 9) and small sections of this wall were scattered along the beach.    Many had pieces of corroding reinforcement protruding from the grout.    Each section was subject to frequent wave action and constant salt water spray.

Figure 8: Veneer wall section from the Howe Developmental Center in Tinley Park, Illinois.

Figure 9: Pieces of wall on the beach in Cayucos, California.

Half cell potential readings from a section of wall still standing (Figure 7) showed that all sections tested had readings greater than 0.50 vs. a $Cu-CuSO_4$ electrode (Figure 10). Such values indicated that greater than 90% of the rebar should have been corroding. The only visible evidence of corrosion appeared on the reinforcement protruding from the grout. ASTM C876 states, "where potentials were more negative than -0.50 Volts, approximately one half of the specimens cracked due to corrosion activity."[21] Only one crack was seen in the area analyzed. Crack analysis revealed that the cause of the crack was not related to corroding rebar expansion because the crack was only 5 centimeters (2 inches) deep, while the reinforcement was embedded 8 centimeters(3 inches) from the surface.

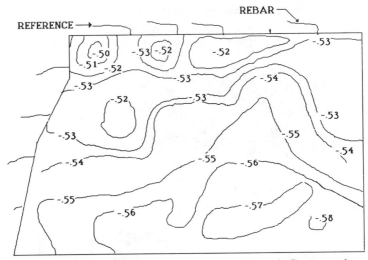

Figure 10: Half-cell potential contour map of Cayucos beach wall shown in Figure 7.

Core samples drilled in the standing wall and in other sections throughout the beach revealed that corrosion of reinforcement embedded in the grout was not occurring (Figure 11). These results contradict the half cell potential results which indicate that greater than 90% of the reinforcement should have been corroding.

Mortar samples drilled from the cores indicated that chlorine was present in the mortar in concentrations near the surface up to 1.09% by weight , decreasing to 0.46% in the grout near the reinforcement (Table 1). Many authorities are of the opinion that a chloride threshold, commonly assumed to be 0.20% to 0.35% by weight of concrete, is necessary for the onset of corrosion[25]. The chloride concentration near the rebar in the samples was much greater than this, and no corrosion was observed.

Figure 11: Core sample showing lack of corrosion on reinforcement removed from wall shown in Figure 7.  Rebar is the  center round section underneath large aggregate at top,   about 1/3 distance from right edge.

Table 1-- Chloride Content and Depth of Penetration

| Chlorides, percent by weight of mortar | Depth of Sample, cm |
|---|---|
| 1.09 | 1.0 |
| 0.81 | 5.6 |
| 0.46 | 8.5 |

Phenolphthalein and pH indicator paper indicated that both the mortar and the grout were carbonated.  pH values of 8.5 to 9.5 were observed.  These values are not surprising since the wall had been in service for at least 7 years, subject to frequent salt spray.

One of the core samples was drilled with an exposed corroding piece of rebar protruding from its surface (Figure 12).  The mortar and grout were removed and the reinforcement examined. Although corrosion was obvious in the exposed area, no corrosion existed just centimeters away, in the previously embedded regions (Figure 13).

SEM and EDX analysis of the embedded section of reinforcement revealed large amounts of both chlorine and calcium (Figure 14).

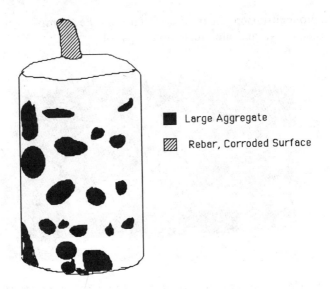

■ Large Aggregate

▨ Rebar, Corroded Surface

Figure 12: Core sample showing exposed, corroding reinforcement removed from wall shown in Figure 7.

Original Raised Pattern

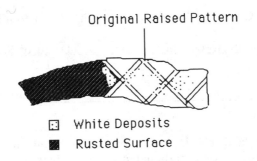

⬚ White Deposits

■ Rusted Surface

Figure 13: Reinforcement previously embedded in core shown in Figure 12.

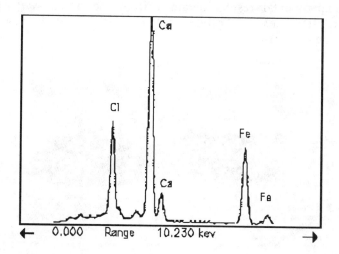

Figure 14: Energy-dispersive X-ray spectrum of embedded reinforcement surface shown on right in Figure 13. Note large amounts of both calcium and chlorine.

The veneer wall from the Howe Developmental Center in Tinley Park, Illinois had been in service for approximately 12 years. At this time much of the wall was replaced because cracking masonry and corroding wall ties had created a unsafe structure. The wall simply spanned one story and was attached by wall ties to metal studs. The section analyzed was approximately one foot square (Figure 8) and had a corroding zinc galvanized wall tie protruding from the rear mortar.

Chloride sampling of mortar samples drilled from the surface, middle and rear of the panel indicated only a maximum of 0.06% chlorides by weight mortar.

Phenolphthalein indicator and pH indicator paper revealed that the mortar was carbonated. pH values ranged from 9.5 near the surface to 10.5 in the center of the mortar.

SEM and EDX analysis of the surface of the wall tie (Figure 15) showed definite corrosion. Large zinc and iron peaks were found with very small chlorine peaks (Figure 16).

**DISCUSSION**

The experimental results obtained from the beach wall seemed to contradict many beliefs taken for granted regarding reinforcement corrosion.

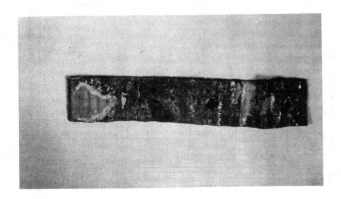

Figure 15: Corroded wall tie removed from wall shown in Figure 8.

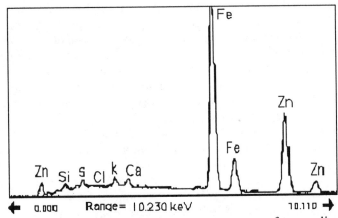

Figure 16: Energy-dispersive X-ray spectrum  of corroding wall tie

Half cell potential values are widely used to pinpoint possible corrosion problems in reinforced concrete and masonry.  Half cell potential readings indicated that large portions of the wall should have been cracked with widespread reinforcement corrosion. Core samples revealed corrosion of steel in the embedded  mortar was not occurring beyond superficial amounts probably present during construction.  Visual examination revealed only the one crack previously-described.    The experimental results obtained in this study indicate that currently accepted potential values are inaccurate for predicting corrosion of reinforcement.

Many authorities assume a chloride threshold of approximately 0.15 to 0.35% by weight chlorides in both mortar and concrete will cause corrosion of reinforcement[19-20,22-23]. The wall analyzed had a much greater chloride contents of 0.46% to 1.09% by weight and yet corrosion had not progressed beyond superficial levels typical of on-site storage of metal prior to construction.

The high chloride content in the mortar was caused by frequent wet-dry cycles of seawater, which has typically 3.5 weight percent salt[26]. Wet- dry cycling occurs when masonry is subject to frequent tidal or splash action. In such an atmosphere, build up of chlorides can be rapid, leading to 0.3-4.0% chlorides by weight concrete in just 10 years.[27]

Wet-dry cycling was also more than likely responsible for carbonation of the mortar. Seawater ions permeated the mortar, reacting with the hydroxides and neutralizing their alkalinity or simply washing them away with the next wetting cycle.

EDX analysis of a sample of reinforcement previously embedded in mortar provided perhaps, the most interesting result of this study. Chlorine was present in the mortar directly adjacent to the reinforcement. Chemical analysis revealed that the chlorine level was 0.46% by weight yet the steel in the mortar was relatively uncorroded. The chlorides had not initiated corrosion. EDX analysis showed that chlorine was present both on the corroding surface of the exposed rebar and on the uncorroded rebar surface embedded in the mortar.

The experimental results obtained from analysis of the Illinois veneer wall section proved that corrosion of wall ties can occur in the absence of chlorides.

Chloride sampling of the mortar indicated only trace levels of chlorides (0.06% by weight mortar). This small amount of chlorides was probably added to the mortar during construction when the mortar was mixed with chlorine treated water. Chlorides in such small amounts do not cause corrosion.

EDX analysis of the wall tie revealed that the galvanized coating had corroded, exposing the steel underneath to corrosion. Corrosion covered approximately 60% of the wall tie. No chlorine was observed with EDX analysis. These results show that chlorides were not responsible for the corrosion of this wall tie.

The only plausible explanation for corrosion of the wall tie was moisture. Little moisture is needed for oxidation-reduction reactions to initiate corrosion. The Building Research Establishment, which investigates a wide variety of masonry problems in England where much of the population lives in masonry buildings, states, "the quantity of moisture necessary for corrosion is less than will cause dampness to be visible to the unaided eye."[11] Such small amounts of moisture will always be present in wall cavities due to rain permeation and condensation. The metal studs attaching the wall ties to the structure of the building were extremely corroded, indicating severe moisture penetration[28].


## CONCLUSIONS


The experimental results obtained from this study indicate that the methods currently used to analyze corrosion in masonry may be inaccurate.  Half cell potential values cannot  predict corrosion of reinforcement in masonry structures.  Chloride presence in mortar in levels above 0.35% does not automatically indicate corrosion of embedded reinforcement.

Additionally, corrosion of wall ties inside wall cavities often occurs due constant moisture penetration in the absence of chlorides. Since moisture will always be present in wall cavities, there is a need for more corrosion resistant wall ties.


## REFERENCES

[1].    Fishburn, C., "Strength and Resistance to Corrosion of Ties for Cavity Walls," National Bureau of Standards Report #BMS101, Washington, D.C., July 1943.

[2].    Stricharchuk, G., Wartzman, R., "Dow Chemical Product Assailed as Causing Brickwork to Collapse," Wall Street Journal, March 21, 1989, p. 1.

[3].    "Owners Reclad Damaged Masonry," Engineering News Record, March 6, 1986, p. 10-11.

[4].    "Dow Wins in Court at Last," Engineering News Record, May18, 1989, pp. 7-8.

[5].    Brown, B., "Leaky Walls Endanger Steel Buildings," Engineering News Record, December 8, 1949, p. 42.

[6].    Plewes, W.G., "Failure of Brick Facing on High-Rise Buildings," Canadian Building Digest, April 1977, p. 185.

[7].   "Facades: Errors can be Expensive," Engineering News Record, January 24, 1980, p. 30-34.

[8].   Beck, B., "Sky Falling? Mice? Just Wind, Bricks," O'Collegian (Oklahoma State University Newspaper), December 30, 1968.

[9].   "Protection Against Corrosion of Reinforcing Steel in Concrete," Building Research Establishment, Watford, U.K., Digest 59.

[10].  Kumar, S.S., Heidersbach, R.H., "The Corrosion of Metal Components in Masonry and Stone Clad Buildings," Proceedings 4th Canadian Masonry Symposium, 1986.

[11].  Moore, J., "The Performance of Cavity Wall Ties," Building Research Establishment Paper CP3/81, April, 1981, p. 15.

[12].  Heidersbach, R.H., Borgard, B., "Corrosion of Metal Components in Masonry Buildings," Proceedings, 4th North American Masonry Conference, UCLA, 1987, pp. 68-4-68-6.

[13].  Fontana, M. G., Corrosion Engineering, 3rd Edition, McGraw Hill, 1986, p. 17, pp. 453-454.

[14].  Treadway, K.W.J., Cox, R.N., "Protection of Ferrous Metals in Brickwork," CP 111- The Next Stage. Symposium at Bloomsbury Centre Hotel, 1979.

[15].  Heidersbach, R.H., Lloyd, J., "Corrosion of Metals in Concrete and Masonry Buildings," Corrosion/85, March 1985,Paper 258, p. 3.

[16].  Randall, F., Panarese, W., "Concrete Masonry Handbook," Portland Cement Association, Skokie, Illinois,1976, p. 107.

[17].  Cornet, I., Ishikawa, T., Bressler, B., Materials Protection, March 1968, pp. 44-47.

[18].  Pourbaix, M., "Atlas of Electrochemical Equilibria in Aqueous Solutions," NACE, Houston, 1974,  pp. 307-321.

[19].  Locke, C. E., "Mechanisms of Corrosion of Steel in Concrete," Paper #2, NACE North Central Region Seminar, NACE, Houston, Tx., September 1982, pp. 2, 4.

[20].  "Corrosion of Metals in Concrete," American Concrete Institute Report #222, Detroit, Michigan, 1985,  pp. R-4, R-12, R-22-24.

[21].  "Standard Test Method for Half-Cell Potentials of Uncoated Reinforcing Steel in Concrete(C 876-87)." American Society for Testing and Materials(ASTM), Philadelphia, Pa.

[22].  Hime, W., Erlin,B, "Chloride-Induced Corrosion," Concrete International, September 1985, p.23.

[23].  West, R.E., Hime, W., "Chloride Profiles in Salty Concrete," Materials Performance, July 1985, p. 29.

[24]. "Standard Methods for Chemical Analysis of Hydralic Cement(C114-87)," American Society for Testing and Materials(ASTM), Philadelphia, Pa.
[25]. Stark, D., "Measurement Techniques and Evaluation of Galvanized Reinforcing Steel in Concrete Structures in Bermuda," Corrosion of Reinforcing Steel in Concrete ASTM STP 713, D.E. Tonini, J.M. Gaidis, Eds., American Society for Testing and Materials, 1980, pp. 132-141.
[26]. Griffin, D., Henry, R., "Effect of Salt in Concrete on Compressive strength, Water Vapor Transmission, and Corrosion of Reinforcing Steel," Technical Report R-217,Civil Engineering Lab NCBC, Port Hueneme, Ca. , November 1962, p. 1052.
[27]. John, D.G., Davies, K.G., " Corrosion of Reinforced Concrete Marine Structures in the Middle East and Options for Control," NACE 1989, Paper #387.
[28]. Kumar, S.S., Heidersbach, R.H., "Corrosion of Steel-Stud Masonry Buildings," Proceedings 4th Canadian Masonry Symposium, 1986.

# DISCUSSION

Questions    (Ian Chin; Wiss, Janney, Elstner Associates)

- The paper includes under "carbonation" the neutralization of hydrated cement by SO2.  Won't this extension of the definition lead to confusion?

- The conclusion that steel in the very negative half-cell readings for the steel in the sea wall (Figure 7) gave a false indication of corrosion are of concern to me.  Such a "low" half cell reading usually indicates chloride-caused corrosion of steel, or the presence of galvanized steel.  Examination of a core (Figure 7) will not often reveal either.  Neither will the EDX pattern (Figure 15) that apparently ends before the zinc peak.  Would the authors please comment?

- The conclusion that "currently accepted potential values are inaccurate for predicting corrosion of reinforcement" appears unwise in the light of the thousands of examples to the contrary, the minimal study of the steel in the sea wall, and the fact that there has never been an understanding that very negative half-cells absolutely indicate corrosion.  ASTM C 876 simply indicates a probability.  Further, half-cell potentials are primarily dictated by chemistry, and not corrosion, and accordingly any half-cell reading should suggest a study of the chemistry of the steel:cementitious material surface.  Would the authors comment?

- The paragraph suggesting that wet-dry cycling leads to carbonation and neutralization is of doubtful accuracy.  Carbonation occurs most rapidly at about 50% R. H., and scarcely at all at 0 or 100 percent.  Sea water is not acid and has little effect on concrete pH except near the surface where soluble hydroxides are leached away.  Indeed such reactions as

$$Na_2SO_4 + Ca(OH)_2 \quad ---- \quad CaSO_4 + NaOH$$

can increase the pH of the concrete.  Would the authors comment?

- The statements that small amounts of chloride -- 0.06 percent by weight of mortar -- are probably added by the mix water and do not cause corrosion are both generally erroneous.  Most mortars contain about 15 percent cement, making the chloride level about 0.4

percent by weight of cement, or twice the corrosion threshold. Most tap waters contain at most 100 ppm chloride, contributing only 0.005 percent chloride by weight of cement. On the other hand, some sands do contain chloride that is bound within the particle matrix and will not contribute to corrosion. Would the authors comment?

AUTHORS' REPLY (R. Heidersbach)

- The point about carbonation is valid and we agree--in the location in question the relative importance of atmospheric gases versus suphur compounds in sea water (or sea spray) cannot be determined. We have used the common practice of stating that cementitious materials that have lost their alkalinity have "carbonated" although it is obvious that other reasons for a loss of alkalinity may have been present.

- We disagree. Our experience is that one cannot determine the presence or absence of corrosion without visual and microscopic examination of the metal surface. The X-ray spectrum referred to by Mr. Chin does not end before the zinc peak. The complete spectrum was recorded and only that portion with peak information was printed. A zinc peak is clearly evident in Figure 17 and would have been shown in Figure 15 if zinc had been detected.

- Mr. Chin questions our phrase "currently accepted potential values are inaccurate for predicting corrosion." We disagree. We have no knowledge of the "thousands" of samples that Mr. Chin refers to, but our experience on concrete structures indicates the same problems we have discussed here for masonry. We chose this extreme example of masonry exposed to sea water to illustrate the point we have observed on numerous occasions on masonry buildings. Unless an independent confirmation of corrosion activity can be made, we can see no justification for using ASTM C876. Mr. Chin correctly suggests that a study of the metal:cementious material interface is necessary. That is why we have reported the visual observations and the SEM and X-ray results discussed above.

We agree that ASTM C876 only states a probability. We disagree with the probability stated in the standard. It should also be noted that ASTM C876 specifically refers to concrete, not masonry. Unfortunately, many forensic consultants use ASTM C876 for masonry.

- We agree that carbonation occurs at intermediate humidities. Since coastal structures dry out between tide and storm cycles, we assume this is when the loss of alkalinity, presumably due to carbonation (see Mr. Chin's first comment) occurs. Whatever the cause, there can be not doubt that the mortar we examined in this study was no longer alkaline, yet the mortar residue protected the underlying metal from corrosion.

- Mr. Chin is correct that aggregates can be a source of chlorides in mortar. He is incorrect in his assumption that there is a corrosion threshhold. Our work clearly shows that corrosion can occur at chloride levels well below the supposed threshold. We have also documented that corrosion need not occur even when the chloride level is several times the supposed threshold.

John H. Matthys and Ranjit Singh

EXPERIMENTAL INVESTIGATION OF MORTAR COMPRESSIVE
STRENGTH USING 5.08 cm. CUBES AND 7.62 cm. X 15.24 cm.
CYLINDERS

---

REFERENCE: Matthys, John H., and Singh,
Ranjit., "Experimental Investigation of
Mortar Compressive Strength Using 5.08 cm.
Cubes and 7.62 cm. x 15.24 cm. Cylinders,"
Masonry: Components to Assemblages: ASTM
STP 1063, J. H. Matthys, Ed., American
Society for Testing and Materials,
Philadelphia, 1990.

ABSTRACT: ASTM C270 "Mortar For Unit
Masonry" lists compressive strength
requirements under the property specification
based on 5.08 cm. cubes. ASTM C780
"Preconstruction and Construction Evaluation
of Mortars For Plain and Reinforced Unit
Masonry" allows the use of 5.08 cm. x 10.16
cm. cylinders, 7.62 cm. x 15.24 cm. cylinders
or 5.08 cm. cubes in determination of mortar
compressive strength. The cylinder
compressive strength may be considered to be
equal to 85% of the cube compressive
strength. An experimental investigation of
the compressive strength relationship between
5.08 cm. cubes and 7.62 cm. x 15.24 cm.
cylinders for (1) laboratory mortars cured in
a moist room, (2) job site mortars cured in
laboratory air, and (3) job site mortars
cured outside has been conducted. The data
includes portland cement lime mortar,
masonry cemment mortar, and ready mix mortar.

KEYWORDS: mortar, compressive strength,
cubes, cylinders.

Dr. John H. Matthys is Professor of Civil
Engineering and Director of the Construction Research
Center at The University of Texas at Arlington, Box
19347, Arlington, Texas 76019-0347.

Ranjit Singh, Graduate Student, University of
Texas at Arlington.

ASTM C270 [1] covers mortars for use in the construction of non-reinforced and reinforced unit masonry structures.  It outlines four mortar types M, S, N and O based on either a proportion or property specification for laboratory prepared mortars.  The property specification requires a minimum compressive strength for each mortar type mixed to a certain flow, cured under appropriate conditions and tested at 28 days.  The compressive strength must be determined in accordance with ASTM C109 [2] which requires use of 5.08 cm. cube specimens.

ASTM C780 [3] is a specification for job site mortars.  This specification allows use of either cube or cylinder specimens for compressive strength.  The cube specimens must be 5.08 cm. x 5.08 cm. x 5.08 cm. The cylinder specimens may be either 5.08 cm. x 10.16 cm. or 7.62 cm. x 15.24 cm. cylinders.  In note 1 of this specification it states "When cube and cylinder test specimens from like mixtures are to compared, the cylinder compressive strength may be considered to be equal to 85% of the cube compressive strength".  The 85% relationship is used regardless of the cylinder size or the mortar composition.  Currently under development in ASTM C12 (Mortar for Unit Masonry) committee is a specification for ready mix mortars. Early drafts of this specification proposed the conversion of cubes to cylinder strengths by using an 85% relationship.

Many testing laboratories and research units are starting to use cylinder molds instead of cube molds for various reasons.  The authors felt that an investigation into the relationship of the compressive strength of cubes to cylinders was in order.  The data presented in this paper is taken from mortar testing on eight masonry cement mortars and two portland cement lime mortars used in a comparative investigation of masonry mortars being conducted at the University of Texas at Arlington and sponsored by National Lime Association.  Additional data on conventional mortars (Portland cement lime and masonry cement) were obtained from both published and unpublished sources.  A limited amount of data for ready mix mortar was obtained using a ready mix mortar supplier.

SCOPE OF TESTING PROGRAM

Mortars of type S and N were made with both masonry cements and combination of portland cement and lime.  Proportions of cementitious material and aggregate for a particular type of mortar used were as per the proportion specification of ASTM C270.  The authors tested and evaluated four type S mortars made with four different brands of masonry cement and one type S mortar made with a combination of portland cement and lime. Also tested were four type N mortars made with four different brands of Type N masonry cement and one type N mortar made with a combination of portland cement and lime . All conventional mortars were mixed in the laboratory to a flow of approximately 115 percent for laboratory mortar and a flow of approximately 135 percent for job site mortar. In addition specimens were made from a type S and type N ready mixed mortars delivered directly to UTA laboratory from a near by mixing plant.

For each type of mortar, there was a single laboratory mix and two job site mixes. This resulted in ten laboratory mixes and twenty job site mixes in all excluding ready mix mortars. For the laboratory mixes, flow, water retention, air content by 400 ml container and cone penetration tests were conducted. Six 5.08 cm. cubes and three 7.62 cm. x 15.24 cm. cylinders were molded for every laboratory mix. These specimens were cured in a moist room for 28 days and tested as per ASTM C109.

Two different batches of job site mortar were evaluated i.e., job site 1 and job site 2. Both job site 1 and job site 2 were mixed to a flow of 135±5 percent. Mortar tests included air content by the 400 mls. container method and cone penetration test. Six 5.08 cm. cubes and three 7.62 cm. x 15.24 cm. cylinders were molded for each of the job site mixes. The only intended difference (other than different batches) between job site 1 and job site 2 mixes was the curing conditions. Job site 1 specimens in their molds were moved outside and covered with polyethylene. After 48 hours, specimens were taken out of their molds and kept covered with polyethylene for 28 days and tested. Job site 2 specimens were kept in the laboratory for the first 48 hours in the molds, demolded and then kept in the laboratory for 28 days before testing. The air temperature of the laboratory room was 73° F ± 2° F with a relative humidity of 55 percent. Ready mix mortar specimens, proportioned and mixed by a commercial ready mix mortar plant, were made every twelve hours for twenty-four hours. Six 5.08 cm. cubes and six 7.62 cm. x 15.24 cm. cylinders were molded from each type "S" and type "N" mortar every twelve hours for curing in the moist room as per ASTM C-109. The same number of ready mix mortar specimens were made for the job site mortar and cured in the laboratory air.


SELECTION OF MATERIALS

Cement: Eight different brands of masonry cement

consisting of four type S and four type N were used.
These masonry cements were obtained across the USA and
are typical of those existing in the market place.
They were numbered A through H.  The specific gravity
of each cement was found to be as follows:

| Designation of Mortar | Type of Masonry Cement | Weight per Sack | Specific Gravity |
|-----------------------|------------------------|-----------------|------------------|
| A | N | 70.0 | 2.95 |
| B | S | 75.0 | 3.05 |
| C | N | 70.0 | 2.92 |
| D | S | 80.0 | 3.06 |
| E | N | 69.5 | 2.91 |
| F | S | 94.0 | 3.02 |
| G | N | 70.0 | 2.98 |
| H | S | 75.0 | 2.99 |

Type I portland cement of specific gravity 3.15
was used for both type S and type N portland cement-
lime mortars in combination with a type S lime of
specific gravity 2.07.  As noted in the table above,
masonry cement F weighted 94 lbs. per bag.  Sources
have indicated that masonry cement F contains fly ash.

Fine Aggregate:  One type of natural sand was used
throughout this investigation.  The maximum size of
sand passed a number 4 sieve and the sand exhibited a
fineness modulus of 2.24.  The sand met the
requirements of ASTM C144 [4].

MIXING AND TESTING

Different mixers were used for the lab mixes and
job site mixes.  For the lab mix, an electrically
operated "STONE" 6 cubic feet mixer was used.  For the
job site mixes, a gasoline engine operated "HURST" 12
cubic feet mixer was used.  Sand was measured by
volume using cubic foot plywood boxes and sacked in
plastic bags before the mixing day.  The volume of a
full sack of masonry cement or portland cement was
considered to be 1 cubic foot.  Whenever less than one
sack of cement was required for a mix, it was weighed
on a "HOMS" beam scale using the printed weight on the
sack as weight of 1 cubic foot of cement.  For the lab
mixes lime was considerd to weigh 18.1 kg. (40 lbs.)
per cubic foot.   For job site mixes lime was
considered to weigh 22.7 kg. (50 lbs. = 1 sack) per
cubic foot.  There were three operations in the mixing
of each mortar type.  Each lab mix used 1.5 cubic feet

of sand. Half of the sand and part of estimated water were added during a two minutes interval while the mixer was running. The cementitious material was then added gradually in a two minutes interval while the mixer was kept running. Finally the entire ingredients were allowed to mix for two minutes and flow was measured via flow table test. Additional water was added if needed. The entire ingredients were mixed for one minute and samples were taken for the mortar specimens. Job site mix 1 and 2 were made using 7.5 cubic feet of sand each. Half the sand plus part of estimated water were added gradually in four minute interval while the mixer was kept running. Cementitious material was added in 2 to 3 minute interval while mixer was kept running. The final part of sand with additional water was added in four minute interval. The flow was checked and water added if needed. A final one minute mixing was conducted before mortar specimens were made. In mixes, which had both portland cement and lime, the lime was added first after addition of half of sand plus part of water and allowed to mix for two minutes and then cement was added and ingredients were allowed to mix for one minute. The remainder of the mixing procedures were the same as explained before.

At 28 days, all the cylinders were capped with sulphur silica compound in a standard 7.62 cm. x 15.24 cm. concrete cylinder capping mold before testing. Compressive load test was performed according to ASTM C780 which recommends that loading be applied within the range of 20 to 50 pounds per square inch per second. The actual loading rate in this study was kept at 25 pounds per square inch per second.


DISCUSSION OF TEST RESULTS

Tables 1 through 5 give all the test data. In table 1, results of laboratory mortars made with masonry cement and cured in moist room are shown. Blank entries indicate the unavailability of data. As seen, the results show a lot of variation in ratio of cylinder strength to cube strength. Table 1 shows a ratio varying from 66.6% to 119% for type N mortars and from 68.7% to 111% for type S mortars made with masonry cement. However, out of all the results shown in Table 1, 62% of the results for type S mortar show a lower ratio when compared to type N mortar. This indicates the ratio of cylinder strength to cube strength appears to decrease with increase in strength. Table 2 shows the results for laboratory mortars made with portland cement-lime and cured in moist room. This data, on an average basis, shows an increase in ratio with increase in strength for type O through type S mortars. Also noted in table 2, (portland cement-lime), the variation in the ratio of cylinder to cube strength is less than the variation table 1 (masonry cement). Table 3 shows results for ready mix mortars cured in moist room. From this table, it appears the ratio of cylinder strength to cube strength may increase with increase in suspension time (i.e., time of molding cubes after initial mixing). Also it is <u>very clear</u> that this ratio is different for different additives. UTA tests used additive 5 which gives the same ratio for type N or type S mortar at same suspension times; noting that the ratio is less than 100%. Correspondingly each additive of multiple mixes indicated generally a narrow range of ratio. Note that additive 2 and 3 indicate a ratio greater than 100%. Thus it appears that the current ASTM draft of the ready mix mortar specification that states "When converting from cylinder strength to cube strength , the relationship shall be documented by the manufacturer with test data" is an appropriate requirement at this time.

Table 4 shows the results for job site mortars cured in the environmentally controlled laboratory air. These results show that ratio of cylinder strength to cube strength is a function of curing condition. The results indicate a conversion factor of more than 100%. The previous results for laboratory mortars (Table 1 and 2) cured in moist room showed a conversion factor less than 100% for most of the results. For the data in laboratory air there is only one masonry cement (i.e., cement F) which gives conversion factor less than 100%. As pointed out earlier this cement may contain fly ash and weighed 94.0 lbs. per sack. Table 4 shows that for mortars made with portland cement-lime the ratio seems to increase with increase in strength. This is also noted in Table 2. For ready mix mortars Table 4 shows a decrease in ratio with increase in mortar strength; also noted is a decrease with increase in suspension time.

Table 5 gives results for job site mortars cured outside. The data suggests that the conversion factor of cylinders to cube is more than 100% for all the mortars, except for mix F. The ratios of cubes to cylinder here, except for portland cement-lime mortar type N, are about the same as for specimens cured in laboratory air.


CONCLUSIONS

Based on previous examined data, the following conclusions have been drawn. (1) For laboratory mortars cured in moist room, the average of all the results of

UTA study shows that compressive strength of 7.62 cm. x 15.24 cm. cylinder is 80.9% of 5.08 cm. cube compressive strength. The average of all the results from other sources gives cylinder strength as 97.7% of 5.08 cm. cube compressive strength. (2) For job site mortars cured in laboratory air, average of all the UTA results gives cylinder strength as 116.3% of 5.08 cm. cube compressive strength. (3) For job site mortars cured outside, the average of all the UTA results gives cylinder strength as 116.5% of 5.08 cm. cube compressive strength.

## RECOMMENDATIONS

It is recommended that 7.62 cm. x 15.24 cm. cylinders be used to check compressive strength of mortars because handling of cylinders is easier than that of cubes and as shown in results, the coefficient of variation of cylinder strength is lower than on cube strength tests. Investigation still needs to be done on other masonry cement mortar such as type M. Future investigations should also consider different gradation of aggregates. There is also a need for thoroughly investigating the relationship between cubes and cylinder for masonry cements containing fly ash. Due to large variability in the cylinder strength versus cube strength data, a statistical analysis needs to be performed to arrive at an appropriate conversion factor. For ready mix mortars, a decision ought to be made regarding the conversion factor based on type of additive rather than using single conversion factor for all ready mix mortars.

## REFERENCES

[1]    "Standard Specification For Mortars For Unit Masonry," ASTM C270-86a, American Society For Testing and Materials, Philadelphia, PA 19103.

[2]    "Standard Test Method For Compressive Strength of Hydraulic Cement Mortar (using 2" or 50mm. cube specimens)," ASTM C109-86, American Society For Testing and Materials Philadelphia, PA 19103.

[3]    "Method For Preconstruction and Construction Evaluation of Mortars For Plain and Reinforced Unit Masonry," ASTM C780-87, American Society For Testing and Materials, Philadelphia, PA 19103.

[4]    "Standard Specification for Aggregate For Masonry Mortar." ASTM C144-87, American Society For Testing and Materials, Philadelphia, PA 19103.

[5]    Data from National Lime Association, Gifford-Hill & Company, Inc., Holmes Testing Laboratory and Construction Research Center.

Table 1:  Masonry Cement Laboratory Mortars

| Mortar Type | Flow (%) | Cube Strength[1] N/mm$^2$ | | Cylinder Strength[1] N/mm$^2$ | | Cylinder Cube |
|---|---|---|---|---|---|---|
| **UTA Study** | (A,C,E,G = Type N Masonry Cement Brands) | | | | | |
| NA | 115.8 | 9.91 | (8.6)* | 8.62 | (4.7)* | 87.0 |
| NC | 115.0 | 8.31 | (7.0)* | 7.07 | (3.3)* | 85.1 |
| NE | 114.0 | 9.01 | (4.5)* | 6.00 | (12.0)* | 66.6 |
| NG | 116.6 | 9.93 | (8.1)* | 8.16 | (7.7)* | 82.2 |
| | | | | | **Average =** | **80.2%** |
| **Other Sources [5]** | | | | | | |
| N | 112.0 | 4.93 | (-)** | 4.87 | | 98.8 |
| N | 108.0 | 6.03 | (-)** | 6.47 | | 107.3 |
| N | 102.0 | 6.05 | (-)** | 7.20 | | 119.0 |
| N | 102.0 | 6.23 | (-)** | 5.64 | | 90.5 |
| N | 111.0 | 10.71 | (-)** | 10.48 | | 97.8 |
| N | 108.0 | 6.60 | (-)** | 6.41 | | 97.1 |
| N | (-)** | 7.47 | (2.6)* | 6.26 | (4.2)* | 83.8 |
| N | (-)** | 6.94 | (23.9)* | 6.05 | (18.6)* | 87.2 |
| N | 125.3 | 5.87 | (10.8)* | 6.07 | (12.4)* | 103.4 |
| N | 102.6 | 6.87 | (7.1)* | 5.67 | (16.6)* | 82.5 |
| N | 115.3 | 9.27 | (6.9)* | 7.20 | (11.2)* | 77.7 |
| N | 120.3 | 4.60 | (10.4)* | 5.03 | (7.2)* | 109.3 |
| | | | | | **Average =** | **96.2%** |
| **UTA Study** | (B,D,F,H = Type S Masonry Cement Brands) | | | | | |
| SB | 112.8 | 16.65 | (3.1)* | 12.32 | (15.4)* | 74.0 |
| SD | 107.8 | 26.0 | (8.1)* | 24.26 | (1.9)* | 93.3 |
| SF | 114.8 | 35.16 | (7.6)* | 35.74 | (1.3)* | 72.4 |
| SH | 115.6 | 13.38 | (6.0)* | 9.69 | (9.3)* | 101.6 |
| | | | | | **Average =** | **85.3%** |
| **Other Sources [5]** | | | | | | |
| S | 116.0 | 12.08 | (-)** | 10.42 | | 86.2 |
| S | 118.0 | 11.60 | (-)** | 11.25 | | 97.0 |
| S | 120.0 | 9.29 | (-)** | 9.93 | | 106.9 |
| S | 112.0 | 9.29 | (-)** | 10.69 | | 94.3 |
| S | 108.0 | 12.11 | (-)** | 9.90 | | 81.7 |
| S | 120.0 | 10.05 | (-)** | 10.40 | | 103.5 |
| S | 124.0 | 8.69 | (-)** | 9.65 | | 111.0 |
| S | 114.0 | 10.03 | (-)** | 9.79 | | 97.6 |
| S | 104.0 | 10.80 | (-)** | 9.16 | | 84.8 |
| S | 104.0 | 8.54 | (-)** | 8.41 | | 98.5 |
| S | 112.0 | 9.25 | (-)** | 8.28 | | 89.5 |
| S | (-)** | 16.04 | (5.1)* | 11.02 | (3.2)* | 68.7 |
| | | | | | **Average =** | **93.3%** |

1   1 N/mm$^2$ = 145 psi
*   Coefficient of Variation
**  Data Unavailable

Table 2 :   Portland Cement & Lime Laboratory Mortars

| Mortar Type | Flow (%) | Cube Strength[1] N/mm$^2$ | | Cylinder Strength[1] N/mm$^2$ | | Cylinder Cube |
|---|---|---|---|---|---|---|
| **Data from Other Sources [5]** | | | | | | |
| O | 117.0 | 4.25 | (-)** | 4.98 | (-)** | 117.2 |
| O | (-)** | 3.96 | (12.8)* | 3.43 | (17.0)* | 86.6 |
| O | (-)** | 2.17 | (4.0)* | 1.51 | (6.6)* | 69.6 |
| O | 114.6 | 4.21 | (11.8)* | 3.94 | (7.0)* | 93.6 |
| O | 110.7 | 3.69 | (9.8)* | 2.92 | (10.2)* | 79.1 |
| | | | | Average = | | 89.2% |
| **UTA Study** | | | | | | |
| N | 116.0 | 8.23 | (3.3)* | 5.94 | (15.3)* | 72.2 |
| **Data from Other Sources   [5]** | | | | | | |
| N | 109.0 | 5.53 | (-)** | 5.27 | (-)** | 95.3 |
| N | (-)** | 9.59 | (5.0)* | 10.61 | (15.4)* | 110.5 |
| N | (-)** | 7.96 | (8.3)* | 6.16 | (7.7)* | 77.4 |
| N | 112.0 | 9.79 | (4.5)* | 12.20 | (3.3)* | 124.6 |
| N | 105.7 | 9.40 | (4.7)* | 9.01 | (2.4)* | 95.8 |
| | | | | Average = | | 100.7% |
| **UTA Study** | | | | | | |
| S | 117.8 | 17.95 | (3.7)* | 15.16 | (7.2)* | 84.4 |
| **Data from Other Sources   [5]** | | | | | | |
| S | 120.0 | 19.77 | (-)** | 21.62 | (-)** | 109.4 |
| S | 120.0 | 21.16 | (-)** | 23.15 | (-)** | 109.4 |
| S | (-)** | 14.95 | (24.3)* | 14.96 | (28.8)* | 100.1 |
| S | 128.0 | 15.50 | (19.6)* | 16.11 | (13.8)* | 103.9 |
| S | 139.3 | 14.68 | (6.7)* | 13.28 | (10.1)* | 90.5 |
| | | | | Average = | | 102.7% |
| **Data from other Sources   [5]** | | | | | | |
| M | 133.0 | 28.90 | (-)** | 27.64 | (-)** | 95.6 |
| M | 132.0 | 24.81 | (-)** | 22.56 | (-)** | 90.9 |
| | | | | Average = | | 93.3% |

1    1N/mm$^2$ = 145 psi
*    Coefficient of Variation
**   Data Unavailable

Table 3 :   Ready Mix Mortars - Moist Room Cured

| Mortar Type | Suspension Time Hrs. | Cube Strength[1] N/mm$^2$ | Cylinder Strength[1] N/mm$^2$ | Cylinder Cube |
|---|---|---|---|---|
| **UTA Study** | (5 | = Ready Mix Additive Brand) | | |
| N-5 | 5 | 8.31  (6.9)* | 6.19  (7.7)* | 74.5 |
| N-5 | 17 | 8.89 (10.4)* | 6.50  (8.4)* | 73.1 |
| N-5 | 29 | 7.36 (29.6)* | 6.94  (9.5)* | 94.3 |
| | | | **Average =** | **80.6%** |
| **Data from Other Sources [5]** | | (1,3,4 = Additive Brand) | | |
| N-1 | (-)** | 9.72  (2.6)* | 8.00  (8.4)* | 82.3 |
| N-3 | 0 | 4.42  (2.7)* | 4.82  (5.8)* | 109.0 |
| N-3 | 12 | 3.81 (11.0)* | 4.75 (10.4)* | 124.7 |
| N-4 | (-)** | 11.95  (1.0)* | 10.11  (4.7)* | 84.6 |
| N-4 | (-)** | 13.29  (2.0)* | 11.19  (6.3)* | 84.2 |
| | | | **Average =** | **97.0%** |
| **UTA Study** | (5 | = Ready Mix Additive Brand) | | |
| S-5 | 5 | 15.15  (8.5)* | 10.75  (9.2)* | 71.0 |
| S-5 | 17 | 17.06  (7.7)* | 12.02 (12.9)* | 70.5 |
| S-5 | 29 | 12.81  (9.0)* | 11.85  (1.8)* | 92.5 |
| | | | **Average =** | **78.0%** |
| **Data from Other Sources [5]** | | (1,2 = Additive Brand) | | |
| S-1 | (-)** | 16.33  (5.2)* | 13.55  (7.0)* | 83.0 |
| S-1 | 0 | 19.79  (2.0)* | 19.79 (16.9)* | 100.0 |
| S-1 | 24 | 19.53  (2.4)* | 20.75  (3.5)* | 106.2 |
| S-1 | 24 | 18.62  (2.6)* | 20.00  (4.6)* | 107.4 |
| S-1 | 25 | 15.24  (1.7)* | 17.35  (2.2)* | 113.8 |
| S-1 | 0 | 15.15  (2.3)* | 18.46  (1.7)* | 121.8 |
| S-1 | 24 | 13.35  (0.9)* | 15.89  (1.6)* | 119.0 |
| S-2 | 48 | 15.48  (1.8)* | 18.04  (1.3)* | 116.5 |
| | | | **Average =** | **108.5%** |

1    1N/mm$^2$ = 145 psi
*    Coefficient of Variation
**   Data Unavailable

Table 4:    Job Site Mortars - Lab Air Cured

| Mortar Type | Flow (%) | Cube Strength$^2$ N/mm$^2$ | | Cylinder Strength$^2$ N/mm$^2$ | | Cylinder Cube |
|---|---|---|---|---|---|---|
| N-MC-A | 139.3 | 6.6 | (7.4)* | 9.82 | (1.9)* | 148.8 |
| N-MC-C | 139.5 | 4.85 | (3.9)* | 5.50 | (1.6)* | 113.4 |
| N-MC-E | 141.3 | 6.47 | (2.9)* | 6.77 | (2.1)* | 104.6 |
| N-MC-G | 141.3 | 4.89 | (6.3)* | 5.85 | (1.4)* | 119.6 |
| | | | | | Average = | 121.6% |
| N-PCL | 132.3 | 5.25 | (1.4)* | 4.54 | (1.6)* | 86.4 |
| N-RM | | | | | | |
| 5-hrs. | (-)** | 6.97 | (7.8)* | 11.00 | (6.8)* | 157.8 |
| 17-hrs. | (-)** | 8.27 | (3.6)* | 11.36 | (7.9)* | 137.4 |
| 29-hrs. | (-)** | 7.73 | (6.9)* | 9.53 | (3.9)* | 123.3 |
| | | | | | Average = | 139.5% |
| S-MC-B | 138.5 | 10.71 | (3.6)* | 13.80 | (3.9)* | 128.8 |
| S-MC-D | 133.3 | 15.86 | (6.3)* | 17.26 | (1.7)* | 108.8 |
| S-MC-F | 142.3 | 20.74 | (5.9)* | 19.61 | (2.1)* | 94.6 |
| S-MC-H | 140.0 | 6.17 | (4.7)* | 7.00 | (2.3)* | 113.4 |
| | | | | | Average = | 111.4% |
| S-PCL | 135.0 | 10.05 | (11.5)* | 10.14 | (1.6)* | 100.9 |
| S-RM | | | | | | |
| 5-hrs. | (-)** | 11.56 | (5.0)* | 13.46 | (3.5)* | 116.4 |
| 17-hrs. | (-)** | 13.24 | (5.1)* | 14.58 | (4.7)* | 110.1 |
| 29-hrs. | (-)** | 13.37 | (5.6)* | 12.87 | (8.8)* | 96.3 |
| | | | | | Average = | 107.6% |

```
 *    Coefficient of Variation
 **   Data Unavailable
 1    N-MC-A = N Mortar, Type N Masonry Cement, Brand A
      N-PCL  = N Mortar, Portland Cement Lime
      N-RM   = N Mortar, Ready Mix
      5 hrs. = Suspension Time - Specimens Made
 2    1N/mm² = 145 psi
```

Table 5 :    Job Site Mortars - Outside Cured

| Mortar Type | Flow (%) | Cube Strength[2] N/mm$^2$ | | Cylinder Strength[2] N/mm$^2$ | | Cylinder Cube |
|---|---|---|---|---|---|---|
| N-MC-A | 139.8 | 7.65 | (17.0)* | 10.01 | (3.2)* | 130.8 |
| N-MC-C | 138.8 | 4.62 | (9.3)* | 4.84 | (4.5)* | 104.8 |
| N-MC-E | 141.0 | 4.52 | (8.6)* | 5.74 | (1.2)* | 127.0 |
| N-MC-G | 141.9 | 3.49 | (9.6)* | 4.46 | (0.5)* | 127.8 |
| | | | | | **Average =** | **122.6%** |
| N-PCL: | 127.8 | 3.52 | (3.4)* | 3.96 | (2.0)* | 112.5% |
| | | | | | | |
| S-MC-B | 140.5 | 11.73 | (7.1)* | 14.69 | (3.4)* | 125.2 |
| S-MC-D | 139.3 | 13.22 | (11.0)* | 15.09 | (0.8)* | 114.1 |
| S-MC-F | 139.8 | 18.79 | (11.9)* | 17.53 | (3.4)* | 93.3 |
| S-MC-H | 140.0 | 5.87 | (6.9)* | 6.97 | (5.4)* | 118.7 |
| | | | | | **Average =** | **112.8%** |
| S-PCL | 132.5 | 9.29 | (3.8)* | 10.31 | (2.4)* | 110.0% |

*    Coefficient of Variation
1    N-MC-A = N Mortar, Type N Masonry Cement, Brand A
     N-PCL$_2$ = N Mortar, Portland Cement Lime
2    1N/mm$^2$ = 145 psi

Weerapun Sriboonlue[1] and Edward M. Wallo[2]

THE EFFECT OF CONSTITUENT PROPORTIONS ON STRESS - STRAIN
CHARACTERISTICS OF PORTLAND CEMENT - LIME MORTAR AND GROUT

---

**REFERENCE:** Sriboonlue, W., and Wallo, E.M., "The Effect
of Constituent Proportions on Stress-Strain
Characteristics of Portland Cement-Lime Mortar and Grout,"
Masonry : Components To Assemblages, ASTM STP 1063 John H.
Matthys, Editor, American Society for Testing and
Materials, Philadelphia, 1990.

**ABSTRACT:** The results of an experimental program designed
to evaluate the influence of constituent proportions on
the stress-strain relationship of Portland cement-lime
mortar and grout are reported.  Grout specimens consisting
of 76 millimeter by 152 millimeter cylinders and mortar
specimens of 50 millimeter cubes were cast and tested.
Mix variations included lime and sand content.  Modulus of
elasticity of both mortar and grout specimens using
different parts of lime and sand are presented.  In
addition, the relationship of the ultimate compressive
strength and modulus of elasticity for various mixes is
presented.  Mortar cube specimens were constructed
following ASTM C270, Standard Specification for Mortar for
Unit Masonry.  The water content used in mortar was
measured as different percentages of flow of fresh mortar.
Grout specimens were made in accordance with the ASTM
C476, Standard Specification for Grout for Masonry.  The
water content was measured as different slumps of the
grout.  The specimens were tested at age twenty-eight
days.  The results define trends, resulting from varying
the constituents, in the stiffness characteristics of
mortar and grout.

**KEYWORDS:** mortar, grout, modulus of elasticity, lime, mix
proportions, compressive strength.

1  Assistant Professor, Civil Engineering Department,
   Villanova University, Villanova, PA  19085
2  Associate Professor, Civil Engineering Department,
   Villanova University, Villanova, PA  19085

The behavior of structural masonry and reinforced masonry systems is strongly influenced by the quality of the mortar or the grout or both.  To better understand such systems one must have sufficient information on the effect of the constituent parts of the system.  This series of tests began to develop some data on the stiffness of both mortar and grout.

A small series of tests was conducted to evaluate the interaction of lime and amount of sand on the mechanical properties of Portland cement-lime mortar and grout.  The mortar tests used standard 50 millimeter cubes.  The grout tests used small cylinders, 76 millimeters in diameter and 152 millimeters in length.  The water content of the mortar mixes was controlled using a flow table.  The water content of the grout was controlled using standard slump tests.  All tests were conducted with test specimens cured in saturated lime solution at room temperature to age twenty eight days.  All test specimens were compression tested and the ultimate compressive strength and compression modulus of elasticity was computed for each. It was expected that this series of tests would identify trends resulting from varying the lime and sand components.

## EXPERIMENTAL PROGRAM

Portland cement-lime mortar is a composite material consisting of Portland cement, lime, sand, and water.  According to ASTM C270, Standard Specification for Mortar for Unit Masonry, Portland cement-lime mortar mix proportions are defined as one part of Portland cement, 0.25 to 2.50 parts of lime, and 2.25 to 3 times the sum of the volumes of the cementitious materials of sand, all by volume [1]. They reflect the four types of mortar M, S, N, and O which are currently used in masonry construction.  Grout is simply a high slump concrete with small percentages of lime.  According to ASTM C476, Standard Specification for Grout for Masonry, coarse grout mix proportions are defined as one part of Portland cement, 0 to 0.1 parts of lime, 2.25 to 3.0 parts of sand, and 1.0 to 2.0 parts of coarse aggregate, by volume [2].  Significant variation of the mix proportions for both mortar and grout may lead to the possibility that the stiffness may also vary.

The test program reported herein involved the testing of 160 - 50 mm mortar cubes and 135-76 mm x 152 mm grout cylinders.  Mortar test specimens were cast following the ASTM C109, Standard Test Method for Compressive Strength of Hydraulic Cement Mortars [3]. Grout mixes were made in accordance with the ASTM C476, Standard Specification for Grout for Masonry.

The proportions of the materials used to prepare different mortar mixes is given in Table 1.  Cement, lime, and sand contents are given in parts by volume.  Water content was measured as different percentages of flow of fresh mortar as defined in ASTM C109, Standard Test Method for Compressive Strength of Hydraulic Cement Mortars.  There were forty sets of test specimens made, each set

contained four specimens, which gave a total of one hundred and sixty
specimens.

TABLE 1 -- Mortar mix proportions.

| Cement | Sand | Lime | Water |
|--------|------|------|-------|
| 1 | 2.25 | 0.00 | 90% flow |
|   | 3.00 | 0.25 | 100% flow |
|   |      | 0.50 | 110% flow |
|   |      | 1.25 | 120% flow |
|   |      | 2.50 |           |

Grout proportions are given in Table 2.  Cement and aggregate
contents are given in parts by volume.  Water content was controlled
by measuring slump of the grout as defined in ASTM C143, Standard
Test Method for Slump of Portland Cement Concrete [4].  There were
twenty seven sets of grout specimens made, each set contained five
specimens, which gave a total of one hundred thirty five specimens.
Sand used for both mortar and grout conformed to ASTM C144, Standard
Specification for Aggregate for Masonry Mortar [5].  Aggregate use in
grout conformed to ASTM C404, Standard Specification for Aggregates
for Masonry Grout, with 9 mm maximum size of aggregate [6].  All test
specimens, both mortar and grout, were removed from the molds after
twenty four hours and stored in a saturated lime bath at room temperature to age twenty-eight days.  This curing method does not
follow recommended curing in ASTM C270.

TABLE 2 -- Grout mix proportions.

| Cement | Lime | Sand | Aggregate | Water |
|--------|------|------|-----------|-------|
| 1.00 | 0.00 | 2.250 | 1.5 | 178 mm slump |
|      | 0.05 | 2.625 |     | 229 mm slump |
|      | 0.10 | 3.000 |     | 279 mm slump |

All specimens were tested at age twenty-eight days using a
universal hydraulic testing machine whose maximum capacity was 534
kN.  Loads and displacements were recorded during the test at suitable intervals of load.  A dial gage was mounted to measure the
change in relative position between the two platens of the testing
machine.  The specimens were loaded to failure and the ultimate load
was recorded.

## RESULTS

The compressive strength and modulus of elasticity were computed
from the load-deformation data.  The modulus of elasticity for each
specimen was determined by taking the slope of the stress-strain
curve.  When interpreting the test results the short initial nonlinear "tail" at the beginning  of each stress-strain diagram was

ignored and assumed to result from "slack" in the test setup.  The
remaining linear portion of the test data was then used in slope
computations.  In all cases the portion of the diagram used for slope
computations was for stresses less than one half of the ultimate
strength.

The modulus of elasticity of mortar was plotted against the
parts by volume of lime used in the mix for sand content of 2.25 and
3.00 as shown in Figure 1.  The relationship between the modulus of
elasticity and the ultimate compressive strength of mortar is shown
in Figure 2.  For grout specimens, the plot of the modulus of
elasticity and the parts by volume of lime used in the mix for sand
content of 2.25, 2.625, and 3.0 parts are shown in Figures 3 through
5, respectively.  All data was plotted and shown in Figure 6.  The
relationship between the modulus of elasticity and the ultimate
compressive strength of grout is shown in Figure 7.  The relationship
between the modulus of elasticity and the square root of the ultimate
compressive strength of grout is shown in Figure 8.

Data in Figure 1 indicates that the amount of lime used in the
mortar mixes affects the modulus of elasticity of mortar.  The plot
shows a trend of increasing in the modulus of elasticity when lime
content is increased to 0.25 parts by volume.  Beyond this point, the
modulus of elasticity decreases when the lime content in the mix was
increased.  This tendency of decreasing modulus for increases in lime
content above 0.25 parts by volume was the same for mixes with
different sand content and appears to be independent of sand content
as shown in Figure 1.  Data in Figure 2 indicates that the modulus of
elasticity increases when the ultimate compressive strength increases.  This effect appears to be the same for both sand contents.

From grout test specimen results, Figures 3 through 6, it
appears that the modulus of elasticity decreases when the lime
content in the mix was increased.  However, there are some results
showing the opposite at lime content of 0.05 parts by volume.  The
change in the modulus of elasticity with variation of lime content is
not sufficient to justify that any trend occurred.  Figures 7 and 8
indicate a trend of an increase in the modulus of elasticity when the
ultimate compressive strength increases.  This effect seems to be the
same for all three sand contents.


## CONCLUSION AND RECOMMENDATIONS


The modulus of elasticity of Portland cement-lime mortar is
increased as lime contents is increased up to about 0.25 part by
volume and then decreases rapidly for any additional increase in lime
content.

The modulus versus strength relationship for both mortar and
grout does not appear to be affected by changes in lime content or
sand content.

The effect of changing aggregate content is not clear and further testing is necessary.

The data to date shows a need for further study of the effect of small amounts of lime, less than 0.5 parts by volume, in the mortar to confirm the skewed bell shaped curve (see Figure 1).

The data for grout show a need for further study of the effects of larger amounts of lime to determine if the same effects as shown for mortar would occur in grout.

## REFERENCES

[1]  ASTM Standard Specification for Mortar for Unit Masonry (C270-87a), American Society for Testing and Materials, Philadelphia, 1988.

[2]  ASTM Standard Specification for Grout for Masonry (C476-83), American Society for Testing and Materials, Philadelphia, 1988.

[3]  ASTM Standard Test Method for Compressive Strength of Hydraulic Cement Mortars (Using 2-in. or 50-mm Cube Specimens) (C109-80), American Society for Testing and Materials, Philadelphia, 1984.

[4]  ASTM Standard Test Method for Slump of Portland Cement Concrete (C143-78), American Society for Testing and Materials, Philadelphia, 1980.

[5]  ASTM Standard Specification for Aggregate for Masonry Mortar (C144-76), American Society for Testing and Materials, Philadelphia, 1980.

[6]  ASTM Standard Specification for Aggregate for Masonry Grout (C404-87), American Society for Testing and Materials, Philadelphia, 1988.

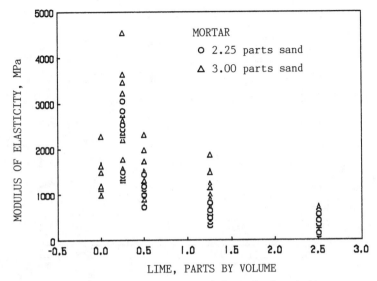

FIG. 1 -- Lime content versus modulus of elasticity.

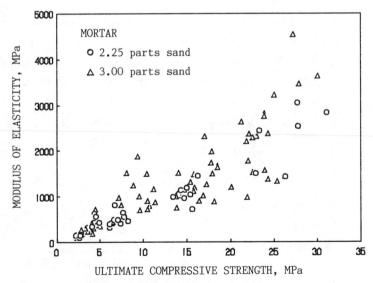

FIG. 2 -- Strength versus modulus.

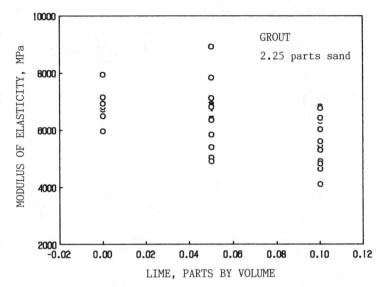

FIG. 3 -- Lime content versus modulus of elasticity.

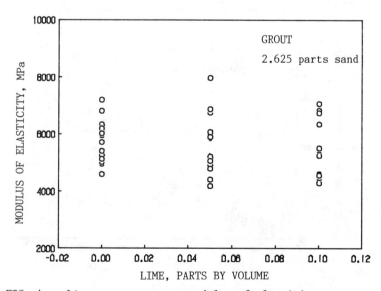

FIG. 4 -- Lime content versus modulus of elasticity.

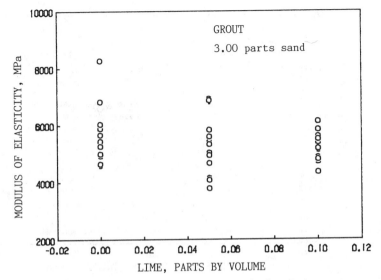

FIG. 5 -- Lime content versus modulus of elasticity.

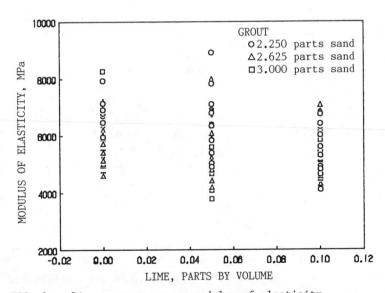

FIG. 6 -- Lime content versus modulus of elasticity.

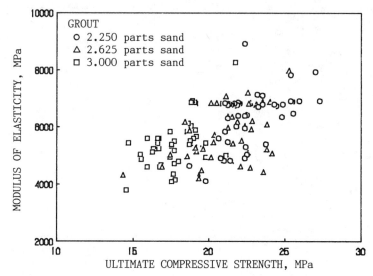

FIG. 7 -- Strength versus modulus.

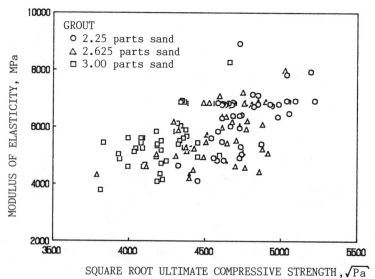

FIG. 8 -- Square root compressive strength versus modulus.

# Assemblages

W. Mark McGinley


IRA AND THE FLEXURAL BOND STRENGTH OF CLAY BRICK MASONRY

---

REFERENCE: McGinley W. M., "IRA and The Flexural Bond Strength of Clay Brick Masonry", MASONRY - "Components to Assemblages", ASTM STP 1063, John H. Matthys, Editor, American Society for Testing and Materials, Philadelphia, 1990.

ABSTRACT: During a recent experimental investigation at the North Carolina A&T State University, the flexural bond strength of a variety of clay bricks and two masonry cement mortars were evaluated. Twenty prisms, comprising ten different brick types and two mortar types, were tested with a bond wrench testing apparatus to determine the flexural bond strengths of each mortar joint of the seven course prisms. Although the investigation only tested a limited number of prisms, the results of these tests indicated that the initial rate of absorption (IRA) of the brick units can have a greater influence on the flexural bond strength of the masonry assembly than is generally accepted by the masonry industry.

The experimental program is described, its results presented and the apparent effects of brick IRA on flexural bond are discussed.

KEYWORDS: masonry, brick, flexural, bond, IRA

INTRODUCTION


In a recent experimental investigation at the North Carolina A & T State University, the flexural tensile bond strengths of a variety of clay brick and two masonry cement mortars were evaluated. Although the investigation only tested a limited number of specimens (six joints were tested for each mortar type and brick type combination), the results of these tests indicated that the initial rate of absorption of the brick (IRA) can have a

---

Dr. McGinley is an assistant professor in the Department of Architectural Engineering at North Carolina A&T State University, Greensboro, NC, 27411

greater influence on the flexural bond strengths of unreinforced clay brick masonry than is generally recognized by the masonry industry.

The following report will summarize the findings of this investigation.

EXPERIMENTAL PROGRAM

It has been found that the flexural bond between mortar and clay brick is affected by a number of factors which include:

1. Mortar Type
2. Workmanship
3. Mortar Consistency
4. Brick IRA
5. Age
6. Curing Conditions [1,2,3,4,5,6,7]

While not specifically designed to evaluate the effects of these factors, the experimental program measured properties of the mortar and the brick that were expected to affect the flexural tensile bond between the two components.

Brick Tests

A randomly chosen sample of ten different lots of brick were tested during this experimental investigation. All brick were extruded, wire cut, clay and shale units with approximately the same surface texture on the bonding faces. Three bricks from each of these lots were subjected to the following tests:

1. Compressive Strength
2. Initial Rate of Absorption (IRA)
3. Absorption

The initial rate of absorption tests were performed on brick units in the "AS LAID" condition, on the same day that the prism specimens were constructed. All tests were performed in accordance to ASTM standard C 67 - 85 [8].

Sand Tests

Four samples of the sand used to mix the mortar were subjected to a sieve analysis as outlined in ASTM Standard C 144 - 84 [9].

Mortar Tests

One batch of type N and one batch of type S prepackaged masonry cement mortars were used to build the brick prism specimens. These mortars were mixed in accordance to the mortar manufacturer's specifications and the flow was adjusted to the satisfaction of the mason.  From each mortar batch, three samples were tested for flow and air content using the procedures outlined in ASTM Standards C 270 [10] and C 780 [11], respectively.  In addition, five mortar cubes were made from each batch, moist cured and then tested for compressive strength.

## Prism Tests

A total of twenty, seven high, stack bonded prisms were constructed. For each of the two mortar types, one prism was constructed for each of the ten lots of brick.  One face of the mortar joints was tooled and the specimens were cured for twenty-eight days.  During curing, the specimens were stored open to the air in an area sheltered from direct sunlight and rain, but otherwise closely approximating an outdoor environment.  At the age of seven days, the specimens were lightly sprayed with water to simulate a rain shower in the field.  This wetting was preformed to ensure that each prism was subjected to similar hydration conditions to those in the field (bond appears to be dependant on the formation of hydration products [6]). The flexural bond strength of the tooled face of each brick prism was determined using a bond wrench testing apparatus and the procedures outlined in ASTM Standard C 1072 [12].

## EXPERIMENTAL RESULTS AND DISCUSSION

The results of the brick tests are summarized in Table 1.  With the exception of Brick #4, all brick satisfied the specified requirements for SW grade brick (ASTM C 67).  It should be noted that ASTM C 67 requires that five brick be tested to classify brick type but only three were tested during this investigation.

The average results of the sand sieve analysis are shown in Figure 1. This plot shows that the sand samples slightly exceeded the ASTM grading limits for masonry sand.

The results of the mortar tests are listed in Table 2.  Both mortars met the compressive strength requirements for laboratory prepared mortar as outlined in ASTM C-270.  As can be seen from the results of the mortar tests, both mortars were stiffer and had lower air contents than those normally used in the field.  Surprisingly, the mason preferred a stiffer mix for building the prism specimens.  As will be discussed later, this low mortar flow affected the bond strengths obtained from the prism joints.

The flexural bond strength of each mortar joint in the twenty prisms is listed in Table 3.  These strengths ranged from 0.0 MPa (0.0 psi) (failed during attachment of lever arm) to 0.610 MPa (87.7 psi) and were highly variable.  However, even allowing for the variability of this small number of tests, a very interesting trend in the data was observed.  Figure 2 shows the variation of the flexural bond of each

Table 1--Brick Properties

| Brick Batch | % Absorption | | I R A | | Compressive Strength | |
|---|---|---|---|---|---|---|
| | 5 hr. | 24 hr. | Kg/m^2/min | g/30in^2/min | (MPa) | (psi) |
| 1 | 8.83 | 7.10 | 0.16 | 3.1 | 81.68 | 11847 |
| | 8.98 | 7.28 | 0.12 | 2.3 | 78.51 | 11387 |
| | 8.92 | 7.24 | 0.14 | 2.7 | 78.07 | 11323 |
| | | Average | 0.14 | 2.7 | 79.42 | 11519 |
| | | COV (%) | 14.3 | | 2.48 | |
| 2 | 8.49 | 6.41 | 0.99 | 19.1 | 103.54 | 15017 |
| | 7.55 | 5.45 | 0.8 | 15.4 | 98.86 | 14338 |
| | 7.05 | 4.69 | 0.66 | 12.7 | 104.14 | 15104 |
| | | Average | 0.82 | 15.8 | 102.18 | 14820 |
| | | COV (%) | 20.7 | | 2.83 | |
| 3 | 7.63 | 5.19 | 1.29 | 24.9 | 74.14 | 10753 |
| | 8.46 | 6.01 | 1.46 | 28.2 | 81.35 | 11799 |
| | 8.42 | 6.00 | 1.56 | 30.1 | 75.52 | 10953 |
| | | Average | 1.44 | 27.8 | 77.01 | 11169 |
| | | COV (%) | 9.72 | | 4.97 | |
| 4 | 10.32 | 8.06 | 1.06 | 20.5 | 81.43 | 11810 |
| | 10.31 | 8.09 | 1.02 | 19.7 | 81.21 | 11779 |
| | 11.01 | 8.81 | 1.07 | 20.7 | 84.71 | 12286 |
| | | Average | 1.05 | 20.3 | 82.45 | 11958 |
| | | COV (%) | 2.89 | | 2.38 | |
| 5 | 11.04 | 8.46 | 1.49 | 28.8 | 85.35 | 12379 |
| | 11.12 | 8.43 | 1.46 | 28.2 | 82.75 | 12002 |
| | 10.82 | 8.4 | 1.55 | 29.9 | 84.91 | 12315 |
| | | Average | 1.5 | 29.0 | 84.34 | 12233 |
| | | COV (%) | 3.33 | | 1.65 | |
| 6 | 10.87 | 8.09 | 1.94 | 37.4 | 56.80 | 8238 |
| | 11.58 | 8.96 | 2.04 | 39.4 | 60.91 | 8834 |
| | 11.00 | 7.91 | 2.13 | 41.1 | 57.76 | 8377 |
| | | Average | 2.00 | 38.6 | 58.49 | 8483 |
| | | COV (%) | 7.50 | | 3.68 | |
| 7 | 6.67 | 4.78 | 0.45 | 8.7 | 98.53 | 14291 |
| | 7.31 | 4.84 | 0.49 | 9.5 | 95.94 | 13915 |
| | 6.11 | 4.29 | 0.46 | 8.9 | 104.01 | 15085 |
| | | Average | 0.47 | 9.1 | 99.50 | 14431 |
| | | COV (%) | 4.26 | | 4.14 | |
| 8 | 4.78 | 3.48 | 0.16 | 3.1 | 104.23 | 15117 |
| | 6.18 | 4.38 | 0.19 | 3.7 | 108.20 | 15693 |
| | 5.81 | 4.03 | 0.28 | 5.4 | 111.28 | 16140 |
| | | Average | 0.21 | 4.1 | 107.90 | 15650 |
| | | COV (%) | 28.60 | | 3.54 | |

Table 1--Brick Properties continued

| Brick Batch | % Absorption | | IRA | | Compressive Strength | |
|---|---|---|---|---|---|---|
| | 5 hr. | 24 hr. | Kg/m^2/Min | g/30in^2.min | (MPa) | (psi) |
| 9 | 5.79 | 5.18 | 0.34 | 6.6 | 103.92 | 15072 |
| | 5.80 | 5.40 | 0.36 | 6.9 | 108.67 | 15761 |
| | 5.65 | 5.25 | 0.38 | 7.3 | 105.17 | 15254 |
| | | | | | | |
| | | Average | 0.36 | 6.9 | 105.92 | 15362 |
| | | COV (%) | 5.56 | | 2.46 | |
| | | | | | | |
| 10 | 5.55 | 4.45 | 0.82 | 15.8 | 117.30 | 17013 |
| | 6.00 | 4.91 | 0.77 | 14.9 | 109.94 | 15945 |
| | 5.58 | 4.51 | 0.74 | 14.3 | 107.33 | 15567 |
| | | | | | | |
| | | Average | 0.78 | 15.1 | 111.52 | 16175 |
| | | COV (%) | 5.13 | | 5.17 | |

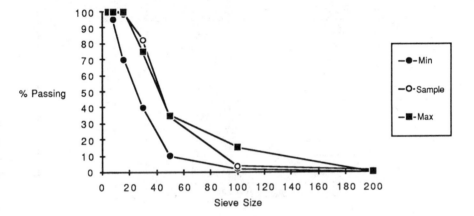

Figure 1
Sieve Analysis of Sand

Table    2-- Mortar Properties

| Properties | Type S Mortar | Type N Mortar |
|---|---|---|
| Average Flow Rate | 100.7 | 111.3 |
| COV (%) | 1.1 | 5.7 |
| | | |
| Average Compressive Strength (MPa) | 26.61 | 14.06 |
| (psi) | 3859 | 2039 |
| COV (%) | 18.0 | 16.4 |
| | | |
| Average Entrained Air (%) | 7.2 | 7.9 |
| COV (%) | 16.0 | 14.3 |

## Table 3--Prism Flexural Bond Strength

| Prism | Mortar Type | Joint | Bond Strength | |
|---|---|---|---|---|
| | | | (MPa) | (psi) |
| 1 | S | 1 | 0.000 | 0 |
| | | 2 | 0.000 | 0.00 |
| | | 3 | 0.034 | 4.93 |
| | | 4 | 0.102 | 14.79 |
| | | 5 | 0.122 | 17.69 |
| | | 6 | 0.154 | 22.34 |
| | | Average | 0.069 | 10.0 |
| | | COV (%) | 69.6 | |
| 1 | N | 1 | 0.034 | 4.93 |
| | | 2 | 0.000 | 0.00 |
| | | 3 | 0.351 | 50.91 |
| | | 4 | 0.028 | 4.06 |
| | | 5 | 0.049 | 7.11 |
| | | 6 | 0.109 | 15.81 |
| | | Average | 0.095 | 13.8 |
| | | COV (%) | 137.3 | |
| 2 | S | 1 | 0.000 | 0.00 |
| | | 2 | 0.384 | 55.69 |
| | | 3 | 0.253 | 36.69 |
| | | 4 | 0.149 | 21.61 |
| | | 5 | 0.199 | 28.86 |
| | | 6 | 0.350 | 50.76 |
| | | Average | 0.223 | 32.3 |
| | | COV (%) | 63.0 | |
| 2 | N | 1 | 0.117 | 16.97 |
| | | 2 | 0.049 | 7.11 |
| | | 3 | 0.093 | 13.49 |
| | | 4 | 0.102 | 14.79 |
| | | 5 | 0.261 | 37.85 |
| | | 6 | 0.141 | 20.45 |
| | | Average | 0.127 | 18.4 |
| | | COV (%) | 56.9 | |
| 3 | S | 1 | 0.057 | 8.27 |
| | | 2 | 0.000 | 0.00 |
| | | 3 | 0.022 | 3.19 |
| | | 4 | 0.000 | 0.00 |
| | | 5 | 0.235 | 34.08 |
| | | 6 | 0.269 | 39.02 |
| | | Average | 0.097 | 14.1 |
| | | COV (%) | 126.0 | |

## Table 3--Prism Flexural Bond Strength continued

| Prism | Mortar Type | Joint | Bond Strength (MPa) | (psi) |
|-------|-------------|-------|---------------------|-------|
| 3 | N | 1 | 0.172 | 24.95 |
|   |   | 2 | 0.000 | 0.00 |
|   |   | 3 | 0.000 | 0.00 |
|   |   | 4 | 0.238 | 34.52 |
|   |   | 5 | 0.000 | 0.00 |
|   |   | 6 | 0.222 | 32.20 |
|   |   | Average | 0.105 | 15.28 |
|   |   | COV (%) | 111.8 | |
| 4 | S | 1 | 0.036 | 5.22 |
|   |   | 2 | 0.067 | 9.72 |
|   |   | 3 | 0.102 | 14.79 |
|   |   | 4 | 0.083 | 12.04 |
|   |   | 5 | 0.000 | 0.00 |
|   |   | 6 | 0.067 | 9.72 |
|   |   | Average | 0.059 | 8.58 |
|   |   | COV (%) | 61.4 | |
| 4 | N | 1 | 0.000 | 0.00 |
|   |   | 2 | 0.030 | 4.35 |
|   |   | 3 | 0.106 | 15.37 |
|   |   | 4 | 0.000 | 0.00 |
|   |   | 5 | 0.085 | 12.33 |
|   |   | 6 | 0.000 | 0.00 |
|   |   | Average | 0.037 | 5.34 |
|   |   | COV (%) | 128.0 | |
| 5 | S | 1 | 0.000 | 0.00 |
|   |   | 2 | 0.000 | 0.00 |
|   |   | 3 | 0.034 | 4.93 |
|   |   | 4 | 0.000 | 0.00 |
|   |   | 5 | 0.044 | 6.38 |
|   |   | 6 | 0.000 | 0.00 |
|   |   | Average | 0.013 | 1.89 |
|   |   | COV (%) | 156.8 | |
| 5 | N | 1 | 0.049 | 7.11 |
|   |   | 2 | 0.000 | 0.00 |
|   |   | 3 | 0.028 | 4.06 |
|   |   | 4 | 0.000 | 0.00 |
|   |   | 5 | 0.067 | 9.72 |
|   |   | 6 | 0.246 | 35.68 |
|   |   | Average | 0.065 | 9.43 |
|   |   | COV (%) | 142.4 | |

## Table 3 Prism Flexural Bond Strength Continued

| Prism | Mortar Type | Joint | Bond Strength | |
|---|---|---|---|---|
| | | | (MPa) | (psi) |
| 6 | S | 1 | 0.031 | 4.50 |
| | | 2 | 0.051 | 7.40 |
| | | 3 | 0.049 | 7.11 |
| | | 4 | 0.091 | 13.20 |
| | | 5 | 0.059 | 8.56 |
| | | 6 | 0.098 | 14.21 |
| | | Average | 0.063 | 9.16 |
| | | COV (%) | 41.3 | |
| 6 | N | 1 | 0.051 | 7.40 |
| | | 2 | 0.190 | 27.56 |
| | | 3 | 0.000 | 0.00 |
| | | 4 | 0.034 | 4.93 |
| | | 5 | 0.049 | 7.11 |
| | | 6 | 0.067 | 9.72 |
| | | Average | 0.065 | 9.45 |
| | | COV (%) | 100.3 | |
| 7 | S | 1 | 0.157 | 22.77 |
| | | 2 | 0.319 | 46.27 |
| | | 3 | 0.157 | 22.77 |
| | | 4 | 0.270 | 39.16 |
| | | 5 | 0.586 | 84.99 |
| | | 6 | 0.558 | |
| | | Average | 0.341 | 43.19 |
| | | COV (%) | 55.7 | |
| 7 | N | 1 | 0.489 | 70.92 |
| | | 2 | 0.156 | 22.63 |
| | | 3 | 0.558 | 80.93 |
| | | 4 | 0.162 | 23.50 |
| | | 5 | 0.422 | 61.21 |
| | | 6 | 0.172 | 24.95 |
| | | Average | 0.327 | 47.35 |
| | | COV (%) | 56.3 | 8169.78 |
| 8 | S | 1 | 0.188 | 27.27 |
| | | 2 | 0.489 | 70.92 |
| | | 3 | 0.385 | 55.84 |
| | | 4 | 0.409 | 59.32 |
| | | 5 | 0.405 | 58.74 |
| | | 6 | 0.308 | 44.67 |
| | | Average | 0.364 | 52.79 |
| | | COV (%) | 28.5 | |

### Table 3--Prism Bond Strength continued

| Prism | Mortar Type | Joint | Bond Strength | |
|---|---|---|---|---|
| | | | (MPa) | (psi) |
| 8 | N | 1 | 0.033 | 4.79 |
| | | 2 | 0.081 | 11.75 |
| | | 3 | 0.157 | 22.77 |
| | | 4 | 0.287 | 41.63 |
| | | 5 | 0.275 | 39.89 |
| | | 5 | 0.162 | 23.50 |
| | | Average | 0.166 | 24.05 |
| | | COV (%) | 61.2 | |
| 9 | S | 1 | 0.472 | 68.46 |
| | | 2 | 0.268 | 38.87 |
| | | 3 | 0.477 | 69.18 |
| | | 4 | 0.527 | 76.44 |
| | | 5 | 0.162 | 23.50 |
| | | 6 | 0.264 | 38.29 |
| | | Average | 0.362 | 52.46 |
| | | COV (%) | 41.2 | |
| 9 | N | 1 | 0.464 | 67.30 |
| | | 2 | 0.287 | 41.63 |
| | | 3 | 0.206 | 29.88 |
| | | 4 | 0.497 | 72.08 |
| | | 5 | 0.303 | 43.95 |
| | | 6 | 0.211 | 30.60 |
| | | Average | 0.328 | 47.57 |
| | | COV (%) | 38.1 | |
| 10 | S | 1 | 0.182 | 26.40 |
| | | 2 | 0.590 | 85.57 |
| | | 3 | 0.235 | 34.08 |
| | | 4 | 0.610 | 88.47 |
| | | 5 | 0.472 | 68.46 |
| | | 6 | 0.203 | 29.44 |
| | | Average | 0.382 | 55.40 |
| | | COV (%) | 52.0 | |
| 10 | N | 1 | 0.101 | 14.65 |
| | | 2 | 0.000 | 0.00 |
| | | 3 | 0.245 | 35.53 |
| | | 4 | 0.206 | 29.88 |
| | | 5 | 0.422 | 61.21 |
| | | 6 | DAMAGED | |
| | | Average | 0.195 | 28.25 |
| | | COV (%) | 81.6 | |

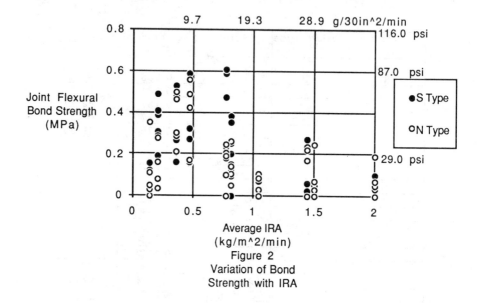

Figure 2
Variation of Bond
Strength with IRA

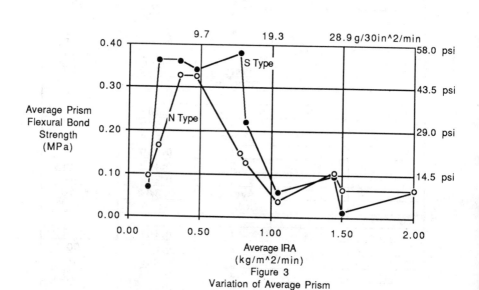

Figure 3
Variation of Average Prism
Strength with Average IRA

mortar joint with average brick IRA.  It appears that flexural bond is
reduced at high and low brick IRA.  This trend is obvious when the
average bond strength obtained from each prism is plotted against the
average brick IRA (see Figure 3).  There is clearly a significant
reduction of bond strength when the brick IRA is outside a certain
range of values.  So severe is this bond strength reduction that, in a
significant number of tests, type S mortar joints constructed using
brick with IRA on either side of this optimum range produced lower bond
strengths than type N mortar joints constructed using bricks with IRA
within this "optimum IRA" range.  For the type and condition of the
brick/mortar combinations tested, this "optimum IRA" range appears to
be approximately 0.26 to 0.52 kg/m$^2$/min (5 to 10 g/30 in$^2$/min) for type
N mortar, and 0.26 to 0.77 kg/m$^2$/min.(5 to 15 g/30 in$^2$/min) for type S
mortar.

   To explain why the IRA of the brick has a variable effect on the
flexural bond strength of the assemblage, one must examine the
mechanism by which mortar bonds to brick.  Lawrence and Cao [6] and
Chase [7] have examined the mechanism of bond at the microscopic level.
Their studies indicate that the bonding between mortar and brick is due
primarily to mechanical interlocking of the cement hydration products
growing in the brick pores and on the brick surface, and those
connected to the mortar paste.  Lawrence and Cao[6] postulate that a
higher IRA will produce higher mortar paste penetration in the brick
and, therefore, higher bond strength.  However, they further postulate
that if the IRA is too high, the bond strength will be reduced because
of an increase in shrinkage induced micro-cracking at the brick/mortar
interface.  An obvious extension of these theories would suggest that
there is low mortar paste penetration in a brick with low IRA and
consequently lower bond strengths.  Although a few researchers have
found the results of their tests to be inconclusive, [2,3] it is
generally recognized that brick IRA reduces flexural bond strength if
either too high or too low [1,4,5,6].

   The above theories are a possible explanation of why the bond
strengths fall off at high and low IRA.  However, the range of "optimum
IRA" shown by the brick/mortar combinations tested during this
investigation is much narrower than has been observed in other testing
programs[1,5], or than is recommended by the B.I.A.[4].  One possible
explanation for this reduced "optimum IRA" range is the low flow of the
mortar.  Most previous research has been done using mortars with flows
of 120 or greater.  The average mortar flow used in this testing
program was 101 for the type S mortar and 111 for the type N mortar.
With a low flow mortar, it is obvious that there will be sufficient
water available at the interface for good bond over a narrower range of
brick IRA.  Mortar flow and initial rate of absorption of the brick
both affect the water available at the interface and, therefore, both
affect the flexural tensile bond strength of the assemblage.  High IRA
brick will require a high flow mortar to provide optimum conditions for
development of flexural bond.  The dependence of the flexural bond on a
combination of mortar flow and brick IRA is, at least in part,
substantiated by the experimental investigation which showed a
significant increase in the bond strength of high IRA brick with
increasing mortar flows[1].

   The degree to which mortar flow and brick IRA combine to affect the flexural tensile bond strength developed between clay brick and different types of mortars is uncertain. The results of this investigation suggest that type S masonry cement mortars are somewhat less sensitive to flow and IRA than type N masonry cement mortar. However, too few specimens and types of mortars were tested during this investigation and further testing is required before a definite relationship can be conclusively determined.

While the mortar flows tested in this investigation were lower than those normally used in the field, the test results reaffirm the importance of using high flow mortars in construction. It is particularly important that masons restrict the length of mortar bed that they lay down before placing the next course of brick. If too long a mortar bed is laid, the last few bricks will be placed into a low flow mortar and consequently have a poorer bond. This becomes particularly important when working in hot weather with high IRA brick. It is important to control both flow and brick IRA to produce maximum flexural bond strengths.


CONCLUSIONS


It has been shown that the initial rate of absorption (IRA) of clay brick masonry has a significant effect on the flexural tensile bond strength developed between brick and masonry cement mortar. The bond strength is reduced if the brick IRA is either too high or too low. It also appears that the sensitivity of flexural bond strength to brick IRA is increased with decreasing mortar flows, although further testing is required to conclusively define this relationship.


REFERENCES

[1]  Engel, G.L., "Factors affecting the Flexural Bond Strength of Brick Masonry", Masonry International: The Journal of the British Masonry Society, Vol. 1, No. 1, Spring, 1987, pp. 16-24.

[2]  Gozzola, E., Bagnariol, D., Toneff, J., and Drysdale, R. G., "Influence of Mortar Materials on the Flexural Tensile Bond Strength of Block and Brick Masonry", Masonry: Research, Application and Problems, ASTM STP 871, J. C. Grogan and J. T. Conway, Eds., American Society of Testing and Materials, Philadelphia, 1985, pp. 15-26.

[3]  Drysdale, R. G. and Gozzola, E., "Influence of Mortar Properties on the Flexural Tensile Bond Strength of Brick Masonry", Proceedings of the 7th International Brick Masonry Conference, Melbourne, Australia, February, 1985, 12 pages.

[4]  Brick Institute of America, "Technical Note 8 Revised", Brick Institute of America, Reston, Va., February, 1987.

[5]  Brick Institute of America, "Background and Commentary", in Recommended Practice for Engineered Brick Masonry, Brick Institute of America, Reston, Va., November, 1969, pp.239-333.

[6]   Lawrence, S. J. and Cao, H. T., "An Experimental Study of the
      Interface Between Brick and Mortar", Proceedings of the
      Fourth North American Masonry Conference, Los Angeles, August,
      1987, pp. 48-1 - 48-14.
[7]   Chase, G. W. "Investigation of the Interface Between Brick and
      Mortar", The Masonry Society Journal, Vol. 3, No. 2, July-
      December, 1984, pp. T-1 - T-9.
[8]   American Society of Testing Materials, "Standard Methods of
      Sampling and Testing Brick and Structural Clay Tile:, ASTM C
      67 85, American Society of Testing and Materials,
      Philadelphia, 1895.
[9]   American Society of Testing and Materials, "Standard
      Specification for Aggregate for Masonry Mortar", ASTM C 144-
      84, American Society of Testing and Materials, Philadelphia,
      1984.
[10]  American Society of Testing and Materials, "Standard
      Specification for Mortar for Unit Masonry", ASTM C 270-86,
      American Society of Testing and Materials, Philadelphia, 1986.
[11]  American Society of Testing and Materials, "Standard Method of
      Preconstruction and Construction Evaluation of Mortars for
      Plain and Reinforced Masonry", ASTM C 780-80, American Society
      of Testing and Materials, Philadelphia, 1980.
[12]  American Society of Testing and Materials, "Standard Method for
      Measurement of Masonry Flexural Bond Strength" ASTM C 1072-86,
      American Society of Testing and Materials Philadelphia, 1986.

ACKNOWLEDGEMENTS


    This program was funded by the Brick Association of North Carolina.
The author would like to specifically thank Marion Cochrane, Tim Conway
and Wright Archer for their assistance during this project.

DISCUSSION

"IRA and the Flexural Bond Strength of Clay Brick Masonry"
W.M. McGinley

Comments   (L.R. Lauersdorf, State of Wisconsin):

According to the ASTM C-270 Appendix, bond is the single most important physical property of hardened mortar, and mortar generally bonds best to masonry units having moderate initial rates of absorption (IRA). Test results from this paper as well as from other sources confirm this statement.

The data in the paper confirms that the IRA of clay brick masonry units can have a significant influence on bond, greater than is generally accepted by the masonry industry. There are actually three facets to bond; namely, strength, extent and durability.

In regard to bond strength, the paper well summarized conclusions on the effects of IRA on bond strength by Figure 3. This graph is similar to that contained in "Factors Affecting Bond Strength and Resistance to Moisture Penetration of Brick Masonry" by T. Ritchie and J.I. Davison, Research Paper No. 192, Division of Building Research, National Research Council, Ottawa Canada, July 1963. This latter paper confirmed a previous similar graph contained in "A Study of the Properties of Mortars and Bricks and Their Relation to Bond" by L.A. Palmer and D.A. Parsons, Research Paper RP683, Bureau of Standards, U.S. Department of Commerce, May 1934. Ritchie and Davison also reported on extent of bond affected by IRA, as measured by leakage rates.

The masonry industry highly recommends that low flow mortars be used with low IRA units in order to increase the bond strength. Even with the extraordinarily low flow mortar which was used as indicated in this paper, the bond strength with both mortar types used was still low with the low IRA masonry units. In fact, the configuration of the bond strength versus IRA plot seemed unchanged when compared with the curvature of the other two references, which utilized a higher, more conventional flow mortar.

In conclusion, the paper along with previous impartial data reported suggests that the existing note in ASTM C-216 relating to IRA should be updated to reflect that both laboratory and field investigations show that strong and watertight joints between mortar and masonry units are not usually achieved by ordinary construction methods when the units as laid have excessively low or high IRA's. Mortar generally bonds best to masonry units having IRA's from 5 to 25 g/min/30 sq. in. (194 sq. cm.) at the time of laying, although adequate bond can be obtained with some units having IRA's less than or greater than these values.

Comments (McGinley)    I appreciate and support the comments put forth by Mr. Lauersdorf.    It should be recognized by the masonry industry that a number of factors affect the bond developed between clay bricks and mortar.    While workmanship and the constituents of the mortar mix have a long recognized and major effect on bond strengths, other factors such as brick IRA, mortar flow and brick surface texture can also have a significant effect on the bond developed between clay bricks and mortar.

DISCUSSION

"IRA and the Flexural Bond Strength of Clay Brick Masonry"
- W. M. McGinley

Question    (J.  H.  Matthys,  University  of  Texas  at
Arlington):

Table 2 on Mortar Properties lists COV on mortar
compressive strengths that appear to be excessively high
particularly for moist cured specimens.  To what factors
do you attribute this high COV?

Table 2 also gives average air entrained air of 7.2% and
7.9% for Type S and Type N prepackaged Masonry Cement.
These values do not seem to be typical levels of air
found in prepackaged masonry cements, i.e., these values
are too low.  I wonder if these products by ASTM C-91
actually qualify as a masonry cement.  Do you know?
What method was used to determine the air content?

For Prism 1, Mortar Type S, the COV on bond strength is
95.6, not 69.6.  The range of COV on bond strength in
Table 3 is from 28.5 to 156.8 with an average COV of
82%.  Do you attribute the high degree of variability to
the test procedure, assemblage, mortar, etc.??

Table 1 on Brick Properties indicates a range of brick
IRA from 2.7 gms./min./30in.$^2$ to 38.6 gms./min./30in.$^2$;
yet the flow of the mortar used was low (100.7 and
111.3) for field mortars.  Do you feel that these flows
are typical of mortars in the field for laying brick?
Do you feel that some of your conclusions might be
significantly altered using a flow of mortar more
closely associated with the IRA demand of the brick?

At the age of seven days you state specimens were
sprayed with water to simulate a rain shower in the
field.  I would suspect that there is a significant
amount of masonry built that is not subjected to applied
moisture at seven days.  There is also a large amount of
masonry built that is never subjected to applied
moisture (interior masonry).  In your opinion would your
bond results in both magnitude and spread in terms of
brick IRA be significantly different if you had not
sprayed the specimens with water?  Would the bond values
be significantly lower without spraying?

Table 3 gives for Type S masonry cement mortar a range
of bond strength from 1.89 psi to 55.40 psi with an
average of 27.98 psi.  Fifty percent of the Type S group
specimens exhibited average bond strengths less than the
allowable flexural tension found in the ACI/ASCE 530
Code.  Type N masonry cement mortar exhibited a range of

bond strength from 5.34 to 47.5 psi with the average
of 21.89 psi.  Thirty percent of the type N group
exhibited average bond strengths less than the
allowable flexural tension found in ACI/ASCE 530
code.  Do you think this should be a cause of concern
or not?  If so, what suggestions would you give.

Answer   (W. M. McGinley) Your questions are
addressed in the following point form:

    1. Since this investigation attempted to
evaluate the flexural bond strengths obtained
using standard field practices, the mason
mixed the mortar batches.  The relatively
high variation in the cube compressive
strengths may have been due to inadequate
mixing of the mortar resulting in larger
variation in the properties of the cubes than
is normally observed for  laboratory mixed
mortar batches.

    2. The values of air entrainment measured for
these masonry cement mortar mixes were low.
They were, in fact, below the minimum allowed
in ASTM C-91 (min 8 %).  The air entrainment
of each batch was obtained using the
procedures in ASTM Standard C 780 and a
roller pressure meter.  However, retesting of
nominally identical mortar batches with the
same procedures and a pump pressure meter
gave air entrainment values of approximately
17 %.  These results suggest that the roller
meter may have given inaccurate readings of
air entrainment.

3. As mentioned previously, this investigation
   was intended as an evaluation of normal field
   practice.  The high variation in flexural
   bond strengths can be attributed primarily to
   a relatively high variation in the mortar
   properties (see 1) and, at best, average
   workmanship on the part of the mason.

4. The flow of the mortar used was low.
   Surprisingly, the mason preffered this low
   mortar flow for laying the stack bonded prism
   specimens.  Higher flow mortars do produce
   better bond strengths [1] and it is expected
   that higher flow mortars will be less
   sensitive to brick IRA.  However, I have
   observed many masons vigorously tapping brick
   units into place, especially the last few
   units laid into a long bed joint of mortar.
   It is likely that these units are laid into

low flow mortar and will result in a poor
bond if the IRA of the brick is high.

5. Since the primary mechanism of mortar to
brick bond appears to be the mechanical
interlocking of cement hydration products at
the interface of the mortar and brick, the
additional water present at the interface
after spray the specimens with water should
affect the bond strength developed.  However,
subsequent testing of similar specimens that
were not subjected to this spraying resulted
in higher bond strength values.  It appears
that the amount of water that was available
at the interface was not greatly affected by
the light spraying.

6. Overall, the bond strengths measured for the
prism specimens tested in this investigation
were low, especially when compared with
applicable allowable values in the ACI/ASCE
530 Code.  If the measured values reflect the
bond strengths present in a wall, there is
indeed a cause for concern.  It has been my
opinion, and that of many others, that there
must be a performance specification for
unreinforced masonry that defines a minimum
flexural bond strength for the brick and
mortar assembly.  Flexural bond is probably
the most important property of the assembly
when used in a non-load bearing application.

Brent A. Gabby


A COMPILATION OF FLEXURAL BOND STRESSES FOR SOLID AND HOLLOW NON-
REINFORCED CLAY MASONRY AND PORTLAND CEMENT-LIME MORTARS


REFERENCE: Gabby, B.A., "A Compilation of Flexural Bond
Stresses for Solid and Hollow Non-reinforced Clay Masonry
and Portland Cement-Lime Mortars," Masonry:  Components to
Assemblages, ASTM STP 1063, J. H. Matthys, Editor, American
Society for Testing and Materials, Philadelphia, PA, 1990.

ABSTRACT:  The purpose of this paper is to make a general
comparison of compiled flexural tensile stress values which
were determined by using one of three standard test methods.
The material in this paper was gathered from seventeen reports which used one or two of the following test methods:
ASTM E 72, ASTM E 518 and ASTM C 1072.  All of the flexural
tensile stress values collected are those normal to the bed
joints for solid and hollow non-reinforced clay masonry and
portland cement-lime mortar.  The reports used contain
specimens which were air cured and broken at or around
twenty-eight days.

KEYWORDS:  mortar, clay masonry, flexural bond strength, test
methods

In the United States there are currently three standard test
methods which can be used to determine the flexural bond strength of
masonry normal to the bed joints.  They are ASTM E 72, ASTM E 518 and
ASTM C 1072.  Both ASTM E 518 and ASTM C 1072 are relatively new
standards when compared to ASTM E 72.  The flexural bond strength
values normal to the bed joints found in the current masonry code are
based on only one method of test, ASTM E 72.  The accuracy of these
values, which were generated nearly twenty years ago, have recently
come under scrutiny.

Until now, no one has made a comprehensive study to determine if
there are any substantial differences between the data which has been
produced by each test method.  This report should help masonry


Brent Gabby is a staff engineer at the Brick Institute of America,
11490 Commerce Park Drive, Reston, VA  22091.

researchers make an educated evaluation of the flexural tensile stress values normal to the bed joints for clay masonry and portland cement-lime mortars as they pertain to allowable flexural tensile values.

REVIEW OF THE TEST METHODS

ASTM E 72

This standard was originally adopted in 1947, and until 1974 was the only good standard available in the United States for testing the flexural strength of masonry. This test method is used for full scale wall specimens only, which are usually 1.2 m (4 ft) wide and 2.4 m (8 ft) high. The specimens can be broken by one of two methods, either by applying a uniformly distributed load of air contained in an air bag (air bag test) or by applying a continuous concentrated load at quarter points along the wall height (quarter point test). See Figures 1 and 2.

The air bag test has been used more widely in the past than the quarter point test method. The air bag test method simulates field conditions much closer, but only one or two joints are subjected to the maximum stress. The joint that fails is usually in the middle of the wall, but this does not necessarily mean it is the weakest joint. On the other hand, the quarter point test method subjects half of the wall to the maximum stress level, so the probability of finding the weakest joint in the wall greatly increases. However, there is a wide spread between bond strength values for similar mortar types when this test is used. This is probably due to the unpredictable stress concentrations which are developed at the points of loading.

A problem with ASTM E 72 is that it is very expensive and unwieldy to conduct. Also, only one value can be determined from a very sizeable specimen. However, this test is useful because it is the only way which composite, masonry tied and metal tied walls can be tested.

ASTM E 518

This standard was adopted in 1974. Like ASTM E 72, the masonry specimens can be broken by one of two methods, either by applying a uniformly distributed load of air contained in an air bag (air bag test) or by applying a concentrated load at third points along the specimen length (third point test). However, unlike ASTM E 72, the scale of the specimens has been greatly reduced. The test specimens consist of stacked bond masonry prisms which are usually four to sixteen courses high. See Figures 3 and 4. Since ASTM E 518 and ASTM E 72 are theoretically similar in scope, the same problems exist between the air bag test method and the concentrated load test method, but only on a smaller scale.

FIGURE 1 -- ASTM E 72 Airbag Test

FIGURE 2 -- ASTM E 72 Quarterpoint Test

FIGURE 3 -- ASTM E 518 Third Point Test

FIGURE 4 -- ASTM E 518 Airbag Test

ASTM C 1072 - Bond Wrench

This standard was adopted in 1986. This test method is a
modification of that which was introduced by Hughes and Zmembery [18]
in 1980. This is a very simple apparatus in which a stacked bond prism
is placed in a stationary frame and a cantilevered arm is clamped to
the brick over the joint to be tested. The free end of the cantilevered arm is then loaded until the clamped brick "wrenches" off the
top of the specimen. See Figure 5. Unlike ASTM E 72 and ASTM E 518,
the bond wrench allows all of the joints in the specimen to be tested.

Analysis of Compiled Data

By test method, Table 1 lists the ranges of both brick and mortar
properties used in each report. Tables 2 and 3 give a summary of the
mean data which was compiled from each report listed in Table 1. The
coefficient of variation in Tables 2 and 3 are based on the variation
of the means which were summarized in each research report. It should
be noted that in Table 3, hollow bonded masonry walls have been
isolated to show that there is a substantial decrease in flexural bond
strength for Type S mortar when this type of wall is tested. The
overall mean of ASTM E 72 air bag test method increased approximately
82.8 kPa (12 psi) and the coefficient of variation decreased
approximately 10 percent when the hollow bonded wall values were
excluded from those calculations. In the current masonry code, the
only requirement for masonry bonded walls is that the masonry headers
be uniformly distributed and the sum of their cross-sectional areas be
at least 4 percent of the wall surface area. The data which was
gathered for this report concerning hollow bonded masonry walls met
this 4 percent code requirement.

Figures 6 and 7 illustrate the mean values for solid and hollow
brick found in Table 2. It should be noted that Figure 8 illustrates a
very familiar profile between the tests conducted with ASTM E 72 air
bag test method and ASTM C 1072. Type S mortar shows a slightly higher
bond strength than Type M and Type M shows a higher bond strength than
Type N. The greatest difference between tests of the three individual
mortar types is only 62.1 kPa (9 psi). However, this comparison is
made with the inclusion of the hollow bonded wall data for Type S
mortar in the compiled ASTM E 72 results.

These two test methods were compared because both have a more
substantial data base than the other test methods. Today, ASTM C 1072
is used almost exclusively for measuring the flexural bond strength of
masonry. Figure 8 illustrates the almost insignificant difference
between the older ASTM E 72 test method, which the current allowable
stresses are based on, and the newer ASTM C 1072 test method.

Figure 7 for hollow brick also exhibits the same profile for all
three mortar types as those shown in Figure 8. Curiously, the greatest
difference between the highest and lowest value between all three
mortars is only 34.5 kPa (5 psi).

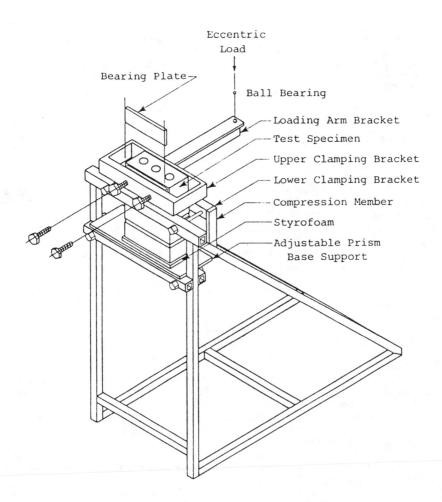

FIGURE 5 -- Schematic Drawing of Bond Wrench
Test Apparatus

CONCLUSION


This paper should give researchers a good taste of the flexural bond data which exists in the United States and Canada for clay masonry and portland cement-lime mortars. Since the validity of all of the flexural bond strengths for all types of masonry and mortars found in the current masonry code are being questioned, it is important that this type of analysis continue.


REFERENCES


[1]  Whittemore, H.L., Stang, A.H., and Parsons, D.E., "Structural Properties of Six Inch Masonry Wall Construction," National Bureau of Standards Report BMS-5, November 1938.
[2]  Monk, C.B., "Transverse Strength of Masonry Walls," Symposium on Methods of Testing Building Constructions, ASTM STP 166, American Society for Testing and Materials, Philadelphia, 1954, pp. 21-50.
[3]  Monk, C.B., "SCR Brick Wall Tests," Research Report No. 1, Structural Clay Products Research Foundation, Geneva, Illinois, Jan. 1965.
[4]  "Compressive and Transverse Test of Five Inch Walls," Research Report No. 8, Structural Clay Products Research Foundation, Geneva, Illinois, May 1965.
[5]  Thompson, J.N., Johnson, F.B., Wheeless, L.A., "Transverse Strength of Masonry Bonded Hollow Walls," Structural Mechanics Research Laboratory, The University of Texas at Austin, Aug. 15, 1965.
[6]  "Compressive, Transverse and Racking Strengths of Four Inch Brick Walls," Research Report No. 9, Structural Clay Products Research Foundation, Geneva, Illinois, Aug. 1965.
[7]  "Transverse Strength Tests of Eight Inch Brick Walls," Research Report No. 10, Structural Clay Products Research Foundation, Geneva, Illinois, Oct. 1966.
[8]  "Compressive and Transverse Strength Tests of Twelve Inch Masonry Bonded Hollow Walls," Structural Clay Products Institute, McLean, Virginia, Aug. 1968.
[9]  "Compressive, Transverse and Shear Strength Tests of Six and Eight Inch Single-Wythe Walls Built with Solid and Heavy Duty Hollow Clay Masonry Units," Research Report No. 16, Structural Clay Products Institute, McLean, Virginia, Sept. 1969.
[10] Gross, J.G., Dikkers, R.D., Grogan, J.C., Recommended Practice for Engineered Brick Masonry, Structural Clay Products Institute, McLean, Virginia, Nov. 1969.
[11] "Effects of Face-Shell Bedding, Unit Strength and Percentage Unit Coring on the Structural Properties of Hollow Brick Masonry," Hollow Brick Research Phase V, Unpublished Report, Brick Institute of America, Reston, Virginia.
[12] Fattal, S.G., and Cattaneo, L.E., "Structural Performance of Masonry Walls Under Compression and Flexure, Building Science Series 73, National Bureau of Standards, Washington, D.C., June

1976.
[13] Palm, B.D., "Flexural Strength of Brick Masonry Using the Bond
     Wrench," A Special Project Presented to the Civil Engineering
     Faculty of Clemson University, In Partial Fulfillment of the
     Requirements for the Degree of Master of Science Civil
     Engineering, Aug. 1982.
[14] Matthys, J.H., and Grimm, C.T., "Flexural Strength of Non-
     Reinforced Brick Masonry with Age," Proceedings of the 5th
     International Brick Masonry Conference, University of Texas at
     Arlington, Arlington, Texas, June 1985.
[15] Gazzola, E., Bagnariol, D., Toneff, V., and Drysdale, R.G.,
     "Influence of Mortar Materials on the Flexural Bond Strength of
     Block and Brick Masonry," Masonry: Research, Application and
     Problems, ASTM STP 871, J.C. Grogan and J.T. Conway, Eds.,
     American Society for Testing and Materials, Philadelphia, 1985,
     pp. 15-26.
[16] Johnson, W.V., "Effect of Mortar Properties on the Flexural Bond
     Strength of Masonry," A Thesis Presented to the Graduate School of
     Clemson University, In Partial Fulfillment of the Requirements for
     the Degree of Master of Science Civil Engineering, May 1986.
[17] Matthys, J.H., "Conventional Masonry Mortar Investigation,"
     National Lime Association, Arlington, Virginia, Aug. 1988.
[18] Hughes, D.M., and Zmembery, S., "A Method of Determining the
     Flexural Bond Strength of Brickwork at Right Angles to the Bed
     Joint," Second Canadian Masonry Conference, Ottawa, Canada, June
     1980, pp. 73-86.

TABLE 1 -- Properties of Mortar and Brick

| Type of Test | Type of Mortar | Range of Mortar Flow(%) | Range of Compressive Strengths 2" Cubes kPa (psi) | Range of Compressive Strengths kPa (psi) | Range of 24 hr cold test g | Range of Sat. coefficent | IRA kg/m²/min (g/30in²/min) |
|---|---|---|---|---|---|---|---|
| | | | | Brick Properties | | | |
| | | Mortar Properties | | | | | |
| E 72[1] Quarterpoint | M,N | 107-113 | 3036-9591 (440-1390) | 18423-121440 (2670-17600) | 1.9-11.3 | .53-.74 | .41-.57 (8.0-11.0) |
| E 72[2] Quarterpoint | M,S,N | 113-130 | 4375-13917 (634-2017) | 68310-76866 (9900-11140) | 4.5-8.8 | .72-.79 | .51-.71 (10.9-13.7) |
| E 72[3] Quarterpoint | M,S,N | 113-120 | 11696-31671 (1695-4590) | 76866 (11140) | 8.8 | .79 | .73 (14.1) |
| E 72[4] Airbag | S | NA | 8715 (1263) | 91908 (13320) | 2.8 | .59 | .41 (7.9) |
| E 72[5] Airbag | S | NA | 8873-11606 (1286-1682) | 39758-44622 (5762-6467) | 8.5-8.7 | .76-.78 | NA |
| E 72[6] Airbag | S | NA | 9260-10716 (1342-1553) | 43511-111042 (6306-16093) | 3.7-9.8 | .74-.90 | .21-1.25 (4.0-24.1) |
| E 72[7] Airbag | S | NA | 10357-10433 (1501-1512) | 81875 (11866) | 6.1 | .72 | .81 (15.6) |
| E 72[8] Airbag | S | NA | 12192 (1767) | 91535 (13266) | 5.0 | .60 | .24 (4.6) |
| E 72[9] Airbag | S | NA | 11523 (1670) | 48162-85560 (6980-12400) | NA | NA | .41-.57 (8.0-10.9) |
| E 72[10] Airbag | M,S,N | NA | NA | NA | NA | NA | NA |

(Table 1 Continued)

| | | | | | | |
|---|---|---|---|---|---|---|
| E 72 [11] Airbag | M,S,N | NA | 7680-29567 (1113-4285) | 58243-91080 (8441-13200) | 1.4-4.8 | .43-.69 | .08 (1.5) |
| E 72 [12] Airbag E 518 3rd Point Airbag | S | 134 | 10357-10440 (1501-1513) | 90287 (13085) | 7.1 | .76 | .38 (21.1) |
| E 518 [13] 3rd Point C 1072 | S | 110-116 | 12123-20810 (1757-3016) | 128892 (18680) | 5.5 | .73 | .52 (10.0) |
| E 518 [14] 3rd Point | M,S,N | 118 | NA | NA | NA | NA | .58-.96 (11.1-18.6) |
| C 1072 [15] | S,N | NA | 5134-12123 (744-1757) | 118459-118963 (17168-17241) | NA | NA | .28-.33 (5.4-6.4) |
| C 1072 [16] | M,S,N | 80-139 | 3623-26393 (525-3825) | 37978-94847 (5504-13746) | 1.8-8.7 | .61-.86 | .10-2.0 (2.0-38.0) |
| C 1072 [17] | S,N | 126-130 | 7832-13365 (1135-1937) | NA | NA | NA | 1.3 (25.0) |

TABLE 2 -- Compiled Data

| Type of Test | E 72 Airbag | | | E 72 Quarterpt. | | | E 518 3rd Point | | | E 518 Airbag | | | ASTM C 1072 | | | E 72 Airbag Hollow Brick | | |
|---|---|---|---|---|---|---|---|---|---|---|---|---|---|---|---|---|---|---|
| Mortar Type | M | S | N | M | S | N | M | S | N | M | S | N | M | S | N | M | S | N |
| Mean kPa | 880 | 903 | 704 | 405 | 618 | 271 | 794 | 731 | 1021 | 0 | 766 | 0 | 818 | 872 | 646 | 353 | 364 | 327 |
| (psi) | (127) | (131) | (102) | (58) | (90) | (39) | (116) | (106) | (148) | 0 | (111) | 0 | (119) | (126) | (94) | (51) | (53) | (47) |
| C.O.V.(%) | 14 | 27 | 22 | 35 | 4 | 20 | 37 | 30 | 0 | 0 | 0 | 0 | 23 | 15 | 29 | 50 | 38 | 34 |
| No. Breaks | 12 | 94 | 11 | 7 | 4 | 10 | 10 | 16 | 5 | 0 | 4 | 0 | 108 | 83 | 191 | 9 | 15 | 9 |
| No. Means | 6 | 27 | 5 | 3 | 2 | 4 | 2 | 4 | 1 | 0 | 1 | 0 | 36 | 19 | 55 | 3 | 5 | 3 |
| No. Reports | 2 | 8 | 2 | 3 | 2 | 3 | 1 | 3 | 1 | 0 | 1 | 0 | 1 | 3 | 3 | 1 | 2 | 1 |

TABLE 3 -- Compiled Data for ASTM E 72 Airbag

| Type of Test | Hollow bonded walls | Without Hollow bonded walls |
|---|---|---|
| Mortar Type | S | S |
| Mean kPa (psi) | 546 (79) | 981 (142) |
| C.O.V.(%) | 43 | 18 |
| No. Breaks | 18 | 79 |
| No. Means | 6 | 22 |
| No. Reports | 2 | 6 |

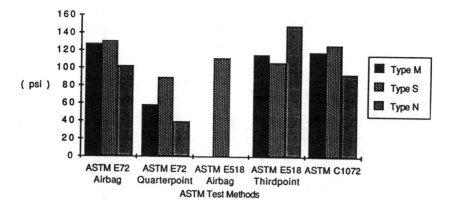

FIGURE 6 -- Compiled Mean Data for Solid Brick

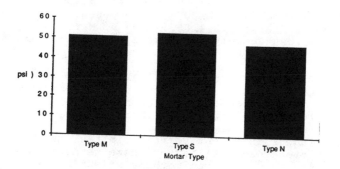

FIGURE 7 -- Compiled Mean Data for Hollow Brick
Using ASTM E 72 Airbag

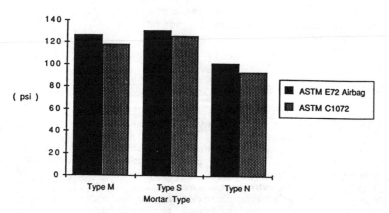

FIGURE 8 -- Comparison Between ASTM E 72
and ASTM C 1072

James L. Noland, Gregory R. Kingsley, and Richard H. Atkinson

NONDESTRUCTIVE EVALUATION OF MASONRY: AN UPDATE

---

REFERENCE: Noland, J.L., Kingsley, G.R., and Atkinson, R.H., "Nondestructive Evaluation of Masonry: An Update," MASONRY: Components to Assemblages, ASTM STP #1063, John H. Matthys, Ed., American Society for Testing and Materials, Philadelphia, 1990.

ABSTRACT: Nondestructive evaluation (NDE) of masonry has been, to date, essentially an adaptation of available methods which have been used for evaluation of concrete. There has been an increasing amount of study and application of these methods in recent years. Limitations on the efficiency and utility of these methods has been observed. Improved methods of NDE of masonry are needed. Recent NDE developments in the medical, aerospace and other fields should be studied for their application to masonry.

KEYWORDS: masonry, nondestructive evaluation, NDE, condition assessment, deterioration, mechanical properties

INTRODUCTION

Although nondestructive evaluation (NDE) methods have been applied to masonry structures for many years, specific recorded instances are few, and, of those, most were exploratory in nature. In the past decade, however, for reasons discussed below, greater attention has been given both in the U.S. and elsewhere to the application of NDE methods to masonry. These efforts revealed that, while useful, present methods are not entirely satisfactory, and

The authors are all engineers with Atkinson-Noland and Associates, 2619 Spruce Street, Boulder, CO 80302.

that improvements are necessary to increase the
effectiveness of NDE of masonry structures.  In this paper,
research and utilization of several prominent methods for
the nondestructive evaluation of masonry are reviewed with
an assessment of their capabilities and limitations.  NDE
methods which have been developed for aerospace materials
evaluation, oil and mineral exploration, physiological
evaluation, nuclear power generating structures evaluation,
manufacturing quality control, and rapid pavement/ground
evaluation are reviewed as candidate NDE methods for
masonry.

Rationale

    There are a great number of existing masonry buildings
throughout the world of various ages and conditions.  The
majority are unreinforced and of natural or cut stone, or
clay brick masonry.  In many cases, these buildings must be
assessed for structural adequacy for purposes of continued
use, change of use, or historical preservation.  A significant component in structural system assessment is the
evaluation of the integrity and properties of the material
used in construction.

    A traditional method of evaluation has been, in addi-
tion to visual inspection, destructive testing of specimens
removed from the structure.  This approach is inherently
limited because the act of specimen removal may be aesthetically and structurally damaging.  Further, the quantity
and quality of data may be poor, because the number of
specimens which may be removed is limited by structural,
cost, and aesthetic reasons, and test results are often inconsistent because of specimen defects.

    Nondestructive evaluation appears to be essential for
the evaluation of masonry condition and quality in a comprehensive and rapid manner, where the term "condition"
refers to the presence of cracks, voids and general deterioration, and "quality" refers to mechanical properties such
as strength, deformation and durability.


REVIEW OF NDE APPLIED TO MASONRY


    Application of nondestructive evaluation techniques to
masonry has traditionally involved a translation of techniques used for the NDE of concrete [1,2,3] and rock to masonry.  Research on NDE of masonry has generally followed a
similar course:  techniques which have proven successful in
the evaluation of concrete and rock have been adapted and
tested on masonry materials [4,5].  Masonry NDE techniques
can be divided into those intended to measure material condition, (i.e. the presence of flaws or deteriorated areas),
and those intended to measure mechanical properties such as

the material compressive strength and deformability.  The
latter can be further divided into two general categories:
(1) "indirect" tests in which masonry mechanical properties
are estimated via correlations to nondestructive measurements, and (2) "direct" physical measurements of mechanical
properties.  The following sections review past and ongoing
research on a number of techniques which have shown some
potential for masonry evaluation, including:

|   |   |   |
|---|---|---|
| 1. | Ultrasonic Pulse | (Indirect) |
| 2. | Mechanical Pulse | (Indirect) |
| 3. | Hardness | (Indirect) |
| 4. | Pull-out | (Indirect) |
| 5. | Neutron Probe | (Indirect) |
| 6. | Flatjack (In-situ Stress) | (Direct) |
| 7. | Flatjack (In-situ Deformability) | (Direct) |
| 8. | In-Place Shear | (Direct) |
| 9. | Vibration | (Indirect) |

## Ultrasonic Pulse

Perhaps the most common application of nondestructive
testing methods to masonry has been with ultrasonic pulse
velocity (UPV) techniques similar to those used for concrete.  The first successful use of UPV techniques for the
detection of flaws and voids in masonry is reported by
Leeper et. al. [6] of Aerojet General laboratories in 1967.
Their study considered clay brick, concrete block and composite construction.  The next published work appeared over
twelve years later, when Wilburn and Associates [7] made
the suggestion that UPV measurements could be correlated to
masonry prism compressive strength.  Noland et. al. conducted extensive correlation studies between UPV measurements and masonry strength parameters (compressive, tensile
and joint shear strength).  They found that UPV correlated
fairly well with masonry compressive strength under controlled conditions (see Figure 1), but that the method was
better suited for flaw detection.  Hobbs [8,9] has also
conducted extensive studies of ultrasonic pulse velocities
in masonry specimens, and reports successful measurement of
variations in compressive strength and the location of
flaws.  Other studies [10,11,12,13] have confirmed the basic conclusions of the earlier studies.

The collected work described above leads to several
common conclusions.  First, UPV measurements can be correlated to masonry prism compressive strength, but best under
carefully controlled conditions.  A generalized relationship between masonry compressive strength and ultrasonic
pulse velocity is not anticipated;  instead, correlations
must be developed for individual structures being assessed.
Second, UPV techniques are better suited to the detection
of flaws than the prediction of compressive strength, but
the signal strength of the ultrasonic pulse deteriorates
rapidly in passage through low density materials hence limiting their usefulness.  Third, many investigators note

that the pulse velocity is only a single descriptive parameter of the pulse transmission, and that attenuation or frequency analysis may reveal more about material condition than the velocity alone.

## Mechanical (Sonic) Pulse

Mechanical pulse tests are similar to ultrasonic tests except that the input pulse is of lower frequency and higher amplitude. The pulse is generated by a hammer blow, and the pulse arrival is recorded with one or more accelerometers (Figure 2). There has been much development of this technique (sometimes referred to as the "pulse-echo" technique or "sonic" testing) for concrete [14]. Because of the high amplitude and long wavelength of the input pulse, the technique is well-suited to masonry, and several researchers have used it for a variety of masonry applications [15,16,17]. Mechanical pulse velocity has been successfully correlated with masonry prism compressive strength, but it is better suited to the detection of flaws and irregularities. A simple device based on the pulse-echo concept has been developed in Japan for the purpose of locating grout flaws in modern reinforced masonry [18].

The detection of flaws and voids in masonry is hindered by several complications revealed by recent research [17]. In general, a series of mechanical pulse velocity tests may be used to develop a contour map of pulse velocities over the area of interest. A single anomalous reading could then be interpreted as the result of a void. However, an area of relatively low pulse velocities may be interpreted as either an area of abnormally low strength masonry or an area of uniformly distributed flaws. The interpretation is further complicated by the fact that a change in the magnitude of the compressive stress in the masonry can have a dramatic effect on the measured pulse velocity, particularly in an area of distributed flaws (Figure 3). Finally, the effect of a void in the center of a wall on the pulse velocity will vary depending on the condition of the surface layers of masonry, as these affect the alternate paths for the pulse as it meets the obstruction to its path.

## Hardness

Penetration tests have shown little application to masonry [4], but the Schmidt Hammer measurement of surface hardness [19] has been shown to have a correlation masonry prism compressive strength (Figure 4) [4]. While the quantitative prediction of masonry compressive strength based on Schmidt Hammer tests is not recommended without companion destructive tests, the method has good potential for the rapid evaluation of material uniformity throughout a structure.

Pull-out

The pull-out test [20] has been adapted from concrete to masonry by decreasing the size of the pull-out specimen and the development of techniques for applying the test to hardened materials. Because the data are extremely variable, there is a limited correlation between the pull-out strength masonry compressive strength. The test is somewhat destructive in that a conical hole is left in the bricks in which the test was performed.

Neutron Probe

The neutron probe [21] represents one of the few truly new developments in NDE of masonry structures in recent years, although it finds applications in many diverse areas of material evaluation. The neutron probe is a specialized use of prompt gamma/neutron activation, a spectroscopic technique that determines elemental composition of a target material. The elements are identified by characteristic gamma rays emitted from the target material while it is being bombarded with neutrons. By moving the probe over a grid of points on a structure such as a wall, it is possible to develop a map of the spatial distribution of the elements in the structure, and from this information important deductions about the condition of the structure can be made.

The neutron probe measures material elemental composition. This function can be useful in several types of building analysis: (1) determination of the composition of building materials, (2) location of contaminants such as water or soluble salts, (3) location of voids, and (4) monitoring of the effectiveness of injection treatments. While the neutron probe does not measure mechanical or structural properties, it can provide a useful compliment to a structural evaluation [22].

Flatjack In-situ Stress Measurement

A flatjack is a thin steel bladder that is pressurized with a fluid to apply a uniform stress over a small area. The well documented use of large flatjacks to determine the in-situ state of stress and deformability of rock was scaled down and adapted for use in masonry structures by Rossi et. al. at ISMES (Istituto Sperimentale Modelli e Strutture) in Bergamo, Italy [5]. Additional studies have been conducted by other researchers [23,24,25,26,27].

Evaluation of in-situ compressive stress is a simple process of stress-relief induced by the removal of a portion of a mortar joint, followed by restoration of the original state of stress by pressurizing a flatjack inserted in the slot. When the mortar is removed from a horizontal joint, the release of the stress across the joint causes the slot to close by a small amount. The magnitude

of this deformation is measured using a removable dial
gauge (such as a DEMEC gauge) between two or more points
located symmetrically on either side of the slot. A flatjack is then inserted in the slot and pressurized until the
original position of the measuring points is restored. At
this point, the pressure in the flatjack -- modified by two
constants to account for the flatjack stiffness and the
area of the slot -- is equivalent to the vertical compressive stress in the masonry. The technique is useful for
verifying analytical models or for determining stress dis-
tributions in masonry walls when conditions of loading or
displacement are unknown or difficult to quantify.

### Flatjack In-situ Deformability Measurement

The deformation properties of masonry may be evaluated
by inserting two parallel flatjacks, one directly above the
other separated by several courses of masonry, and pressurizing them equally, thus imposing a compressive load on the
intervening masonry [5]. The deformations of the masonry
between the flatjacks are then measured for several increments of load, and used to calculate the masonry stress-
strain curve (Figure 5) and deformability modulus. If some
damage to the masonry is acceptable, the test may be carried out to ultimate stress. This technique is useful when
an estimate of material deformability is needed for stress
analysis or deflection calculations.

The two-flatjack test, like the single-flatjack test
was developed by Rossi [5] and others [24,25,26,27]. In
both cases, the tests may be considered nondestructive, because the mortar may be replaced leaving no evidence of
disturbance.

### In-place Shear (Push) Test

The in-place shear test is designed to measure the in-
situ bed joint shear resistance of masonry walls. It requires the removal of a single masonry unit and a head
joint on either side of a test unit. The test unit is then
displaced horizontally relative to the surrounding masonry
using a hydraulic jack, and the horizontal force required
to cause first movement of the test unit is recorded. The
test may be considered nondestructive, because the removed
unit and mortar joints may be replaced and/or repaired to
their former appearance.

The in-place shear test was first developed by Kariotis
et. al. [28] in response to a need to measure the available
shear capacity of existing unreinforced masonry structures
in areas of high seismic risk. The test procedure has been
modified by Noland et. al. [27] to address some of the in-
accuracies in the existing test. In the modified test, the
vertical stress in the wall at the test unit is measured
directly using the single flatjack test, and the normal
stress on the test unit during the test is controlled by

flatjacks above and below the unit. The test is then conducted on the same unit for several levels of normal
stress, so the friction angle $\phi$ is measured directly.
LVDTs are used to monitor the movement of the unit continuously during the test (Figure 6).

## Vibration

Vibration of a 20-story masonry building due to ambient
disturbances was used by Medearis [29] to perform a system-
level assessment of the structure in conjunction with the
finite element technique. Noland et. al. [4] attempted to
correlate natural period and logarithmic decrement to masonry wall condition and quality with marginal results.
The former application can reveal general response information, but because of the very low amplitude of vibration,
may not be an indication of large amplitude behavior. The
correlations between vibration parameters and component
quality were not good and, in general, would be affected by
component boundary conditions.

## Combination of Existing NDE Techniques

The nondestructive evaluation of masonry for mechanical
properties and condition is quite difficult, because the
heterogeneous and highly variable nature of the material
hinders the simple analysis and interpretation of test results. At the current level of development, the best application of NDE to masonry would make use of a number of
complimentary techniques. For example, rapid methods such
as the Schmidt Hammer might be used to assess the condition
of the entire structure, and pulse velocity methods would
then be used to map the variation in material condition in
critical areas of the structure. Direct mechanical measurements of material deformability and joint shear
strength might then be made in locations defined during the
ultrasonic or sonic pulse velocity mapping. In-situ stress
measurements might be made in areas where more information
was needed to interpret pulse velocity measurements, or in
locations defined by building analysis needs. In general,
the procedure and methods used would vary depending on individual building requirements. In all cases, considerable
experience and judgement would be required for the accurate
interpretation of results.


## FUTURE DIRECTIONS IN MASONRY NDE

In recent years, there has been a broadly based advance
in technology and application of NDE in many diverse
fields. Among the most widely publicized are the NDE developments and applications in medicine which include computerized tomography (CAT scans), nuclear magnetic resonance (NMR) and doppler ultrasonics. Other fields in which

advancements in the use of NDE have been made are seismic
exploration, aerospace structures and manufacturing.   Il-
lustrative examples are listed below.

Seismic Exploration.

Advanced detection and signal processing technologies
are being used in the search for oil and mineral deposits
at greater depths and in more complex formations.  These
technologies are based on reflection seismology using multiple receivers and energy sources and utilize two and
three dimensional digital analysis methods to describe the
under ground structure [30,31].

Aerospace Structures.

The use of embedded optical fibers as sensors is receiving considerable study as part of an effort to create
"smart" materials, i.e., materials which can act as sensors
to report current stress or strain, for example.  Optical
fibers when embedded in composite materials can serve as
dynamic strain sensors or as real-time fracture sensors.
Full-field holography is now being used to determine deformations fo structures as large as aircraft wings [32].

Medical.

The introduction of digital technology and the computed
tomography algorithm in the early 1970's allowed data from
transmission X-ray absorption to be processed to give a
two-dimensional cross-section of the body [33].  Other applications of digital imaging technology in medicine are
ultrasound, gamma and positron detection sensors for ra-
dioactive isotopes, and magnetic resonance imaging.

Manufacturing Techniques.

Manufacturing techniques are being developed for real
time inspection of critical components using X-ray and
cobalt source topography methods [34].

Nuclear Power.

NDE methods are being developed to detect flaws within
thick sections.  Included are ultrasonic surveys employing
advanced computerized data analysis and presentation [35]
and acoustic emission monitoring under service and test
loadings.

Civil Engineering.

Several variations of the ground penetrating radar
method exist which utilize both time and frequency domain
signal analysis and various frequency contents of the
transmitted signal [36].  These techniques have been applied as a high-speed NDE method to detect voids, deterio-

rated joints and delaminations in pavements [37]  The techniques have also been used to detect defects in an embankment dam and in mining to measure seam thickness and to detect subsurface anomalies [38].


CONCLUSIONS


"Conventional" NDE methods previously described (including NDE mechanical methods such as the flatjack) may be used to develop an impression and provide some physical data on the condition and properties of masonry.  Generally, it is a process of building a circumstantial case and requires evaluation by an engineer.  Destructive testing of specimens removed from a building is a seriously limited means of building evaluation.  It is concluded, therefore, that "advanced" NDE technologies such as those described herein and others should be considered as possibilities for masonry evaluation under the general criteria of:

- ability to detect flaws and material degradation
- ability to directly or indirectly measure material properties needed for structural assessment.
- ability to be used by technician-level personnel.
- ability to perform NDE of large and small areas of a structure.


ACKNOWLEDGEMENT

Preparation of the paper and the work referenced herein by the authors [4,17,22,25,28] was supported by the National Science Foundation, Dr. J.B. Scalzi - Program Manager.  Opinions and conclusions are those of the authors and not necessarily of the National Science Foundation.

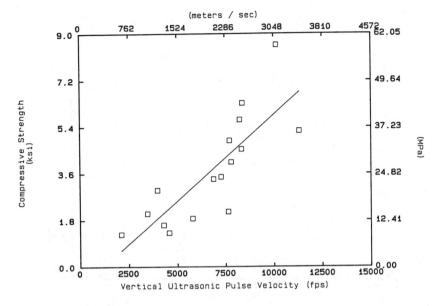

Figure 1.  Correlation of Ultrasonic Pulse Velocity with Masonry Prism Compressive Strength.

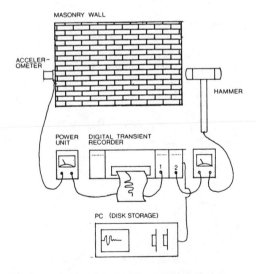

Figure 2.  Mechanical Pulse Equipment Set-up Schematic

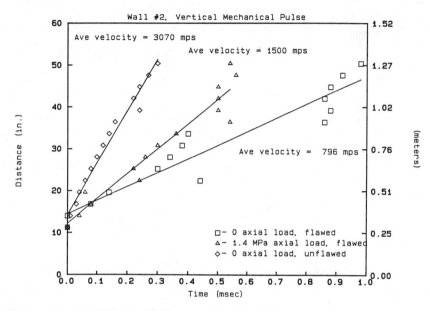

Figure 3.   The Effect of Flaws and Compressive Stress on Mechanical
            Pulse Velocity.

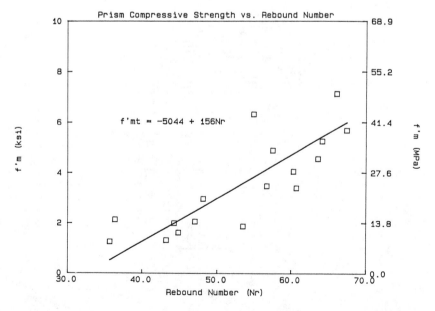

Figure 4.   Correlation of Rebound Number (Schmidt Hammer Reading) with
            Masonry Prism Compressive Strength.

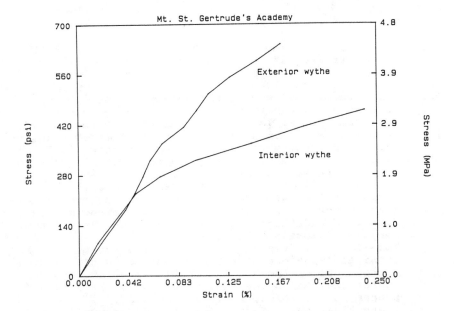

Figure 5.   Example Stress-strain Curves Obtained from Flatjack Tests
on an Existing Masonry Structure.

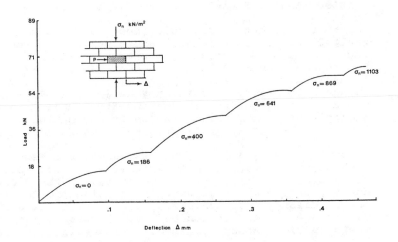

Figure 6.   Typical Results from a Modified In-place Shear Test.

**REFERENCES**

[1]    DeVekey, R.C., "Non-destructive Test Methods for Ma-
       sonry Structures", Proceedings of the 8th Interna-
       tional Brick/Block Masonry Conference, Dublin, Ire-
       land, September 1988.

[2]    Malhotra, V.M., Testing Hardened Concrete:
       Nondestructive Methods, American Concrete Institute
       Monograph no. 9, Iowa State  University Press, 1976.

[3]    Zoldners, N.G., Soles, J.A., "An Annotated Bibliogra-
       phy on  Nondestructive Testing of Concrete, 1975-
       1983," ACI SP 82-39,  In-Situ Nondestructive Testing
       of Concrete, p. 745, 1984.

[4]    Noland, J.L., Atkinson, R.H., Baur, J.C., "An
       Investigation into  Methods of Nondestructive Evalua-
       tion of Masonry Structures,"  Report to the National
       Science Foundation, Atkinson-Noland and  Associates,
       Boulder, Colorado, USA, 1982.

[5]    Rossi, P. P., "Flat-jack Test for the Analysis of Me-
       chanical Behaviour of Brick Masonry Structures,"
       Proceedings of the 7th International Brick Masonry
       Conference, Melbourne, Australia, Vol. 1, 1985.

[6]    Leeper, C.K., Eales, E.P., Brown, S.J.,
       "Investigation of Sonic  Testing of Masonry Walls,"
       Aerojet General Corporation,  Sacramento Plant, 1967.

[7]    Victor Wilburn Associates, "Nondestructive Test
       Procedures for Brick, Block, and  Mortar," HUD Con-
       tract #H2540, July  1979.

[8]    Hobbs, B., Wright, S.J., "An Assessment of Ultrasonic
       Testing for Structural Masonry", Proceedings of the
       British Masonry Society, No. 2, H.W.H. West, Ed.,
       April 1988.

[9]    Hobbs, B., Wright, S.J., "Ultrasonic Testing for
       Fault Detection in Brickwork and Blockwork", Proceed-
       ings of the International Conference on Structural
       Faults and Repair, Volume 1, London, July, 1987.

[10]   Binda Maier, L., Rossi, P.P., Sacchi Landriani, G.,
       "Diagnostic Analysis of Masonry Buildings", Interna-
       tional Association for Bridge and Structural
       Engineering (IABSE) Symposium on Strengthening of
       Building Structures -- Diagnosis and Therapy, Venice,
       Italy, 1983.

[11]   Senbu, O., Baba, A., Abe, M., Tomosawa, F., Mat-
       sushima, Y.,  "Effect of Admixtures on Compactibility
       and Properties of  Grout," Proceedings of the Second
       Meeting of the Joint Technical  Coordinating Commit-
       tee on Masonry Research, Keystone Colorado,  1986.

[12]   Calvi, G.M., "Correlation between Ultrasonic and Load
       Tests on Old Masonry Specimens", Proceedings of the
       8th International Brick/Block Masonry Conference,
       Dublin, Ireland, September 1988.

[13]    Berra, M., Binda, L., Baronia, B., Fatticcioni, A.,
        "Ultrasonic Pulse Transmission: A Proposal to Evalu-
        ate the Efficiency of Masonry Strengthening by Grout-
        ing", Proceedings of the Second Joint USA-Italy Work-
        shop on Evaluation and Retrofit of Masonry Struc-
        tures, Binda and Noland, Eds., August 1987.
[14]    Carino, N.J., Sansalone, M., Hsu, N.N., "A Point
        Source-Point  Receiver, Pulse-Echo Technique for Flaw
        Detection in Concrete,"  ACI Journal, March 1986.
[15]    Komeyli, F., Forde, M.C., and Whittington, H.W.,
        "Sonic  Investigation of Shear Failed Reinforced
        Brick Masonry," Masonry  International, November
        1984.
[16]    Forde, M.C., and Batchelor, A.J., "Low Frequency NDT
        Testing of  Historic Structures," 3rd European
        Conference on NDT Testing,  Florence, Italy, 1984.
[17]    Kingsley, G.R., Noland, J.L., Atkinson, R.H.,
        "Nondestructive Evaluation of Masonry Structures Us-
        ing Sonic and Ultrasonic Pulse Velocity Techniques,"
        Proceedings of the Fourth North American Masonry
        Conference, Univ. of Calif. Los Angeles, August,
        1987.
[18]    BEAT WALL: Defect Detecting Device for Grout Concrete
        / Mortar of Masonry Structures, Operation Manual,
        Taisei Corporation, January 1988.
[19]    American Society for Testing Materials, "Tentative
        Test Method for Rebound Number of Hardened Concrete,"
        ASTM C805-75.
[20]    American Society for Testing Materials, "Standard
        Test Method for Pullout Strength of Hardened Con-
        crete," ASTM C900-82.
[21]    Livingston, R.A., Chang, L., Evans, L.S., Trombka,
        J.I., "The Application of the Neutron Probe to the
        Nondestructive Evaluation of Building Condition",
        Proceedings of the International Workshop on
        Nondestructive Evaluation for Performance of Civil
        Structures, USC, February 1988.
[22]    Kingsley, G.R., Livingston, R.A., Taylor, T.H.,
        Noland, J.L., "Nondestructive Evaluation of a Nine-
        teenth Century Brick Smokehouse in Colonial Williams-
        burg, Virginia", Proceedings of the First Interna-
        tional Conference on Structural Studies, Repairs and
        Maintenance of Historical Buildings, Florence, Italy,
        April 1989.
[23]    Sacchi-Landriani, G., Taliercio, A., "Numerical
        analysis of the  flat jack test on masonry walls,"
        Journal de Mecanique Theorique  et Appliquee, Vol. 5,
        No. 3, 1986, p. 313-339.
[24]    Wang, Q., Wang, X., "Evaluation of Compressive
        Strength of Brick Masonry in Situ", Proceedings of
        the 8th International Brick/Block Masonry Conference,
        Dublin, Ireland, September 1988.

[25]  Kingsley, G.R., Noland, J.L., "A Note on Obtaining
      In-situ Load-deformation Properties of Unreinforced
      Brick Masonry in the United States Using Flatjacks",
      Final Report of the Joint USA-Italy Workshop on Eval-
      uation and Retrofit of Masonry Structures, 1988.
[26]  Abdunur, C., "Stress and Deformability in Concrete
      and Masonry", International Association for Bridge
      and Structural Engineering (IABSE) Symposium on
      Strengthening of Building Structures -- Diagnosis and
      Therapy, Venice, Italy, 1983.
[27]  Noland, J.L., Kingsley, G.R., and Atkinson, R.H.,
      "Utilization of Nondestructive Techniques in the
      Evaluation of Masonry," Proceedings of the 8th Inter-
      national Brick/Block Masonry Conference, Dublin, Ire-
      land, October, 1988.
[28]  ABK, A Joint Venture, "Methodology for Mitigation of
      Seismic Hazards in Existing Unreinforced Masonry
      Buildings: Diaphragm Testing," Topical Report 03,
      Dec. 1981.
[29]  Medearis, K., "An Investigation of the Dynamic Re-
      sponse of the Park Lane Towers to Earthquake Load-
      ings", Proceedings of the First North American Ma-
      sonry Conference, Boulder, CO, 1978.
[30]  Robinson, E.A., and Durrani, T.S., Geophysical Signal
      Processing, Prentice-Hall, 1986.
[31]  Nelson, H.R., New Techniques in Exploration Geo-
      physics, Gulf Publishing Company, Houston, 1983.
[32]  Dube, C.M., et. al., "Laboratory Feasibility Study of
      a Composite Embedded Fiber Optic Sensor for Measure-
      ment of Structural Vibrations," Dynamics Technology,
      Inc. Report DT 8723-01, Feb. 1988.
[33]  Digital Imaging in Health Care, United Nations
      Publication No. ECE/ENG.AUT/25, 1987.
[34]  Engel, H.P., "NASA's Computed Tomography System,"
      ASTM Standardization News, March 1989, pp. 34-37.
[35]  Hollar, P., Visit by Dr. R.H. Atkinson to the NDE In-
      stitute, University of Saarbruchen, West Germany,
      Oct. 1988.
[36]  Pittman, W.E., Church, R.H., Webb, W.E., and Mclen-
      don, "Ground Penetrating Radar - A Review of its Ap-
      plication in the Mining Industry,"  U.S. Bureau of
      Mines Information Circular - IC-8964, 1984.
[37]  Holt, F.B., and Earles, J.W., "Nondestructive Evalua-
      tion of Pavements," ACI Concrete International, June,
      1987, pp. 41-45.
[38]  Church, R.H., and Webb, W.E., "Evaluation of a Ground
      Penetrating Radar System for Detecting Subsurface
      Anomalies," U.S. Bureau of Mines Report of Investiga-
      tion RI-9004, 1986.

Russell H. Brown and J. Gregg Borchelt

COMPRESSION TESTS OF HOLLOW BRICK UNITS AND PRISMS

---

REFERENCE: Brown, R.H. and Borchelt, J.G., "Compression
Tests of Hollow Brick Units and Prisms", Masonry:
Components to Assemblages, ASTM STP 1063, J.H. Matthys,
Editor, American Society for Testing and Materials,
Philadelphia, 1990.

ABSTRACT: It is customary to relate the compressive
strength of the masonry to that of its components: mortar
and units. The correlation between solid unit compressive
strength, mortar type and assemblage compressive strength
is well documented, and is generally independent of unit
coring. The relationships of prism components and prism
dimensions to assembly compressive strength and modulus
of elasticity for hollow brick are presented in this paper.

Hollow brick meeting the requirements of ASTM C 652 in four
widths were selected. Portland cement-lime mortars, Type
N, S, M by proportions, were used. Both face-shell and
full mortar bedding were used to construct prisms with
height to thickness ratios from 2 to 5. Mortar, brick and
prisms were tested in compression. Deformations measured
during prism testing permitted the determination of modulus
of elasticity in compression.

Hollow unit masonry prisms fail by tensile splitting of the
crosswebs, regardless of their aspect ratio. Prism compressive
strength and modulus of elasticity are influenced least by fully
bedded, stack bond prims. Compressive strengths of prisms made
from hollow brick have similar relations to h/t and mortar
strength as prisms of solid brick, but less dependence on unit
strength.

KEYWORDS: hollow brick, mortar, prisms, compressive
strength, modulus of elasticity

Dr. Brown is Professor and Head of the Department of Civil
Engineering, Lowry Hall, Clemson University, Clemson, SC 29631; Mr.
Borchelt is Director of Engineering and Research for the Brick
Institute of America, 11490 Commerce Park Drive, Reston, VA 22091.

INTRODUCTION

Hollow clay masonry units have been used for many years under a variety of material specifications and names.  In 1970, the American Society for Testing and Materials published a standard for hollow brick, ASTM C 652-70 [1].  This document, revised periodically, provides the manufacturer and specifier with guidelines for producing and specifying high-quality hollow clay brick units.  The current version is ASTM C 652-87a [2].

Model building codes use the specified compressive strength of masonry to determine the allowable stresses in engineered masonry design.  This important parameter must be verified by the contractor prior to the construction process.  Verification is often accomplished by combining the compressive strength of the brick unit with the type of mortar used and correlating that combination to the compressive strength of the masonry.

A test program to investigate the compressive strength of hollow brick masonry was conducted in two phases at the laboratories of the Brick Institute of America.  The first phase of the program was completed in 1972, the second phase in 1976.  The results of those tests have been combined and the relationship of prism components and prism height to thickness (h/t) ratios to assembly compressive strength and modulus of elasticity are discussed in this paper.


SCOPE OF TEST PROGRAM

Both phases of the program included compressive tests of the hollow brick, mortar and prisms.  Prisms were made with both full and face shell mortar bedding with four aspect ratios:  h/t of approximately 2, 3, 4 and 5.  Five specimens were constructed for each combination of variables.

All hollow brick used in phase 1 were from a single manufacturer.  Nominal unit widths of 100 mm (4 in.), 150 mm (6 in.), 200 mm (8 in.) and 250 mm (10 in.) were included.  Three types of portland cement-lime mortar, Types N, S and M by volume proportions, were used.  A total of 480 prisms were tested, with the units laid in stack bond.

Hollow brick used in phase 2 were 200 mm (8 in.) nominal thickness from three manufacturers.  Only Type S portland cement-lime mortar by volume proportions was used.  Prisms were built with the units laid in running bond for all h/t ratios and in stack bond with an h/t of 5.  A total of 150 prisms were tested.

Axial strain was measured in selected prisms during the compression testing.  In phase 1, each of the 5 specimens having an h/t ratio of 5 for each thickness was selected.  In phase 2, one prism for each combination of brick and mortar bedding with h/t ratios of 3, 4 and 5 was monitored for axial strain.  Modulus of elasticity was determined from these data.

MATERIALS

Brick

All hollow brick units for phase 1 were obtained from one manufacturer and were formed from the same raw material and fired under as identical firing schedules as was possible.  Hollow brick from three manufacturers were used in phase 2, representing a broad range of strength.  The units were in compliance with ASTM C 652-70 and C 652-87a and were closer to the requirements of Class H40V than H60V.  Table 1 includes unit dimensions, areas and compressive strengths.  Coring percentages ranged from 29.5% to 41.8% and are solid shell hollow brick.  All of the units had one large cell at each end and a single, smaller center cell.  All units have nominal face dimensions of 100 mm x 300 mm (4 in. x 12 in).

Mortar

All mortars were mixtures of portland cement, lime and sand batched by equivalent weight to the proportions specification of ASTM C 270. Type I portland cement and Type S hydrated lime were used.  Sand was a natural sand procured near the laboratory.

The mortar was mixed in a 0.071 m$^3$ (2.5 cu. ft.) paddle type mixer with enough water to produce a workability satisfactory to the mason. The size of each batch was such that the mason could use it within 2 hours.  Retempering the mortar to replace water lost by evaporation was permitted.

TEST SPECIMENS

Mortar Cubes

Representative batches of mortar were sampled daily and three 50 mm (2 in.) compressive strength cubes molded from each sample.  The samples were taken from the mortar board immediately after discharging from the mixer and before any retempering had occurred.

Each mold was kept covered with polyethylene film for 24 hours to prevent rapid evaporation of water.  After that time the cubes were removed from their molds and permitted to age in laboratory air for 27 days, or until the day the corresponding prisms were tested.

Masonry Prisms

Masonry prisms were fabricated under inspection by a journeyman mason in a laboratory.  Masonry prisms were constructed in accordance with ASTM E 447-72, Method A, except that some prisms were laid in running bond.  The height-to-thickness ratio (h/t) varied from a low of 2.04 to a high of 5.61.  Prisms built with full mortar bedding had mortar spread on all crosswebs and both face shells.  The net area of

the unit was therefore considered to be the minimum net bedded area.
Prisms built with face shell bedding had mortar spread only on the face
shells. The minimum net bedded area used in calculating compressive
stresses was the combined area of only the two face shells. Geometric
properties of prisms are given in Table 2.

All prisms were cured in laboratory air ranging in temperature from
4.4°C (40°F) to 25.6°C (78°F) and in relative humidity from 4% to 88%.
All prisms were cured at least 28 days before testing. None were
tested at ages greater than 31 days.

## Equipment and Procedure

All masonry compressive prisms in phase 1 were tested in a 4.45 MN
(one-million lb) capacity hydraulic testing machine. Phase 2 prisms
were tested in compression on a 2.67 MN (600,000 lb) capacity Forney
testing machine. All prisms were capped using high-strength gypsum
capping compound at least 24 hours prior to testing. Capping compound
was applied to the full surface area of all prisms. The load was
applied to the prism through an upper spherical head to accommodate
nonparallel upper and lower bearing surfaces. Loads were applied
continuously except where dial gage readings were taken.

Compressometers were attached to prisms in order to monitor axial
deformation. Micrometer dial gages with 0.025 mm (0.001 in.)
increments were used. The gage length of the compressometer was 305 mm
(12 in.) for h/t values of 3 and 4. For h/t values of 5, the gage
length was 432 mm (17 in.). The instrumentation was removed at
approximately 75% of the anticipated maximum load to prevent damage to
the dial gages. The hollow brick and the mortar cubes were tested in a
0.534 MN (120,000 lb) capacity universal testing machine.

## TEST RESULTS

### General

Average results of prism compressive tests are tabulated in Tables
3 to 6 for phase 1 and include compressive strength, coefficient of
variation, mortar cube compressive strength and modulus of elasticity.
The results for phase 2 are found in Tables 7 to 9. All stresses are
based on mimimum net bedded area, whether the specimen was constructed
using full bedding or face shell bedding. The modulus of elasticity in
compression was determined by the secant method, measuring from 0.05 to
0.33 of the prism compressive strength.

The influence of experimental parameters are examined by comparison
in this discussion. Comparison of test results are computed by
dividing the tested value of one prism by the corresponding value of
another prism and looking at the ratios. Thus, all variables but one
are constant when calculating the ratio.

TABLE 1 -- Hollow Brick Properties^a

| Brick Type | Actual Dimensions | | | | Area | | | Unit Compressive Strength^c | | | | |
|---|---|---|---|---|---|---|---|---|---|---|---|---|
| | width mm (in.) | height mm (in.) | length mm (in.) | FST^b mm (in.) | Gross mm² (sq in) | Net mm² (sq in) | % Solid | Gross Area f_u, kPa (psi) | s, kPa (psi) | Net Area f_u, kPa (psi) | s, kPa (psi) | v % |
| 1-4 | 88.9 (3.5) | 88.9 (3.5) | 295.3 (11.63) | 23.5 (0.92) | 26260 (40.7) | 17820 (27.2) | 66.9 | 74190 (10760) | 2880 (418) | 110940 (16090) | 4310 (625) | 3.8 |
| 1-6 | 144.8 (5.7) | 91.4 (3.6) | 298.5 (11.75) | 35.6 (1.40) | 43230 (67.0) | 24800 (44.5) | 66.4 | 44670 (6478) | 2861 (415) | 67190 (9750) | 4300 (623) | 6.4 |
| 1-8 | 193.8 (7.63) | 88.9 (3.5) | 295.9 (11.65) | 35.6 (1.40) | 57350 (88.9) | 35100 (54.4) | 61.2 | 38030 (5515) | 2040 (296) | 62140 (9012) | 3340 (484) | 5.3 |
| 1-10 | 242.6 (9.55) | 88.9 (3.5) | 295.9 (11.65) | 41.9 (1.65) | 71810 (111.3) | 44770 (69.4) | 62.4 | 40030 (5806) | 2450 (356) | 64200 (9311) | 3940 (571) | 6.1 |
| 2-1 | 190.5 (7.5) | 88.9 (3.5) | 292.1 (11.5) | 45.7 (1.80) | 55650 (86.3) | 39230 (60.8) | 70.5 | 91010 (13200) | 7830 (1135) | 129110 (18730) | 3550 (515) | 8.6 |
| 2-2 | 190.5 (7.5) | 88.9 (3.5) | 292.1 (11.5) | 30.5 (1.20) | 55270 (85.7) | 32140 (49.8) | 58.6 | 58200 (8440) | 3080 (447) | 99320 (14405) | 5260 (763) | 5.3 |
| 2-3 | 193.0 (7.6) | 91.4 (3.6) | 320.0 (11.6) | 35.6 (1.40) | 56880 (88.2) | 34270 (53.1) | 60.3 | 76580 (11110) | 8040 (1166) | 131820 (19120) | 13840 (2007) | 10.5 |

^a Average of 5 specimens
^b FST = average thickness of one face shell
^c $f_u$ = unit compressive strength, s = standard deviation, v = coefficient of variation, based on applied loads.

TABLE 2 -- Prism Geometric Properties[a]

| Brick Type | thickness mm (in.) | Dimensions length mm(in.) | height mm(in.) | h/t | height mm(in.) | h/t | height mm(in.) | h/t | height mm(in.) | h/t | Net Area Full Bedding (FB) mm²(in.²) | Face Shell (FS) mm²(in.²) |
|---|---|---|---|---|---|---|---|---|---|---|---|---|
| 1-4 | 88.9 (3.5) | 295.3 (11.63) | 190.5 (7.5) | 2.14 | 292.1 (11.5) | 3.28 | 393.7 (15.5) | 4.43 | 495.3 (19.5) | 5.57 | 17820 (27.2) | 13810 (21.4) |
| 1-6 | 144.8 (5.7) | 298.5 (11.75) | 295.7 (11.6) | 2.04 | 499.1 (19.6) | 3.44 | 592.4 (23.6) | 4.14 | 812.8 (32) | 5.61 | 24800 (44.5) | 21230 (32.9) |
| 1-8 | 193.8 (7.63) | 295.9 (11.65) | 396.2 (15.6) | 2.04 | 602.0 (23.7) | 3.11 | 805.2 (31.7) | 4.15 | 1010.9 (39.8) | 5.18 | 35100 (54.4) | 21030 (32.6) |
| 1-10 | 242.2 (9.55) | 295.9 (11.65) | 499.1 (19.6) | 2.05 | 805.2 (31.7) | 3.32 | 1008.4 (39.7) | 4.16 | 1315.7 (51.8) | 5.4 | 44770 (69.4) | 24770 (38.4) |
| 2-1 | 190.5 (7.5) | 292.1 (11.5) | 393.7 (15.5) | 2.07 | 604.5 (23.8) | 3.17 | 812.8 (32.0) | 4.27 | 1010.9 (39.8) | 5.31 | 39230 (60.8) | 26710 (41.4) |
| 2-2 | 190.5 (7.5) | 292.1 (11.5) | 401.3 (15.8) | 2.10 | 604.5 (23.8) | 3.17 | 810.3 (31.9) | 4.25 | 1005.8 (39.6) | 5.28 | 32140 (49.8) | 17810 (27.6) |
| 2-3 | 193.0 (7.6) | 294.6 (11.6) | 406.4 (16.0) | 2.11 | 609.6 (24.0) | 3.16 | 812.8 (32.0) | 4.21 | 1016 (40.0) | 5.26 | 34270 (53.1) | 20950 (32.5) |

[a] Average of 5

TABLE 3 -- Prism Compressive Test Results, Brick 1-4, Stack Bond

| Prism | Mortar | | | Prism Compressive Strength | | | $E^c$ x $10^6$ |
|---|---|---|---|---|---|---|---|
| h/t | Type | Comp. Str.[a] kPa (psi) | Area[b] | Avg[c], kPa (psi) | s, kPa (psi) | v % | kPa (psi) |
| 2.14 | N | 8736 (1267) | FB | 31260 (4533) | 3110 (451) | 9.9 | ... |
| 2.14 | N | 8736 (1267) | FS | 36710 (5324) | 4260 (618) | 11.6 | ... |
| 3.28 | N | 8736 (1267) | FB | 29790 (4320) | 3160 (458) | 10.5 | ... |
| 3.28 | N | 8736 (1267) | FS | 32350 (4692) | 3330 (484) | 10.3 | ... |
| 4.43 | N | 8736 (1267) | FB | 25680 (3724) | 2050 (297) | 7.9 | ... |
| 4.43 | N | 8736 (1267) | FS | 26340 (3821) | 2120 (308) | 8.0 | ... |
| 5.57 | N | 8736 (1267) | FB | 24550 (3560) | 1500 (218) | 6.1 | 13.9 (2.01) |
| 5.57 | N | 8736 (1267) | FS | 26450 (3837) | 2150 (269) | 6.9 | 17.6 (2.55) |
| 2.14 | S | 13980 (2028) | FB | 35350 (5127) | 2990 (434) | 8.4 | ... |
| 2.14 | S | 13980 (2028) | FS | 40170 (5830) | 3260 (472) | 8.0 | ... |
| 3.28 | S | 13980 (2028) | FB | 39780 (5770) | 2280 (331) | 5.7 | ... |
| 3.28 | S | 13980 (2028) | FS | 43260 (6274) | 3130 (454) | 7.2 | ... |
| 4.43 | S | 13980 (2028) | FB | 36310 (5266) | 3810 (553) | 10.5 | ... |
| 4.43 | S | 13980 (2028) | FS | 37710 (5469) | 2219 (322) | 5.8 | ... |
| 5.57 | S | 13980 (2028) | FB | 33450 (4852) | 1990 (288) | 5.9 | 16.0 (2.32) |
| 5.57 | S | 13980 (2028) | FS | 33970 (4927) | 328 ( 48) | 0.9 | 16.3 (2.36) |
| 2.14 | M | 26440 (3835) | FB | 54990 (7976) | 2360 (343) | 4.3 | ... |
| 2.14 | M | 26440 (3835) | FS | 56250 (8158) | 8765 (1269) | 15.5 | ... |
| 3.28 | M | 17400 (2523) | FB | 40120 (5819) | 5400 (783) | 13.4 | ... |
| 3.28 | M | 17400 (2523) | FS | 43410 (6296) | 3662 (531) | 8.4 | ... |
| 4.43 | M | 26440 (3835) | FB | 45620 (6617) | 1990 (289) | 4.3 | ... |
| 4.43 | M | 26440 (3835) | FS | 43010 (6238) | 3585 (520) | 8.3 | ... |
| 5.57 | M | 17400 (2523) | FB | 31580 (4580) | 4520 (656) | 14.3 | 17.3 (2.51) |
| 5.57 | M | 17400 (2523) | FS | 33140 (4831) | 1148 (166) | 3.4 | 20.5 (2.98) |

[a] Average of 3
[b] FB = full mortar bedding, FS = face shell mortar bedding
[c] Average of 5

TABLE 4 -- Prism Compressive Test Results, Brick 1-6, Stack Bond

| Prism h/t | Type | Mortar Comp. Str.[a] kPa (psi) | Area[b] | Avg[c], kPa (psi) | Prism Compressive Strength s, kPa (psi) | v % | $E^c$ x 10[6] kPa (psi) |
|---|---|---|---|---|---|---|---|
| 2.04 | N | 9430 (1367) | FB | 31380 (4552) | 1409 (204) | 4.4 | ... |
| 2.04 | N | 9430 (1367) | FS | 29030 (4210) | 2920 (423) | 10.0 | ... |
| 3.44 | N | 9430 (1367) | FB | 27210 (3946) | 1493 (217) | 5.5 | ... |
| 3.44 | N | 9430 (1367) | FS | 25380 (3681) | 1150 (167) | 4.5 | ... |
| 4.14 | N | 8860 (1285) | FB | 31260 (4533) | 1883 (273) | 6.0 | ... |
| 4.14 | N | 8860 (1285) | FS | 28060 (4070) | 2270 (329) | 8.0 | ... |
| 5.61 | N | 8860 (1285) | FB | 32340 (4691) | 1420 (206) | 4.4 | 15.9 (2.30) |
| 5.61 | N | 8860 (1285) | FS | 29580 (4290) | 1120 (162) | 3.7 | 18.3 (2.65) |
| 2.04 | S | 12820 (1860) | FB | 36190 (5248) | 1294 (187) | 3.5 | ... |
| 2.04 | S | 12820 (1860) | FS | 40680 (5900) | 1990 (162) | 2.7 | ... |
| 3.44 | S | 12820 (1860) | FB | 36690 (5322) | 2997 (435) | 8.1 | ... |
| 3.44 | S | 12820 (1860) | FS | 34890 (5060) | 2330 (338) | 6.6 | ... |
| 4.14 | S | 12410 (1800) | FB | 34810 (5048) | 1136 (164) | 3.2 | ... |
| 4.14 | S | 12410 (1800) | FS | 33920 (4920) | 1120 (163) | 3.3 | ... |
| 5.61 | S | 12410 (1800) | FB | 34220 (4968) | 891 (129) | 2.6 | 14.1 (2.08) |
| 5.61 | S | 12410 (1800) | FS | 33580 (4870) | 1061 (154) | 3.1 | 16.4 (2.38) |
| 2.04 | M | 19310 (2800) | FB | 34980 (5073) | 6331 (918) | 18.0 | ... |
| 2.04 | M | 19310 (2800) | FS | 37080 (5378) | 3310 (480) | 8.9 | ... |
| 3.44 | M | 25370 (3680) | FB | 43487 (6306) | 2187 (316) | 5.0 | ... |
| 3.44 | M | 25370 (3680) | FS | 38320 (5558) | 4100 (595) | 10.7 | ... |
| 4.14 | M | 19310 (2800) | FB | 42035 (6096) | 4942 (716) | 11.7 | ... |
| 4.14 | M | 19310 (2800) | FS | 41050 (5954) | 2082 (302) | 5.0 | ... |
| 5.61 | M | 19310 (2800) | FB | 43360 (6289) | 696 (101) | 1.6 | 13.7 (1.98) |
| 5.61 | M | 19310 (2800) | FS | 38820 (5630) | 2760 (400) | 7.0 | 14.5 (2.11) |

[a] Average of 3
[b] FB = full mortar bedding, FS = face shell mortar bedding
[c] Average of 5

TABLE 5 -- Prism Compressive Test Results, Brick 1-8, Stack Bond

| Prism h/t | Mortar Type | Mortar Comp. Str.[a] kPa (psi) | Area[b] | Prism Compressive Strength Avg[c], kPa, (psi) | s, kPa (psi) | v % | E[c] x 10[6] kPa (psi) |
|---|---|---|---|---|---|---|---|
| 2.04 | N | 9450 (1370) | FB | 29170 (4230) | 2760 (400) | 9.4 | ... |
| 2.04 | N | 9450 (1370) | FS | 32060 (4650) | 2650 (385) | 8.2 | ... |
| 3.11 | N | 9450 (1370) | FB | 28680 (4160) | 1360 (197) | 4.7 | ... |
| 3.11 | N | 9450 (1370) | FS | 32060 (4650) | 3280 (475) | 10.2 | ... |
| 4.15 | N | 9450 (1370) | FB | 29860 (4330) | 2250 (326) | 7.5 | ... |
| 4.15 | N | 9860 (1430) | FS | 33720 (4890) | 3550 (515) | 10.5 | ... |
| 5.18 | N | 9860 (1430) | FB | 25720 (3730) | 1970 (285) | 7.6 | 12.6 (1.80) |
| 5.18 | N | 9860 (1430) | FS | 24820 (3600) | 2900 (420) | 11.6 | 13.3 (1.93) |
| 2.04 | S | 15820 (2295) | FB | 35100 (5090) | 1590 (230) | 4.5 | ... |
| 2.04 | S | 15820 (2295) | FS | 41920 (6080) | 4140 (600) | 9.8 | ... |
| 3.11 | S | 15820 (2295) | FB | 38060 (5520) | 2200 (319) | 5.7 | ... |
| 3.11 | S | 15820 (2295) | FS | 30650 (5750) | 1990 (289) | 5.0 | ... |
| 4.15 | S | 15820 (2295) | FB | 33440 (4850) | 2250 (327) | 6.7 | ... |
| 4.15 | S | 11890 (1725) | FS | 36890 (5350) | 2930 (425) | 7.9 | ... |
| 5.18 | S | 11890 (1725) | FB | 37850 (5490) | 1680 (243) | 4.4 | 16.1 (2.34) |
| 5.18 | S | 11890 (1725) | FS | 34540 (5010) | 2530 (367) | 7.3 | 17.5 (2.54) |
| 2.04 | M | 23100 (3350) | FB | 46130 (6690) | 1280 (185) | 2.7 | ... |
| 2.04 | M | 23100 (3350) | FS | 45580 (6610) | 3160 (458) | 6.9 | ... |
| 3.11 | M | 23100 (3350) | FB | 42960 (6230) | 1770 (257) | 4.1 | ... |
| 3.11 | M | 23100 (3350) | FS | 42610 (6180) | 2410 (350) | 5.6 | ... |
| 4.15 | M | 20930 (3035) | FB | 41920 (6080) | 3900 (565) | 9.3 | ... |
| 4.15 | M | 20930 (3035) | FS | 43020 (6240) | 1500 (217) | 3.4 | ... |
| 5.18 | M | 20930 (3035) | FB | 37230 (5400) | 9670 (1402) | 25.9 | 16.5 (2.40) |
| 5.18 | M | 20930 (3035) | FS | 39030 (5660) | 2110 (306) | 5.4 | 19.9 (2.87) |

[a] Average of 3
[b] FB = full mortar bedding, FS = face shell mortar bedding
[c] Average of 5

TABLE 6 -- Prism Compressive Test Results, Brick 1-10, Stack Bond

| Prism | Mortar | | | Prism Compressive Strength | | | $E^c$ x $10^6$ |
|-------|--------|--------|--------|--------|--------|--------|--------|
| h/t | Type | Comp. Str.[a] kPa (psi) | Area[b] | Avg[c], kPa (psi) | s, kPa (psi) | v % | kPa (psi) |
| 2.05 | N | 6000 (870) | FB | 39300 (5700) | 2070 (300) | 5.2 | ... |
| 2.05 | N | 6000 (870) | FS | 36750 (5330) | 2790 (405) | 7.5 | ... |
| 3.32 | N | 6000 (870) | FB | 34960 (5070) | 2980 (432) | 8.5 | ... |
| 3.32 | N | 6000 (870) | FS | 35920 (5210) | 910 (132) | 2.5 | ... |
| 4.16 | N | 7580 (1100) | FB | 37920 (5500) | 1950 (283) | 5.1 | ... |
| 4.16 | N | 7580 (1100) | FS | 35790 (5190) | 1370 (198) | 3.8 | ... |
| 5.40 | N | 7580 (1100) | FB | 33920 (4920) | 1680 (243) | 4.9 | 15.1 (2.19) |
| 5.40 | N | 7100 (1030) | FS | 27990 (4060) | 1720 (249) | 6.1 | 22.2 (3.22) |
| 2.05 | S | 10480 (1520) | FB | 32340 (4690) | 1790 (259) | 5.5 | ... |
| 2.05 | S | 10480 (1520) | FS | 39090 (5670) | 2360 (343) | 6.0 | ... |
| 3.32 | S | 10480 (1520) | FB | 34060 (4940) | 1190 (172) | 3.4 | ... |
| 3.32 | S | 9720 (1410) | FS | 39720 (5760) | 2300 (333) | 5.7 | ... |
| 4.16 | S | 9720 (1410) | FB | 38060 (5520) | 1330 (193) | 3.4 | ... |
| 4.16 | S | 9720 (1410) | FS | 38540 (5590) | 890 (129) | 2.3 | ... |
| 5.40 | S | 9720 (1410) | FB | 36610 (5310) | 820 (119) | 2.2 | 15.2 (2.20) |
| 5.40 | S | 9830 (1425) | FS | 39830 (5776) | 3670 (532) | 9.2 | 18.5 (2.68) |
| 2.05 | M | 21580 (3130) | FB | 48130 (6980) | 1100 (160) | 2.2 | ... |
| 2.05 | M | 21580 (3130) | FS | 53440 (7750) | 1600 (232) | 2.9 | ... |
| 3.32 | M | 21580 (3130) | FB | 45710 (6630) | 1370 (199) | 3.0 | ... |
| 3.32 | M | 21580 (3130) | FS | 46060 (6680) | 2930 (425) | 6.3 | ... |
| 4.16 | M | 18270 (2650) | FB | 42680 (6190) | 1450 (210) | 3.3 | ... |
| 4.16 | M | 15860 (2300) | FS | 45710 (6630) | 2940 (427) | 6.4 | ... |
| 5.40 | M | 15860 (2300) | FB | 44680 (6480) | 1390 ( 201) | 3.0 | 15.0 (2.17) |
| 5.40 | M | 15860 (2300) | FS | 45580 (6610) | 4060 (589) | 8.9 | 18.5 (2.68) |

[a] Average of 3
[b] FB = full mortar bedding, FS = face shell mortar bedding
[c] Average of 5

## Mode of Failure

In all cases the mode of failure was vertical splitting of the crosswebs, regardless of unit strength, mortar type or bedding, and prism bond pattern or h/t.

TABLE 7 -- Prism Compressive Test Results,
Type S Mortar, Brick 2-1

| Bond[a] | h/t | Mortar Comp.Strength[b] kPa (psi) | Area[c] | Prism Compressive Strength Average[b] kPa (psi) | s kPa (psi) | v % | E[d] x 10^6 kPa (psi) |
|------|------|------|------|------|------|------|------|
| R | 2.07 | 12960 (1880) | FB | 39730 (5762) | 1190 (173) | 3.0 | ... |
| R | 2.07 | 12960 (1880) | FS | 48780 (7077) | 2540 (368) | 5.2 | ... |
| R | 3.17 | 12960 (1880) | FB | 31570 (4578) | 1540 (224) | 4.9 | 23.6 (3.42) |
| R | 3.17 | 12960 (1880) | FS | 39740 (5764) | 2030 (294) | 5.1 | 24.0 (3.48) |
| R | 4.27 | 14960 (2170) | FB | 34670 (5029) | 520 ( 75) | 1.5 | 22.8 (3.31) |
| R | 4.27 | 14960 (2170) | FS | 36470 (5290) | 2700 (391) | 7.4 | 23.0 (3.34) |
| R | 5.30 | 11160 (1618) | FB | 32530 (4718) | 1140 (165) | 3.5 | 22.9 (3.32) |
| R | 5.30 | 11720 (1700) | FS | 38770 (5623) | 1360 (197) | 3.5 | 26.2 (3.80) |
| S | 5.30 | 11160 (1618) | FB | 35490 (5147) | 2730 (396) | 7.7 | 23.6 (3.43) |
| S | 5.30 | 11720 (1700) | FS | 36310 (5266) | 1560 (226) | 8.8 | 27.1 (3.93) |

[a] R = running bond, S = stack bond
[b] Average of 5
[c] FB = full mortar bedding, FS = face shell mortar bedding
[d] Single value

## Effect of Mortar Bedding and Bond Pattern

There was no significant difference between the minimum net area strengths of prisms built with either type of mortar bedding when stack bond prisms were tested. More full bedded prisms had lower strengths than those with face shell bedding, 32 versus 18. The overall average of full bedded prism strength divided by face shell bedded prism strength was 0.997 for the prisms laid in stack bond. Part of this data can be found in Table 10, column IV. In the running bond prisms, all of the full bedded prisms had lower strengths than the corresponding face shell bedded prisms. The overall average of full bedded prism strength divided by face shell bedded prism strength was 0.771.

Strength of prisms laid in running bond divided by strength of prisms laid in stack bond can be determined from phase 2 data at an h/t of approximately 5.3. This ratio is less for full bedded prisms, averaging 0.848, than for face shell bedded prisms which averaged 1.022.


Effect of Mortar Type

Using Type S mortar as the reference (relative strength = 1.00), the average relative prism strengths are:

Type N mortar    0.81
Type S mortar    1.00
Type M mortar    1.14

The rank order and relative strength do not differ materially from the results of earlier work with solid masonry prisms [3], although the value for Type M mortar is somewhat lower.


TABLE 8 -- Prism Compressive Test Results,
Type S Mortar, Brick 2-2

| Bond[a] | | Mortar | | Prism Compressive Strength | | | E[d] |
|---|---|---|---|---|---|---|---|
| | h/t | Comp.Strength[b] kPa (psi) | Area[c] | Average[b] kPa (psi) | s kPa (psi) | v % | x 10^6 kPa (psi) |
| R | 2.10 | 13340 (1935) | FB | 22240 (3226) | 1070 (155) | 4.8 | ... |
| R | 2.10 | 13340 (1935) | FS | 25570 (3709) | 4190 (608) | 16.4 | ... |
| R | 3.17 | 13340 (1935) | FB | 21710 (3149) | 2320 (337) | 10.7 | 18.0 (2.61) |
| R | 3.17 | 13340 (1935) | FS | 31150 (4518) | 2050 (298) | 6.6 | 18.6 (2.70) |
| R | 4.25 | 18400 (2668) | FB | 18320 (2657) | 2360 (343) | 12.9 | 18.3 (2.65) |
| R | 4.25 | 18400 (2668) | FS | 27580 (4000) | 1960 (284) | 7.1 | 25.0 (3.63) |
| R | 5.28 | 12810 (1858) | FB | 20700 (3002) | 910 (132) | 4.4 | 15.3 (2.22) |
| R | 5.28 | 13310 (1930) | FS | 26650 (3865) | 29940 (329) | 8.5 | 24.5 (3.55) |
| S | 5.28 | 12810 (1858) | FB | 23110 (3352) | 1570 (228) | 6.8 | 19.7 (2.85) |
| S | 5.28 | 13310 (1930) | FS | 25840 (3747) | 1760 (255) | 6.8 | 20.4 (2.96) |

[a] R = running bond, S = stack bond
[b] Average of 5
[c] FB = full mortar bedding, FS = face shell mortar bedding
[d] Single value

Mortar strength does influence prism compressive strength. For a given brick strength, and the same h/t, prisms with higher compressive strength mortar generally attained a greater percentage of the brick compressive strength. Tables 3 through 6 can be used to verify this fact.

TABLE 9 -- Prism Compressive Test Results,
Type S Mortar, Brick 2-3

| Bond | h/t | Mortar Comp.Strength[b] kPa (psi) | Area[c] | Prism Compressive Strength Average[b] kPa (psi) | s kPa (psi) | v % | E[d] x 10^6 kPa (psi) |
|---|---|---|---|---|---|---|---|
| R | 2.11 | 19740 (2863) | FB | 28800 (4177) | 2500 (363) | 8.7 | ... |
| R | 2.11 | 19740 (2863) | FS | 48590 (7047) | 4910 (712) | 10.1 | ... |
| R | 3.16 | 17220 (2497) | FB | 31370 (4549) | 2320 (337) | 7.4 | 26.2 (3.80) |
| R | 3.16 | 19740 (2863) | FS | 39890 (5785) | 3630 (526) | 9.1 | 26.9 (3.90) |
| R | 4.21 | 17220 (2497) | FB | 27300 (3960) | 2240 (325) | 8.2 | 27.9 (4.04) |
| R | 4.21 | 15930 (2310) | FS | 34370 (4985) | 2720 (394) | 7.9 | 31.9 (4.62) |
| R | 5.26 | 12740 (1848) | FB | 26840 (3893) | 1150 (167) | 4.3 | 31.0 (4.50) |
| R | 5.26 | 20980 (3043) | FS | 33250 (4822) | 2590 (376) | 7.8 | 36.3 (5.26) |
| S | 5.26 | 21310 (3090) | FB | 36750 (5330) | 1510 (219) | 4.1 | 29.0 (4.21) |
| S | 5.26 | 18220 (2643) | FS | 34320 (4978) | 2340 (339) | 6.8 | 35.2 (5.11) |

[a] R = running bond, S = stack bond
[b] Average of 5
[c] FB = full mortar bedding, FS = face shell mortar bedding
[d] Single value

Effect of Brick Compressive Strength

The ratio of prism compressive strength to net area compressive strength of the hollow brick decreases as unit strength increases for both full bedded and face shell bedded prisms. The comparison for prisms with Type S mortar, laid in stack bond and with an h/t of approximately 5.3 are given in Table 10.

There is no apparent relationship between net area unit strength and prism compressive strength.

TABLE 10 -- Effect of Brick Compressive Strength[a]

| Brick Type | Net Area Unit Strength kPa,(psi) | (I) | (II) | (III) | (IV) | (V) |
|---|---|---|---|---|---|---|
| 1-8 | 62140 (9012) | 0.609 | 0.556 | 1.096 | 259.7 | 281.8 |
| 1-10 | 64200 (9311) | 0.570 | 0.620 | 0.919 | 236.3 | 287.8 |
| 1-6 | 67190 (9750) | 0.510 | 0.500 | 1.020 | 213.3 | 244.1 |
| 2-2 | 99320 (14405) | 0.233 | 0.260 | 0.895 | 197.8 | 205.5 |
| 1-4 | 110940 (16090) | 0.302 | 0.306 | 0.985 | 144.2 | 146.7 |
| 2-1 | 129110 (18730) | 0.275 | 0.281 | 0.977 | 183.1 | 209.8 |
| 2-3 | 131820 (19120) | 0.279 | 0.260 | 1.071 | 220.2 | 267.2 |

| | |
|---|---|
| (I) | Full Bedded Prism Strength/Net Area Unit Strength |
| (II) | Face Shell Bedded Prism Strength/Net Area Unit Strength |
| (III) | Full Bedded Prism Strength/Face Shell Bedded Prism Strength |
| (IV) | Full Bedded Modulus of Elasticity/Net Area Unit Strength |
| (V) | Face Shell Bedded Modulus of Elasticity/Net Area Unit Strength |

[a]Prisms laid in stack bond with Type S mortar and h/t of 5.3

## Effect of Prism h/t

Increasing the aspect ratio of the prism decreases the compressive strength of the masonry. This effect is more pronounced with the 100 mm (4 in) nominal brick than with the thicker units. The change in prism h/t ratio has less of an effect on the strength of hollow brick prisms than on prisms built of solid brick [3]. This is not surprising since the hollow units have their area distributed nearer the perimeter of the cross-section.

The height to thickness ratio had no effect on the mode of failure.

## Modulus of Elasticity Comparisons

The phase 2 tests indicate an increase in modulus of elasticity with an increase in the height to thickness ratio of prisms laid in running bond with face shell mortar bedding. The ratio ranged from 1.09 to 1.35, with an average of 1.25, comparing h/t of 3 and 5. Only one of the three brick exhibited this phenomenon when the prisms were laid with full mortar bedding. This increase in modulus of elasticity with h/t was unexpected.

The modulus of elasticity is less for prisms laid with full mortar bedding than with face shell bedding. This ratio ranged from 0.625 to 0.991, with an average of 0.794 for prisms in running bond. For stack bond prisms the range was 0.680 to 0.983 with an average of 0.864.

There does not appear to be a defined trend between modulus of elasticity and mortar type with a given unit strength. Unit strength alone does have an effect on modulus of elasticity, E generally increases with unit strength. However, there is significant scatter to the data. Combining mortar type and unit compressive strength shows an increase in modulus of elasticity as mortar strength and unit strength increase.

The ratio of modulus of elasticity to unit compressive strength generally decreases with increasing unit strength. Prisms laid with full mortar bedding have a more narrow range than face shell bedded prisms. See Table 10, columns (IV) and (V). The ratio of E to unit strength has less scatter than E to prism strength.

The ratio of modulus of elasticity to prism compressive strength also varies widely. In phase 1, the value ranged from 314 to 793, with an average of 469. In phase 2, the ratio ranged from 598 to 1161, with an average of 810.

SUMMARY

An experimental program was performed to evaluate the compressive strength of prisms built with hollow clay brick. Four thicknesses of units, four prism h/t ratios, three mortar types, two bond patterns and two types of mortar bedding were considered. Five replications of each combination resulted in testing of 790 prisms. Observations which result from evaluation of test data are as follows:

1. The mode of failure was always vertical splitting of the crosswebs for prisms built with hollow brick units.
2. Stack bond prisms show negligible influence of mortar bedding on prism compressive strength and the modulus of elasticity is influenced less by mortar bedding in stack bond prisms than in running bond prisms.
3. Full bedded prisms develop lower compressive strengths when laid in running bond than when laid in stack bond.
4. Compressive strength of prisms made with a given hollow brick increases with increased mortar strength.
5. A greater percentage of the brick compressive strength is achieved in the prisms with lower strength brick.
6. Prism h/t ratio has less effect on prism compressive strength of hollow units than on those of solid units. Moderate strength reductions are produced by increasing h/t ratios.
7. The modulus of elasticity is less for full bedded prisms than for face shell bedded prisms.
8. The modulus of elasticity increases with unit strength and with a combined increase in mortar and unit strength.

REFERENCES

[1]  C 652-70, Standard Specification for Hollow Brick (Hollow Masonry
     Units Made From Clay or Shale), Annual Book of ASTM Standards,
     Part 12, American Society for Testing and Materials,
     Philadelphia, PA, 1971, pp. 464-467.
[2]  C 652-87a, Standard Specification for Hollow Brick (Hollow
     Masonry Units Made From Clay of Shale), Annual Book of Standards,
     Vol. 04.05, American Society for Testing and Materials,
     Philadelphia, PA, 1988, pp. 356-359.
[3]  Recommended Practice for Engineered Brick Masonry, Brick
     Institute of America, Reston, VA, 1969.

J.E. Amrhein[1] and R.H. Hatch[2] and M.W. Merrigan[3]

ANCHOR CONNECTIONS OF STONE SLABS

REFERENCE: Amrhein, J.E., Hatch, R.H., and
Merrigan, M.W. "Anchor Connections of Stone
Slabs," Masonry: Components to Assemblages,
ASTM STP 1063, American Society for Testing and
Materials, Philadelphia, 1989.

ABSTRACT: This paper presents the results of
testing anchorage systems of stone panels onto
buildings. It includes various typical anchor
systems for the two most common stone thicknesses
and five different classes of stones. The anchor
systems and their ultimate loads allow comparisons
with code requirements and factors of safety.

KEYWORDS: Anchors, Kerfs, Pull-Outs, Stone, Wire
Ties.

Anchored stone slab veneer systems have been in
use for centuries; even the Ancient Romans used lead
anchors to secure stone to the structure. Experiences
from the past provided the guidelines for the
development of anchoring procedures which have been
included in building codes as empirical requirements.

For thin stone slabs under 1.9 square meter (20
square feet) in area using 9-gauge copper or brass wire
every 0.61 meters (24 inches) around the perimeter is
an acceptable method of securing stones to buildings.
However, the strength requirements of these anchors or
ties are not defined.

This test program investigated the strength of
common anchor systems and provided information on
various anchor systems. Test results revealed the
range of safety factors of these connections for the
type of anchor and stone tested.

[1]Executive Director, Masonry Institute of America
[2]DBM/Hatch, Inc.
[3]Engineer, Masonry Institute of America

## TYPES AND THICKNESS OF STONE

Stones have variable strength characteristics and perform differently under load and stress. Since different type of stones are used in construction, the more common ones were selected for this test program.

The stones were selected based on their current availability and usage.

1.- Spanish Pink Granite, Rosa Povina, with a polished face; 2cm and 3cm thick. Quartz; Alkalie; Feldspar.
2.- Italian Green Marble with a polished face; 2 cm thick. Serpentine; Group A.
3.- Indian Red Sandstone with a cleft face; 3cm thick nominal. Quartz, Feldspar; Group A.
4.- Portuguese Limestone with a honed face; 2cm and 3cm thick. Calcite, Dolmite; Group A.
5.- Italian Limestone with a honed face; 2cm and 3cm thick. Calcite, Dolmite; Group A.
6.- French Limestone with a polished face; 2cm and 3cm. Calcite, Dolmite; Group A.
7.- Italian Travertine with a polished face; 2 cm thick. Calcite, Dolmite; Group A.

## ANCHOR SYSTEMS

The most commonly used and available anchor tie systems were selected for testing.

1.- Hand set wire tie of 3.8mm (9-gauge) copper or brass wire inserted in drilled 4.8mm (3/16") diameter holes 25mm (1") deep in the edge of the stones. (Tension condition for wall and soffit use.)

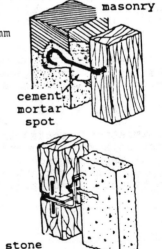

masonry

cement mortar spot

2.- Strap and dowel anchors; 4.8mm (3/16") diameter dowels inserted in 6.4mm (1/4") diameter drilled holes 25mm (1") deep in the edge of the stones. (Tension condition for wall and soffit use.)

stone

concrete

3.- Bent bar anchor bolts 6.4mm
(1/4") and 9.5mm (3/8")
diameter set in the back of
the stones.
(Shear condition for vertical
support and tension condition
for wall use.)

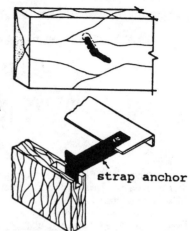

4.- Strap anchors in kerfs
along the edge of the stone.
(Tension condition for wall
use.)

strap anchor

TESTING PROGRAM

The testing technique was dependent upon the type
of anchor system and was conducted as follows:

Wire tie anchor system

The hand set wire tie anchor system is a standard
installation technique for stone slab veneer.

The test procedure was as follows:

The 9-gauge wires were set into 4.8mm (3/16")
diameter 25mm (1") deep drilled holes in the
middle third of the edge of the stones using three
different standard practices.

1.- Wood plugs

2.- Cement

3.- Wood plug with a crimp

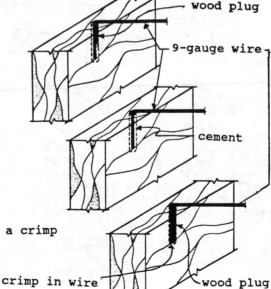

wood plug

9-gauge wire

cement

crimp in wire          wood plug

### Kerf and strap anchor systems

The kerf and strap anchor system was as follows:

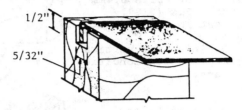

The kerfs were tested using both 25mm (1") wide straps and continuous anchor straps 76mm (3") wide in the sawn kerfs.

### Dowel and Strap anchor system

The dowel anchor system was as follows:

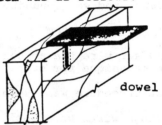

### Bent Bolt anchor systems

The bent bolt anchor system was as follows:

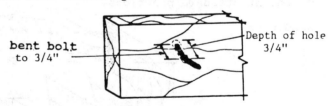

## TEST SET UP AND PROCEDURE

All specimens were tested using a universal testing machine with 545 kilogram (1,000 pound) capacity in 0.9 kilogram (2 pound) increments, 1,135 kilogram (2,500 pound) capacity in 2.3 kilogram (5 pound) increments and 2,270 kilogram (5,000 pound) capacity in 4.5 kilogram (10 pound) increments. All loads were applied uniformly and monotonically

Support for the specimens was provided so as not to affect the test results for the anchor systems. Restraint supports for the specimens were arranged to be as far from the anchor as feasible without creating a primary flexural failure mode in the specimen. Spacing of restraint devices was based upon an assumed failure cone of 15 degrees from the plane of the face of the stone specimen.

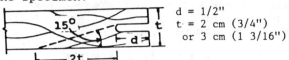

d = 1/2"
t = 2 cm (3/4")
   or 3 cm (1 3/16")

Hand set wire ties and dowel anchors were tested using similar set ups:

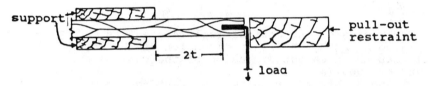

Bent bolt anchor were tested for pull-out and for shear loading conditions:

Inside diameter of restraining ring is 4".

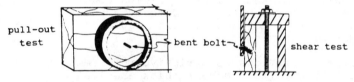

Bent strap anchors were tested with the anchors applying loads perpendicular to the plane of the specimen:

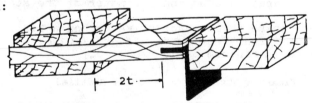

## TEST RESULTS

The test results for the various anchor systems and stone types demonstrated that for similar loading conditions the same basic type of failures occurred. Generally the greater the flexural strength of the stone and the greater its thickness then the greater its load capability before failure.

In tension the wire tie, dowel and strap, kerf and strap and bent bolt had the loads applied perpendicular to the plane of the specimen and they exhibited either pull-out of the anchor, conical spalling, splitting of the stone, flexural failure at the anchor or support or jp6a combination of these.

In shear the bent bolt anchors tested exhibited either bending of the bolts, bolt shear, crushing of the stone, spalling of the stone, pull-out of the anchor or a combination of these.

An attempt was made to calculate the surface areas of the failures cones from the pull-out tests.   The area of failure was based upn the equation:

$$A = r\pi\sqrt{r^2 + h^2}.$$

The radius of the conical failure area was calculated as being the least dimension of the conical spall plus the greatest dimension of the conical spall divided by two ((a+b)/2).
Table 2 and 3 show the results from testing the various anchors and stone.

ANALYSIS AND COMPARISON OF TEST RESULTS

The 1988 edition of the Uniform Building Code requires that all anchored veneer and its attachments be "designed to resist a horizontal force equal to twice the weight of the veneer."

The U.B.C. also provides prescriptive requirements limiting slabs to 5 cm. (two-inches) thick and 1.86 square meters (20 square feet) in area and that ties be dowels or wire type and spaced not more than 0.61 meters (24") apart around the perimeter of the stone.

According to this prescriptive method a 1.2 meters by 1.5 meters (four feet by five feet) (20 sq. ft.) slab requires a minimum of 10 ties.

**Table 1:  Flexural Strength (ASTM C880) Modified[1]**

|  | (2.0 cm) | | (3.0 cm) | |
|---|---|---|---|---|
|  | KPA | (PSI) | KPA | (PSI) |
| Italian Limestone | 6670 | ( 967) | 5300 | ( 768) |
| Portuguese Limestone | 7540 | (1094)(2 samples tested) | 5230 | ( 759) |
| French Limestone | 12,480 | (1810) | 10900 | (1581) |
| Travertine | 10,470 | (1518) | | |
| Green Marble | 12,640 | (1833)(2509 psi if sample that broke along a vein is disregarded) | | |
| Spanish Pink Granite | 11,780 | (1709) | 9670 | (1402) |
| Indian Red Sandstone | - | - | 13270 | (1925) |

TABLE 2:    Test Results - 2.0 cm Stones

| Anchor Test Stone[1] | Avg[2] Max. Load (Kgs) | Area of failure (cm[2]) | Failure[3] Strength (Kilo Pascals) | Load per Cm of Kerf (Kgs/Cm) | Avg. Max. Shear (kgs) | Type(s)[4] of Failure(s) |
|---|---|---|---|---|---|---|
| **Wire Ties** | | | | | | |
| S.P.G.[6] | 106 | 13.8 | 731 | - | - | a |
| Trav.[6] | 75 | 6.8 | 1082 | - | - | a,a&b,b |
| G.M.[6] | 113 | 9.9 | 1117[5] | - | - | a,b |
| I.L.[7] | 69 | 10.8 | 627 | - | - | a |
| P.L.[7] | 49 | 7.0 | 689 | - | - | a |
| F.L.[7] | 129 | 6.3 | 2006[5] | - | - | a,a&b,b |
| **Dowels** | | | | | | |
| S.P.G. | 118 | 15.6 | 738 | - | - | a |
| Trav. | 95 | 10.7 | 869 | - | - | a |
| G.M. | 146 | 19.0 | 752 | - | - | a |
| I.L. | 59 | 10.8 | 531 | - | - | a |
| P.L. | 65 | 12.8 | 503 | - | - | a |
| F.L. | 102 | 10.6 | 945 | - | - | a |
| **Strap & Kerf** | | | | | | |
| S.P.G. | 70 | 2.5 | - | 16 | - | c |
| Trav. | 48 | 6.1 | 772 | 11 | - | c,e |
| G.M. | 92 | - | - | 19 | - | c,f |
| I.L. | 33 | 4.9 | 662 | 8 | - | c,d,e |
| P.L. | 34 | 3.6 | 938 | 8 | - | c,d |
| F.L. | 54 | 3.1 | 1710 | 12 | - | c,d |
| **Bent Bolts** | | | t | | | |
| S.P.G. | 380 | 23.9 | - | - | 732 | a,f |
| Trav. | 353 | 23.4 | - | - | 587 | a,f |
| G.M. | 453 | 12.8 | - | - | 649 | a,f |
| I.L. | 74 | 8.9 | - | - | 154 | a,b |
| P.L. | 138 | 14.3 | - | - | 203 | a |
| F.L. | 320 | 16.3 | - | - | 401 | b |

[1]S.P.G.- Spanish Pink Granite, Trav.-Travertine, G.M.-Green Marble, I.L. Italian Limestone, P.L.-Portuguese Limestone, F.L.-French Limestone.
[2]The average value of the sum of the maximum load at failure.
[3]Maximum Failure load divided by the area of failed surface.
[4]a=conical, b=pull-out, c=kerf, d=corner of kerf, e=at support, f=splitting along the vein.
[5]Some samples failed entirely by pulling out of the anchor with no damage to the stone and therefore had no surface area of failure for calculating failure strength in kilo newtons.
[6]9-gauge copper wire 276,480 kilo pascals (40100 psi) tension load at failure.
[7]9-gauge brass wire 442,370 kilo pascals (64160 psi) tension load at failure.

TABLE 3:          Test Results - 3.0 cm Stones

| Anchor Test Stone[1] | Avg[2] Max. Load (Kgs) | Area of failure (cm$^2$) | Failure[3] Strength (Kilo pascals) | Load per Cm of Kerf (Kgs/Cm) | Avg. Max. Shear (kgs) | Type(s)[4] of Failure(s) |
|---|---|---|---|---|---|---|
| **Wire Ties** | | | | | | |
| S.P.G.[6] | 168 | 25.9 | 634[5] | - | - | a&b,b |
| I.L.[7] | 111 | 18.1 | 600[5] | - | - | a,a&b,b |
| P.L.[7] | 135 | 26.6 | 496[5] | - | - | a,b |
| F.L.[7] | 166 | 7.7 | 2100[5] | - | - | b,a&b |
| I.R.S.[6] | 154 | 32.6 | 462 | - | - | a |
| **Dowels** | | | | | | |
| S.P.G. | 193 | 30.9 | 614 | - | - | a |
| I.L. | 102 | 22.6 | 441 | - | - | a |
| P.L. | 127 | 29.9 | 427 | - | - | a |
| F.L. | 184 | 20.0 | 903 | - | - | a |
| I.R.S. | 188 | 34.7 | 531 | - | - | a |
| **Strap & Kerf** | | | | | | |
| S.P.G. | 63 | 6.0 | 1020 | 13 | - | c |
| I.L. | 50 | 7.7 | 641 | 9 | - | c,d,e |
| P.L. | 45 | 6.6 | 676 | 11 | - | c,d,e |
| F.L. | 111 | 6.6 | 1655 | 26 | - | c,d |
| I.R.S. | - | - | - | - | - | - |
| **Bent Bolts** | | | | | | |
| S.P.G. | 542 | 24.3 | - | - | 773 | c,g,h,i |
| I.L. | 220 | 24.3 | - | - | 602 | b&c,c,g,h,i |
| P.L. | 193 | 25.6 | - | - | 445 | b&c,c,f,g |
| F.L. | 563 | 35.9 | - | - | 1022 | b,c,g,h,i |
| I.R.S. | 335 | 29.5 | - | - | 654 | c,g |

[1]S.P.G.- Spanish Pink Granite, Trav.-Travertine, G.M.-Green Marble, I.L.-Italian Limestone, P.L.-Portuguese Limestone, F.L.-French Limestone, I.R.S.-Indian Red Sandstone.
[2]The average value of the sum of the maximum load at failure.
[3]Maximum Failure load divided by the area of failed surface.
[4]a=conical, b=pull-out, c=kerf, d=corner of kerf, e=at support, f=splitting along the vein.
[5]Some samples failed entirely by pulling out of the anchor with no damage to the stone and therefore had no surface area of failure for calculating failure strength in kilo newtons.
[6]9-gauge copper wire 276,480 kilo pascals (40100 psi) tension load at failure.
[7]9-gauge brass wire 442,370 kilo pascals (64160 psi) tension load at failure.

A review of the test data for wire ties and dowel type anchors indicates the safety of such anchor systems.

Table 4: Weight of masonry veneer per square meter

| Weight per Sq.meter For Slab Stone Thicknesses | | | | |
|---|---|---|---|---|
| Type of Stone | Density, (max) | 2.0 cm (psf) | 3.0 cm (psf) | 5.0 cm (psf) |
| | $Kg/m^3$ (pcf) | $Kg/m^2$ (psf) | $Kg/m^2$ (psf) | $Kg/m^2$ (psf) |
| Granite | 3040 (190) | 61 (12.5) | 91 (18.7) | 155 (31.7) |
| Limestone | 2960 (185) | 59 (12.1) | 89 (18.2) | 150 (30.8) |
| Marble | 2950 (184) | 59 (12.1) | 88 (18.1) | 150 (30.7) |
| Sandstone | 2720 (170) | 54 (11.1) | 82 (16.7) | 138 (28.3) |
| Travertine | 2560 (160) | 51 (10.5) | 77 (15.7) | 130 (26.7) |

## DESIGN LOAD

A 1.9 sq.meter (20 sq.ft.) slab of 3cm granite, with a maximum weight of 170kgs (374 lbs), requires the lateral connection be capable of resisting twice this load in a horizontal direction perpendicular to the plane of the slab.

Therefore a 1.2 meter by 1.5 meter (4-feet by 5-feet) slab 1.9 sq.meters (20sq.ft.) of granite with 10 wire anchors must resist 340 kgs (748 lbs) or 34 kgs (74.8 lbs) per anchor.

Test results on wire ties in granite were between 134 and 200 kgs (296 and 438 lbs) per anchor 168 kgs average (370.5 lbs). This results in a factor of safety 5 times the code requirement.

Similarly a 2.0cm travertine slab would weigh a maximum of 95 kgs. (210 lbs)(for 1.9 sq.m. slab). The 10 anchors must resist 191 kgs (420 lbs). Test results were between 36 and 117 kgs (80 and 258 lbs) per anchor (average = 75 kgs (166 lbs) each). The factor of safety would be between 2 and 6 times the UBC requirements.

Table No. 5 lists the relative factors of safety against failure for anchor systems tested in each stone type for 2.0cm and 3.0cm thicknesses. These values are based on one anchor per 0.19 square meters (2 sq.ft.) of stone.

Table 5: Relative Factors of Safety, Tension or Pull-Out

| ANCHOR & STONES | 2.0cm Thick Stone | | | 3.0cm Thick Stone | | |
|---|---|---|---|---|---|---|
| | Low | Avg. | High | Low | Avg | High |
| **WIRE TIES** | | | | | | |
| Travertine | 3.8 | 7.9 | 12.3 | - | - | - |
| Green Marble | 9.5 | 10.2 | 11.5 | - | - | - |
| Spanish Pink Granite | 7.7 | 9.0 | 10.5 | 7.9 | 9.9 | 11.7 |
| Italian Limestone | 4.0 | 6.3 | 10.7 | 5.8 | 6.7 | 8.4 |
| Portuguese Limestone | 3.6 | 4.5 | 5.4 | 6.9 | 8.2 | 10.6 |
| French Limestone | 8.8 | 11.7 | 14.3 | 8.1 | 10.0 | 12.9 |
| Indian Red Sandstone | - | - | - | 6.5 | 10.2 | 14.2 |
| **DOWELS** | | | | | | |
| Travertine | 5.3 | 9.9 | 12.8 | - | - | - |
| Green Marble | 11.8 | 13.3 | 14.3 | - | - | - |
| Spanish Pink Granite | 8.6 | 10.4 | 13.0 | 7.9 | 11.4 | 14.2 |
| Italian Limestone | 3.6 | 4.7 | 6.8 | 4.6 | 6.2 | 7.4 |
| Portuguese Limestone | 4.5 | 5.9 | 8.6 | 5.1 | 7.7 | 11.4 |
| French Limestone | 7.8 | 9.3 | 12.2 | 9.3 | 11.2 | 12.7 |
| Indian Red Sandstone | - | - | - | 9.5 | 12.4 | 15.4 |
| **Straps & Kerf** | | | | | | |
| Travertine | 1.9 | 3.0 | 6.0 | - | - | - |
| Green Marble | 3.0 | 4.3 | 7.1 | - | - | - |
| Spanish Pink Granite | 2.7 | 3.8 | 5.0 | - | - | - |
| Italian Limestone | 0.7 | 1.9 | 3.8 | 0.7 | 1.4 | 1.9 |
| Portuguese Limestone | 0.6 | 1.5 | 2.7 | 0.6 | 1.7 | 2.9 |
| French Limestone | 2.2 | 2.8 | 4.5 | 2.5 | 4.0 | 5.9 |
| Indian Red Sandstone | - | - | - | 1.9 | 2.3 | 3.2 |
| **Bent Bolts** | | | | | | |
| Travertine | 10.6 | 14.8 | 17.1 | - | - | - |
| Green Marble | 13.2 | 16.5 | 21.5 | - | - | - |
| Spanish Pink Granite | 12.4 | 13.4 | 14.5 | 11.8 | 12.8 | 13.6 |
| Italian Limestone | 1.3 | 2.7 | 4.1 | 3.6 | 5.3 | 7.1 |
| Portuguese Limestone | 3.4 | 5.0 | 7.6 | 2.9 | 4.7 | 6.9 |
| French Limestone | 10.0 | 11.6 | 14.0 | 12.4 | 13.6 | 14.3 |
| Indian Red Sandstone | - | - | - | 8.6 | 8.8 | 9.0 |

NOTES FOR TABLE No.5

1.-  The limestone factors of safety were based upon the
     maximum density limestones (the heaviest possible
     load).  The tested stones were of lighter weight and
     would have had a higher factors of safety.
2.-  The strap and kerf anchor factors of the safety are
     based on an assumed 2.5cm (one inch) strap placed into
     a sawn kerf space approximately 60cm (24") around the
     perimeter of the 1.9 sq.m.(20 sq.ft.) slab.
3.-  The bent bar anchor bold factors of safety are based on
     the assumption that only four bolts support the stone.
     This is a typical anchoring pattern when bent bolts are
     used.

CONCLUSION

When test results were compared to the empirical requirements of the Uniform Building Code the factors of safety were relatively high.

The average factor of safety for wire-ties in drilled holes was 8.6 and for dowels in drilled holes it was 9.3. For straps in kerfs, the average was 2.7 and bent bolts averaged 9.9.

These results should be carefully evaluated since they are based on the assumption the slabs will be of the heaviest possible stone of each type. Realistically, the values would be higher if compared to the weights of the actual stone tested. For comparison purposes, the factors of safety were calculated in the most conservative fashion.

The samples for this testing program were supplied by three stone contractors and their methods of installation were not always identical.

For example, the Indian Red Sandstone specimens with wire ties had a 180 degree bend with a 1.0cm return at the end. This practice was developed by the contractor just for this particular stone. All other wire anchors were installed using straight wires in cement or wood plugs or both. The sample size was too small to determine a reliable difference between the various methods.

The 4.8mm diameter 25mm long dowel anchors (3/16"Ø x 1") were tested in 6mm (1/4")diameter holes drilled in the middle of the edge of the stones. The dowels were taped with a single layer of masking tape and tightly fitted into the drilled holes.

Not all of the drilled holes were perfectly aligned. The testing rig in the laboratory was adjustable and the misaligned holes did not affect the results. It would be interesting, however, to specifically test the effects of varying hole alignments on anchor strengths.

When testing the same kerf stones with strap anchors, there were several instances of just the corner failing. This only occurred for the wider 76mm (3-inch) samples and was due to uneven loading by the wide strap. No testing was done on the effects of a narrow strap in a wide kerf (a common practice).

The results of the strap and kerf testing are given as kilograms of load per meter (pounds of load per inch) of kerf. If the kerf was wider than the strap, there would be a large area of stone resisting the load and higher anchor strengths would result.

Higher loads for kerf type anchors can be expected if an improved seating method is used as a cushion between the edge of the kerf and the anchor strap.  For testing purposes a double layer of 1.6mm (1/16") foam strips were used as a cushion.  Silicon or some other more resilient material will distribute loads evenly, give higher results and there would be less tendency for the kerf corners to fail.

Figures A and B in the Appendix are graphical representations of the range of anchor strengths achieved using dowel type and wire type anchors.  There is an apparent relationship between maximum loads and conical failure areas.  Additional testing and analysis is required for a valid numerical analysis procedure to predict strengths of anchor systems.

Acknowledgements

This testing program was made possible by the full support and cooperation of the mason contractors who made the test specimens.

The individuals and companies that contributed to this test program were:

Robert Hatch of DBM/Hatch Inc.
Lou Carnevale of Carnevale & Lohr, Inc.
Fred Cordova of Carrara Marble Company of America.

All testing was by Smith Emery Company, Los Angeles, Ca
Dowel anchor hardware was by Hohmann & Barnard, Inc.

INFORMATIONAL PUBLICATIONS.

Marble and Stone Slab Veneer, 2nd Edition, Amrhein and Merrigan, published by the Masonry Institute of America, 1989.

American Society for Testing and Materials, Standard Test Method for Flexural Strength of Natural Building Stone, C880-89.

1988 Edition of the Uniform Building Code, Published by International Conference of Building Officials, Whittier, Ca.

Comparisons of maximum loads (lbs) to failure
loads per conical failure areas (psi)

Diagram A.  Wire Anchor Tests - 2.0 cm

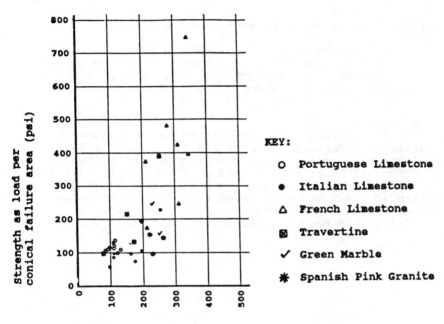

KEY:

O   Portuguese Limestone

●   Italian Limestone

△   French Limestone

▣   Travertine

✓   Green Marble

✳   Spanish Pink Granite

Maximum load at failure (lbs)

Diagram B.   Dowel Anchor Tests - 2.0 cm

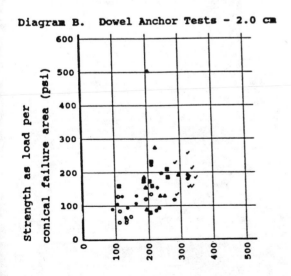

Maximum  load at failure (lbs)

DISCUSSION

ANCHOR CONNECTIONS OF STONE SLABS; J.E.Amrhein, R.H.Hatch and M.W. Merrigan.

Question: (Edward O. Benovengo, Skidmore, Owings & Merrill, New York, NY)
   Though it appears that the code requirement for lateral loads applied to a stone panel, it is also a requirement for the stone to resist a sustained lateral wind pressure.  Pending on stone face size and quantity of anchors, this load stress can exceed the twice gravity load values.

Answer:
Amrhein - Chapter 30 of the Uniform Building Code requires that "anchored veneer and its attachments shall be designed to resist a horizontal force equal to twice the weight of the veneer" as a minimum force.  If the actual forces are greater than this due to wind suction at peaks and corners of a building, the actual forces must be used in the design.  Information provided in the paper provides this opportunity.

DISCUSSION

ANCHOR CONNECTIONS OF STONE SLABS; J.E.Amrhein, R.H.Hatch and
M.W. Merrigan.

Question: (Edward O. Benovengo, Skidmore, Owings & Merrill,
   New York, NY)
      Stresses in stone anchor pockets and cutouts can change due to
the profile of the cut cross section, radiused transitions have lower
stress build-up than sharp corners.  Did the stone specimens have all
radiused or all sharp edged cutouts?  If both were used, did the test
results vary between these two types?

Answer:
   Amrhein - There were no slots with radius ends.  The
   kerfs were continuous across the specimen, the
   specimens were either one inch wide or three inches
   wide.  If the kerf slot was a circular cut this would
   be stronger than a continuous horizontal cut in the
   stone.

DISCUSSION

ANCHOR CONNECTIONS OF STONE SLABS; J.E.Amrhein, R.H.Hatch and M.W. Merrigan.

Question : (Edward O. Benovengo, Skidmore, Owings & Merrill, New York, NY)
    Do the tests described follow a specific standardized testing method? If so, please identify. If not, then it would be desirable to describe the procedure more completely.

Answer:
Amrhein - We are not aware of any specific ASTM test procedures for this connection test. What the test consisted of was uniform application of load as per ASTM load application, and loaded to failure. Tests set ups were arranged so that there would be direct tension or shear on the connection.

DISCUSSION

<u>ANCHOR CONNECTIONS OF STONE SLABS</u>; J.E.Amrhein, R.H.Hatch and
M.W. Merrigan.


<u>Question</u>;   (Edward O. Benovengo, Skidmore, Owings & Merrill,
    New York, NY)
    It would be desirable to state the tolerance or oversize
dimensions for the anchor holes and/or cutouts.  Were the anchors all
snug fit, or where were the anchors loose versus snug fitting.
<u>Answer</u>:
<u>Amrhein</u> -  The revised paper provides the information
on tolerances.  All anchors has a loose fit or may have
been wedged in after they have been placed in the slot
or hole.

DISCUSSION

ANCHOR CONNECTIONS OF STONE SLABS; J.E.Amrhein, R.H.Hatch and
M.W. Merrigan.


Question:   (Edward A. Benovengo, Skidmore, Owings & Merrill, New York,
    NY)
     The stones used in the testing are identified by the common trade
names, it would be more informative (and accurate) to also identify
the quarry source for all stones.  Only then would the actual results
have full meaning and accuracy. It is a common practice in the stone
industry to use a single name for many different sources of a given
stone, and it has been shown that different quarries will have stone
of varying properties.

Answer:
Amrhein - To identify the quarry source would not
provide much additional information as stones from
different locations in a quarry may have significant
variation in properties.  In the revised paper we
included the mineralogical names and classified the
specimens by group.  All specimens were Group A which
would be marble and stones unimpaired by weakened
planes with very little variation in fabrication
quality.  Stone are selected based upon their quality
and thus are grouped.

DISCUSSION

ANCHOR CONNECTIONS OF STONE SLABS; J.E.Amrhein, R.H.Hatch and
M.W. Merrigan.


Question :    (Edward A. Benovengo, Skidmore, Owings & Merrill,
    New York, NY)
    Is "Monotonically" used in the proper context?  How can a load be
applied incrementally and simultaneously be a constant or "Monotonically".
Answer:
Amrhein -  The load was applied in increments and in
one direction only.  No reversal and no dynamic forces
were applied.

DISCUSSION

ANCHOR CONNECTIONS OF STONE SLABS; J.E.Amrhein, R.H.Hatch and M.W. Merrigan.

Question: (Edward A. Benovengo, Skidmore, Owings & Merrill, New York, NY)
    Where the bent bar (dowel) type anchor bolts set in a cement or epoxy?  If so please indicate.  By using the term "Bolt" do you mean to say that the anchors were threaded?

Answer:
Amrhein -  The bent bar is actually a bent threaded rod.  And it was set in epoxy.  We have changed reference from Bent Bar to Bent Bolt in the paper.

Maureen T. Brown

A CRITICAL REVIEW OF FIELD ADAPTING ASTM E 514 WATER PERMEABILITY
TEST METHOD

---

REFERENCE: Brown, M.T., "A Critical Review of Field Adapting
ASTM E 514 Water Permeability Test Method," Masonry: Components
to Assemblages, ASTM STP 1063, John H. Matthys, Editor,
American Society for Testing and Materials, Philadelphia, 1990.

ABSTRACT: This paper investigates the use of a field adapted
ASTM E 514 water permeability test which is currently used by
masonry consultants to document moisture penetration problems
in brick masonry. There is a growing debate of whether this
adapted test method can appropriately be used to decide if a
masonry wall is considered a failure regarding moisture
penetration.

It is the intent of this adapted test method to follow the
requirements of ASTM's E 514-74 test method except the test
is used on in-place walls. However, additional differences
between the actual and adapted test methods significantly
alter the results and interpretation of the test. Information on the differences between the two test methods is
needed by engineers, architects and building owners so that
informative decisions can be made on the appropriateness of
the field adapted ASTM E 514 water permeability test method.

KEYWORDS: water permeance, masonry, field test, unit masonry,
exterior wall surface, cavity wall, drainage wall, leakage

A test method which classifies in-place masonry walls as those
which resist damage due to water penetration to various degrees would
be a great asset to the masonry industry. Uncontrolled water
penetration through masonry walls can reduce the durability of brick
masonry and damage the interior wall. It is the intention of some who
use the Field Adapted ASTM E 514, Water Permeability Test Method [1],
to make such a classification.

Maureen Brown is a staff engineer at Penkiunas Associates in
Lanham, MD, formerly at the Brick Institute of America, 11490 Commerce
Park Drive, Reston, VA 22091.

If it were possible to detect an in-place wall which will later experience water related damage, the wall could be corrected before the damage occurred and great savings could result. However, to be cost effective, the test must not erroneously classify in-place walls. If unwarranted repair is made due to the test method, needless cost could occur.

The Field Adapted ASTM E 514 Test Method has been proposed as a test which adapts ASTM E 514-74, Standard Test Method for Water Permeance of Masonry [2], to allow the procedure to be performed on in-place walls. The following discussion presumes that the reader is familiar with the ASTM E 514-74 Test Method, the 1986 revision ASTM E 514-86 Test Method [3] and the Field Adapted ASTM E 514 Test method, herein called Standard E 514-74, Standard E 514-86 and Field E 514, respectively. When no clarification is made regarding the specific dates of the Standard ASTM E 514 Test Method, it will be referenced as Standard E 514.

The following discussion addresses the significant differences between Standard E 514 and Field E 514. The purpose, apparatus, procedure, record of observations, and interpretation of results of the tests are compared. Standard E 514-74 will be used as the basis for comparison with Field E 514 although Standard E 514-86 will also be given consideration when it differs from Standard E 514-74.

The current Standard E 514 is E 514-86. E 514-86 declares that it is a laboratory test procedure only. This declaration questions the validity of adapting Standard E 514-86 Test Method as a field test. This non-agreement will not be pursued in this discussion since Field E 514 was written before E 514-86 was adopted.


PURPOSE


Standard E 514's purpose is to determine unit masonry's resistance to wind-driven rain. The test is intended to produce data which will help evaluate the effect different materials, coatings, construction details and workmanship have on the water penetration resistance of a unit masonry wall.

Field E 514's intent is to alter Standard E 514 for field use. Field E 514 is claimed to determine the masonry exterior face's resistance to wind-driven rain. Field E 514 is also suggested by some to evaluate a brick wall's water permeability, measuring the quality and compatability of materials and workmanship.

The intent of Standard E 514 and Field E 514 is not equivalent. The water penetration resistance of unit masonry as defined in Standard E 514, is not determined by the amount of water which passes into the exterior face of a masonry wall. The water penetration resistance of unit masonry in Standard E 514 is determined from observed moisture or water on the interior of the wall, water collected from behind the exterior wythe or water collected at the interior side of the wall. The water penetration resistance of unit masonry is a measure of water which must first pass through the exterior wythe of masonry. The amount of water which passes the exterior face of the masonry does not necessarily pass through the exterior wythe of masonry. A consistent relationship between the purpose of Standard E 514 and Field E 514 is therefore not established.


APPARATUS


The basic apparatus of Standard E 514 includes a test chamber with a minimum opening of 1.1 $m^2$ (12 $ft^2$), an air line with manometer, a water line with valves and water manometer, a water overflow pipe at the bottom of the chamber and a refillable reservoir. Standard E 514 does not make an attempt to account for the total amount of water which leaves the refillable reservoir. The total amount of water which enters the exterior wall surface is not labeled as leakage water. Figure 1 is a schematic of the diagram.

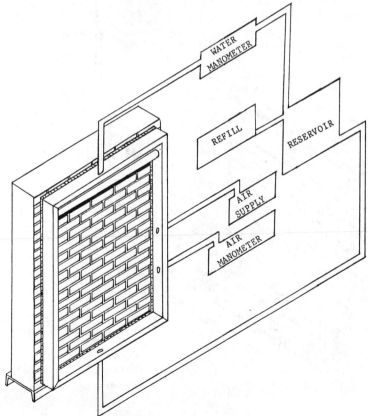

FIGURE 1 -- Standard E-514*
*Drawing adapted from ASTM, Designation E 514, Figure 1

The basic apparatus of Field E 514 is the same as Standard E 514 except the reservoir is placed on a scale and is not refillable. Field E 514 measures the total amount of water which leaves the non-refillable reservoir and labels it as leakage water. Figure 2 is a diagram of the apparatus.

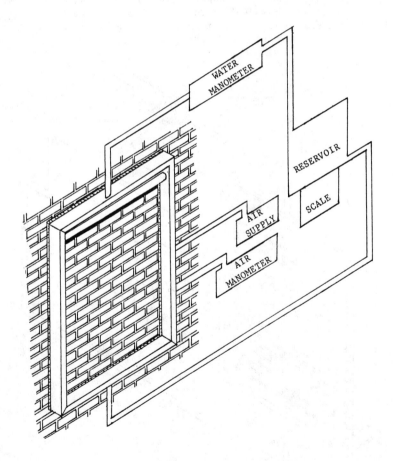

FIGURE 2 -- Field E 514*
*Drawing adapted from ASTM, Designation E 514, Figure 1)

No significance is given in Standard E 514 to water which enters the exterior wall surface unless it passes through the exterior brick wythe. It would be a simple adjustment to Standard E 514 to require a measure of the amount of water which enters the exterior wall surface as is done in Field E 514. However, this adjustment was not made in Standard E 514 since water passing the exterior wall surface, but not through the exterior wythe, is not considered a significant threat to a masonry wall. Water which does not pass through the exterior wythe cannot damage the interior wall system.

PROCEDURE

Mounting Chamber

Standard E 514 requires the test chamber to be attached to the test specimen and sealed around the edges to inhibit water leakage between the chamber and exterior wall. The test specimen is isolated vertically and horizontally by the ends of the test panel. Because the test panel is isolated, it is reasonable to assume that the amount of water which migrates vertically and horizontally beyond the test chamber can be calculated with limited error.

Field E 514 also requires the test chamber to be attached to the test specimen and sealed around the edges to inhibit water leakage between the chamber and exterior wall. However, Field E 514 is run on in-place walls and the test area is not laterally or vertically isolated from the remaining wall. Since the test area is not isolated, it is not known how far water may migrate beyond the test chamber. The amount of water which migrates beyond the test chamber is dependent on the physical properties, size and coring of the brick units, the mortar type, the quality of workmanship, the wall design and configuration, the field dampness of the wall and the climatic conditions during the test. This new variable may result in an inability to provide a rational precision statement and does not permit comparison of results from different tests.

Water Rate and Air Pressure

After the test chamber is in place, Standard E 514 test water is supplied to the chamber at a rate of 138 L/m$^2$/h (3.4 gal/ft$^2$/h) and the air pressure in the chamber is held constant at 500 Pa (10 lbf/ft$^2$) unless otherwise specified. These exposure conditions simulate 13.97 cm (5 1/2 in) of rain per hour with a 100 km/h (62.5 mph) wind velocity which is rarely exceeded in nature for time periods longer than one hour.

Field E 514 does not designate a specifice water flow rate or air pressure. Field E 514 suggests that local climatological data be used for determining water flow rates and air pressures used. Comparative results are not produced when different water flows or air pressures are used.

## Preconditioning

The required preconditioning of Standard E 514-74 and Standard E 514-86 are specified. Standard E 514-74 requires preconditioning of the test specimen for one full day. Standard E 514-86 requires no preconditioning. Field E 514 does not give specific guidance for preconditioning. Comparative results are not produced when different preconditioning times are used.

## RECORD OF OBSERVATIONS

Standard E 514 requires the recording of the time dampness and also the time water first appears on the interior face of the specimen. Standard E 514-74 also requires the recording of the time leakage begins to flow from the flashing, the maximum rate of leakage from the flashing and the time which the maximum occurred. Standard E 514-86 also requires the recording of the area dampness on the interior side of the specimen at the end of 4 h and the total water collected during a 4 h test period.

Field E 514 requires the recording of the rate of water passing the masonry exterior wall face. This rate is recorded by placing the non-refillable water reservoir on a scale. The change in weight of the reservoir is used to calculate the rate of water passing the exterior wall face.

Field E 514 does not differentiate between water which migrates directly behind the test chamber and water which migrates beyond the test chamber. The amount of water which migrates vertically and horizontally beyond the test chamber affects the amount of water which enters the exterior wall surface. Field E 514 does not account for this effect.

## INTERPRETATION OF RESULTS

Standard E 514-74 gives five classifications to rate the results based on the recorded observations; Standard E 514-86 gives no rating system. The classes for Standard E 514-74 are as follows:

Class E:  No water visible on the interior wall surface. No more than 25% of the interior wall surface area damp at the end of 3 days.

Class G:  No water visible on the interior wall surface at the end of 1 day. Less than 50% of the interior wall surface area damp at the end of 3 days. No leaks through the wall at the end of 1 day.

Class F:  Water visible on the interior of the wall surface in more than 3 and less than 24 h.  Rate of leakage less than 1 L/h (.264 gal/h) at the end of 1 day.

Class P:  Water visible on the interior of the wall surface in 3 h or less and a rate of leakage less than 5 L/h (1.32 gal/h) at the end of 1 day.  Water visible on the interior of the wall surface in more than 3 and less than 24 h and a rate of leakage more than 1 L/h (.264 gal) and less than 5 L/h (1.32 gal/h) at the end of 1 day.

Class L:  A rate of leakage through the wall equal to or greater than 5 L/h (1.32 gal/h) at the end of 1 day.

Field E 514-86 gives a tentative suggested rating system as follows:

Expected:  Less than 1 L/h (.264 gal/h)

Significant:  Between 1 (.264) and 5 L/h (1.32 gal/h)

Excessive:  Greater than 5 L/h (1.32 gal/h)

The tentative suggested ratings of Field E 514 go beyond the scope of Standard E 514.  Standard E 514-74 uses a rating system of arbitrary letters with no implication as to which walls perform satisfactorily.  Standard E 514-86 gives no rating system.  Field E 514 which rates a wall with the terms expected, significant and excessive implies the acceptability of the wall's performance.  For example, excessive is defined in Webster's Dictionary as going beyond a normal or acceptable limit.  A rating of excessive therefore, may be construed as an indication of failure which may inspire some sort of corrective action.

No persuasive evidence is available which concludes that walls which pass more than 5 L/h (1.32 gal/h) of water through the exterior face are beyond any acceptable limit.  Due to the many variables inherent in a masonry wall and the complexity of their combination, no single leakage rate can be used as an acceptance criteria for all masonry walls.  In broad terms, these variables are the same as those discussed under Procedure.  This is largely the reason that Standard E 514 dropped the 1974 rating system in its 1986 revision.  For example, it is erroneous to assume that the two following examples, with variable design and workmanship, will have the same likelihood of failure given identical Field E 514 ratings.

A test method must yield reproducible results under similar test conditions, procedures and specimens before the test can be used for comparative studies.  Field E 514 has not been proven reproducible by ASTM standard procedure.

Example 1

A brick and brick wall with a 5.08 cm (2 in) clean and clear airspace between the two brick wythes, properly installed copper flashing at the base and top of the wall with open head joint weepholes spaced 60.96 cm (24 in) o.c. directly above the base flashing.

Example 2

A brick and brick wall with a 1.27 cm (1/2 in) obstructed airspace between the two brick wythes, no flashing or weepholes installed.

It is correct to assume that example 1 wall will dry out faster than example 2 wall and that example 1 wall is less likely to allow water to migrate to the backup wall than example 2 wall. Given these two correct assumptions, it is also correct to assume that example 1 wall is less likely to experience water related damage than exmaple 2 wall even though they have the same Field E 514 rating.


CONCLUSIONS

There are significant differences between Standard E 514 and Field E 514. These differences persuade the author to conclude that Field E 514 is not an adaptation of Standard E 514 but a completely different test warranting a title disassociating it with Standard        E 514. Significant alterations to the purpose, apparatus, procedure, record of observations, the interpretation of results and new variables introduced in Field E 514 lead to this conclusion.

The tentative suggested rating system included in Field E 514 is of particular concern since it may erroneously classify walls. Properly constructed drainage walls, such as cavity walls, are especially susceptible to erroneous classification by Field E 514. The following is quoted from NBS Report BMS 82 by Cyrus C. Fishburn [4]. "Cavity walls are extensively used in England, where they have given excellent protection from wind-driven rains when properly flashed over wall openings and at the bottom so as to divert possible leakage to the outside. When so constructed, the penetration of water through the facings has been of minor importance..." Standard E 514 is based on the test method described in NBS Report BMS 7, co-authored by Cyrus C. Fishburn [5], and the above quoted BMS 82, authored by Cyrus C. Fishburn.

The author is persuaded to believe that the suggested rating system may lead to unwarranted repair of walls which would otherwise never experience significant moisture related damage. An excepted relationship between a specific amount of water which passes through the exterior face of a masonry wall and the likelihood of whether a masonry wall will experience moisture related damage has not yet been established. Until such an excepted relationship is found, it is the author's belief that Field E 514 alone cannot be used to distinguish between in-place masonry walls which will resist damage due to water penetration and those which will not.

REFERENCES

[1]  Monk, C.B., Jr., "Adaptations and Additions to ASTM Test Method
     E 514 (Water Permeance of Masonry) for Field Conditions," Masonry:
     Materials, Properties and Performance, ASTM STP 778, J.G.
     Borchelt, Ed., American Society for Testing and Materials, 1982,
     pp. 237-244.
[2]  E 514-74, Standard Test Method for Water Permeance of Masonry,
     Annual Book of ASTM Standards, Part 18, American Society for
     Testing and Materials, Philadelphia, PA, 1975.
[3]  E 514-86, Standard Test Method for Water Penetration and Leakage
     through Masonry, Annual Book of ASTM Standards, Vol. 04.07,
     American Society for Testing and Materials, Philadelphia, PA,
     1988, pp. 386-390.
[4]  Fishburn, C.C., "Water Permeability of Walls Built of Masonry
     Units", BMS 82, National Bureau of Standards, Washington, DC,
     1942.
[5]  Fishburn, C.C., Watstein, D., and Parsons, D.E., "Water
     Permeability of Masonry Walls", BMS 7, National Bureau of
     Standards, Washington, DC, 18 October 1938.

DISCUSSION

"A Critical Review of Field Adapting ASTM E 514 Water Permeability Test Method" - Maureen T. Brown

QUESTION(Justin Henshell, Justin Henshell Architects) The author's point is well taken. The need for appropriate field tests is important. The operative word is "appropriate".

In order to develop an appropriate test it is necessary to determine its significance and use. What is the purpose of the test?

E 514 states that the test is to obtain information to aid in evaluating "quality of materials, coatings, wall design and workmanship". The test method may be refined to aid in evaluating masonry assemblies in the field provided the caveats in par 3.3 are applied.

The attempt to modify E 514 for use as a field test met with resistance not only because the users tried to apply the test results to evaluate the water resistance of the wall, but because it was used to investigate the cause of reported leakage.

These are two different purposes. A field test that measures the amount of water absorbed in a composite wall has no meaning when applied to a cavity wall or a concrete masonry wall.

Tests that are conducted to determine if a wall leaks, i.e. permits water applied on one face to appear on the other, will help locate the source of leaking but cannot be used to characterize the workmanship of the balance of the wall unless a consistent pattern of leaking is developed.

The developers of an appropriate field test should first determine its purpose - significance and use - and restrict it to that. Two tests are required - one to evaluate workmanship and one to investigate leaking. Then each will become useful tools. It is inappropriate to drive a nail with a screwdriver.


ANSWER(Maureen Brown) Your question and comments help define the appropriateness of the field test. Test results cannot be used for quantitative results unless there is an absolute value which can be used for comparison. None exists for E 514 conducted in the laboratory so none exists for a field adaptation. Separate tests with specific purposes, as you suggest, should be developed.

Robert W. Crooks, Frederick A. Herget

A DISCUSSION OF THE ABUSE OF SOME COMMON MASONRY INDUSTRY PRACTICES

---

REFERENCE: Crooks, R.W. and Herget, F.A., "A Discussion of
the Abuse of Some Common Masonry Industry Practices,"
MASONRY: Components to Assemblages, ASTM STP 1063, John H.
Matthys, Editor, American Society for Testing and Materials,
Philadelphia, 1990.
ABSTRACT: Theory and practical experience have been
combined to develop standard practices for the design and
construction of exterior masonry walls. Such practices have
been presented in the literature and have stood the test of
time. Strict adherence to these practices is critical with
todays thinner masonry walls where abuse often results in
costly repair or replacement of the masonry. Examples of
some of the more typical abuses the authors have observed and
subsequent repairs in the areas of workmanship, moisture
control and differential movement are presented.

KEYWORDS: Differential movement, workmanship, moisture
control, abuses, repairs.

INTRODUCTION

Prior to the early part of the twentieth century the design of
buildings consisted mainly of low-rise buildings framed by wood,
concrete or steel floor systems resting on load bearing masonry
walls. Where the interior of the building was opened up using
columns and beams in lieu of load bearing walls, the exterior of the
building was still constructed of thick, monolithic masonry walls.

As taller and taller buildings were designed the building code
criteria demanded thicker and thicker walls. The construction of
these thick walls (under the existing codes) became prohibitive from
a cost consideration and from a space consideration.

Mr. Crooks is the President of ARSEE Engineers, 14500 E. 136th
St., Noblesville, Indiana 46060; Mr. Herget is a Professional
Engineer with ARSEE Engineers.

Designers then turned to full skeleton frame buildings in which not only the interior of the building was constructed using columns and beams but also the exterior. Exterior walls then only had to span one story in height and could be thinner. Not only could they be thinner they no longer had to be monolithic; that is, they could be constructed using a thin veneer for aesthetic qualities and a more economical back-up material for support.

The skeleton frame buildings were also more flexible since they no longer had the lateral support furnished by the stiff diaphragm consisting of the monolithic exterior masonry walls. In fact, the thinner exterior walls were being supported by the frame rather than the frame being supported by the walls. This support by the frame meant that the walls had to be tied to the frame yet (as we shall see later) be independent of the frame.

Other changes were occurring which affected the performance of the exterior walls. Temperature differentials between the outside face of the exterior walls and its back-up material and/or its supporting framework were becoming greater.

Thin veneers heat up and cool off more quickly than thicker, monolithic walls. In addition, the temperature of the framework to which the thin veneer is tied stays close to the interior temperature of building while the temperature of the brick fluctuates with exterior conditions. This problem was further exacerbated by the introduction and increased use of air conditioning and thermal insulation in walls starting in the 1930's.

The thin, vulnerable masonry veneer must be firmly supported to resist lateral loads, yet these supports must be flexible enough to allow for differential movement between the veneer and the building frame. Thinner walls are less forgiving than the old thicker walls to problems associated with movement, workmanship and moisture control. Proper consideration of these aspects of construction present a challenge to the designer and the craftsman.

The concepts necessary to address such challenges have long been presented in the literature by such well known authorities as the SCPI/BIA, the MIA and the National Bureau of Standards to name a few. These publications have been directed to all aspects of the masonry industry ranging from building codes and textbooks to apprentice training manuals.

Theory and practical experience have been combined in the literature to develop recommended masonry practices. Unfortunately these practices are often abused resulting in the unsatisfactory performance of masonry. This in turn has led to poor performance of the wall system in general, costly repairs and even total replacement of the masonry in some instances.

Three basic areas where such abuses often occur will be addressed. These are: differential movement, workmanship and moisture control.

DIFFERENTIAL MOVEMENT

Differential movement between the exterior brick masonry and the building frame occurs when either the masonry or the frame move relative to the other.  Three major causes of differential movement are:

1) Thermal expansion and contraction of the brick.
2) Moisture expansion of the brick.
3) Lateral and vertical movement of the building frame.

Provisions must be made for such movements or stresses will be induced into the masonry.  If these stresses are large enough, they will crack, crush or spall the masonry.  This in turn allows moisture to enter the wall which can cause portions of the wall to become unstable.

The thermal and moisture expansion characteristics of clay brick have long been documented.  Moisture expansion of brick was reported in the late 1800's [1].  A coefficient of thermal expansion has been reported for brick at least as early as 1930 [2].

Concern by the industry over the potential damage to masonry as a result of frame movements dates back to at least the 1920's.  As early as 1924 training manuals for masons proposed a dry "vibration joint" for use in single wythe brick veneer over wood stud construction to prevent cracks caused by the vibration of the wood frame during a wind storm [3].  This type of construction was primarily limited to residential construction of the time.

Regarding larger structures with thicker masonry walls, Plummer identified the potential problems associated with not providing for expansion of masonry facades in 1939 [4].  He reported that, although theoretically applicable to all portions of a masonry clad structure, problems observed in the field were more generally confined to the parapet walls.  In 1942 Mulligan attributed cracking in tall masonry structures to differential thermal expansion between the brick and frame and to frame movements [5].  In 1954 [6] McBurney attributed cracking in several structures clad with structural clay tile to moisture expansion.

As masonry construction changed and walls became thinner, more instances of cracking were reported in the literature.  In 1956 Myhre attributed cracking in several industrial plant buildings to differential movement and in particular to thermal and moisture expansion of the masonry.  He offered several suggestions for the prevention of such cracking including provision for flexible anchorage, expansion joints and separation of the facade from the building frame [7].

Plummer discussed differential movement at length in 1962 [8] and SCPI expanded this work into the Tech Note 18 series in 1963 [9].  These are excellent resources for the designer and contractor.  They still have not been improved upon.  Adherence to the recommendations in these publications would have prevented all of the problems related to differential movement which the authors have observed to date.

These and other similar references [10 thru 14] discuss in depth the concept of how to provide for differential movements. This generally takes the form of including horizontal and vertical expansion joints in the masonry and the provision for an anchorage which will resist lateral loads on the face of the masonry but is flexible enough to permit movement of the masonry parallel to the plane of the wall.

Examples of abuses of these concepts which the authors have seen include:

1) Buildings where no expansion joints have been detailed or constructed.
2) Expansion joints which are left full or partially full of mortar, raked out, and caulked.
3) Laying of brick copings in mortar beds such that mortar is forced up into the coping expansion joints.
4) Intrusion of debris or other incompressible materials in the expansion joints.
5) Relief angles which completely fill the joint defeating the provision for vertical expansion of the masonry relative to itself and the building frame.
6) Use of inflexible anchorages to support the masonry.

These abuses have resulted in pushing off of the masonry at the corners of the building, cracking, spalling and bulging brick. This in turn allows water to infiltrate and cause further damage to the wall system. Examples of typical distress are shown in Figures 1, 2 and 3.

The authors have been involved in several types of remedial work involved with these types of distress. These have included both total and partial replacement of the masonry. Unstable portions of masonry have been stabilized by drilling a hole thru the veneer and into the back-up and gluing a stainless steel rod to reattach the exterior masonry to the original back-up. An illustration of this is presented in Figure 4.

One of the most common forms of distress associated with differential movement is cracking of the masonry. Tuckpointing, caulking and epoxy injection are methods of repair which the authors have used to repair such distress. Brick dust may be added to epoxy in aesthetically sensitive areas to help hide crack repairs in the brick.

Each structure must be analyzed to determine the potential movements of the masonry and provisions made to accommodate such movement. Inoperative expansion joints must often be opened up and recaulked. New joints can be cut out with the aid of a track mounted saw. These joints must extend completely through the thickness of the veneer to be effective.

FIG. 1 - Bulging brick at building corner where no expansion
joints were provided.

FIG. 2 - Cracking caused by inoperative expansion joints.

FIG. 3 – Cracking caused by lack of provision for differential movement.

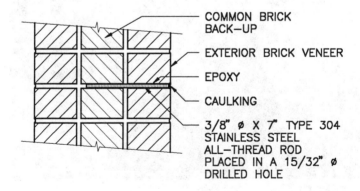

COMMON BRICK BACK–UP

EXTERIOR BRICK VENEER

EPOXY

CAULKING

3/8" ⌀ X 7" TYPE 304 STAINLESS STEEL ALL–THREAD ROD PLACED IN A 15/32" ⌀ DRILLED HOLE

FIG. 4 – An example of reanchoring brick veneer to back-up masonry.

WORKMANSHIP

The term workmanship covers a broad spectrum of activities during construction. Applied to brick masonry workmanship is associated with the construction of the mortar joints and the installation of the metal embedments as well as with the aesthetic qualities of the masonry.

Good workmanship has been emphasized in all forms of masonry literature including building codes, design manuals, technical bulletins and reports, apprentice training manuals, inspectors manuals and even "how to" books.

Since prior to the turn of the century [15] the message has remained the same:  for exterior brick masonry to be strong and durable there must be good workmanship.  Good workmanship includes complete filling and proper tooling of the outer wythe mortar joints.  Furthermore where metal embedments are included they must be completely encased in mortar without voids and without contact between the metal and the brick.

Tests conducted by the National Bureau of Standards in the 1930's and 40's concluded that workmanship was the most important aspect affecting the permeability of masonry walls and that walls with incomplete or open head joints were highly permeable and leaked excessively [16, 17].  More recent publications by prominent research and testing organizations continue to reference these reports [18, 19].

Other excellent publications have stressed not only the necessity of complete mortar joints but also the complete encasement of metal embedments as well [20 thru 24].  The encasement of metal embedments is critical to masonry performance from both the standpoint of anchorage of the metal in the mortar and for corrosion protection of the metal.  These and other references have further emphasized the need to properly tool the mortar joints to seal or compress the mortar against the edges of the brick.

Contrary to such good advice the authors have often observed inadequate workmanship in the field.  Some of the more common deficiencies include:

1) Mortar joints (particularly head joints) left unfilled.
2) Deeply furrowed bed joints.
3) Mortar joints improperly tooled resulting in mortar/brick separations or leaving checks, voids or ledges where moisture can penetrate the wall.
4) Methods of jointing which do not seal the joint and/or do not readily shed water.
5) Incompletely encased metal embedments (voids in contact with the brick).
6) Metal embedments installed too close to the front or back face of the brick.
7) Wall ties improperly spaced, unused or omitted.

These deficiencies have resulted in excessive water penetration of the wall and decreased masonry strength. These in turn can and have led to deterioration of the brick or mortar and to corrosion of the metal embedments leading to instability of the masonry.

Remedial work where such deficiencies exist often takes the form of tuckpointing or waterproofing the exterior of the masonry. These practices are of some benefit but they cannot correct poor workmanship practices. Where wall ties are inadequate, supplemental ties may be installed. A variety of methods to tie back brick veneer exist, unfortunately they are typically labor intensive and therefore expensive. The few dollars saved using inadequate workmanship cannot equal these repair costs.


MOISTURE CONTROL


There are three principle sources of moisture infiltration in an exterior brick masonry wall. These are rain and snow penetration, condensation and capillary action from the ground. Each of these sources must be considered to insure the satisfactory performance of the masonry.

Moisture infiltration can have a deleterious effect on the wall system in general. It can result in deterioration of the brick and/or the mortar. Moisture will lead to corrosion of the metal embedments and the masonry support system. It can cause excessive staining of the masonry and may cause damage to interior finishes if not properly directed back out of the wall.

Workmanship, design and proper maintenance collectively affect the moisture permeance of the masonry. All three are equally important.

In the solid, monolithic brick walls of the late 1800's and early 1900's moisture control depended upon constructing masonry which was impervious to moisture [25]. Such practices included the complete filling of all mortar joints particularly those in the exterior wythe of the wall. In some cases these joints were then raked out and pointed with a Portland cement or high quality lime mortar. Copings, sills and other projections were made of stone, hard-burned brick, terra cotta or vitrified clay tile which provided a barrier to moisture penetration. Damp-proof courses, typically made of hot asphalt or coal tar, were installed just above grade and above the roof line where parapet walls were constructed.

Review of the literature reveals the industry's concern over moisture penetration of exterior brick masonry. With the advent of thin veneer walls which not only provided a shorter path for moisture to penetrate but were more flexible and vulnerable to cracking a second line of defense was conceived to control moisture penetration. This defense was the installation of flashings. In 1939, Plummer and Reardon devoted an entire chapter in their handbook to the subject of moisture control [26]. In it they discussed the

purpose and location of flashings in the wall.  Recommendations
concentrated largely on the more vulnerable portions of the masonry
including parapets, copings, sills, lintels and spandrel beams in the
wall.  In these and other similar publications of the day [27], both
through-wall and concealed flashings were shown.  The use of weep
holes was not mentioned.

Similar recommendations were presented in the 1962 SCPI/BIA
Technical Notes on flashing clay masonry [28, 29].  A note was added
concerning the necessity for the installation of weep holes above all
flashings.  Reference was made to tests at the National Bureau of
Standards which indicated that concealed flashings in tooled mortar
joints are not self-draining without weep holes.

The BIA Technical Note Series 7 was revised and reissued in the
early 1980's [30].  Recommendations are now made that all flashings
should extend beyond the face of the masonry and turn down to form a
drip.  Concealed flashings are referred to as dangerous and are not
recommended.

It is apparent that there has been a gradual evolution in the
approach toward moisture control in clay brick masonry.

While the methods used to control moisture have changed somewhat,
the basic concepts have long been known.  Still problems associated
with moisture penetration continue to plague the masonry industry.
It is the authors' opinion that the majority of these problems are a
result of the abuse of these basic concepts.  Examples which the
authors have observed include:

1) Flashings never installed at all.
2) Cap flashings improperly designed or installed.
   - Flashing stops short of the face of the brick.
   - Flashing does not extend a sufficient distance down over
     the face of the brick.
   - Flashing extends to far out from the face of the brick
     permitting wind blown rain to blow up underneath the
     flashing.
3) Continued use of concealed flashings.
   - Many of which are stopped over the brick cores.
4) Seams of flashings left unsealed allowing water to circumvent
   the flashing.
5) Ends of flashings not turned up or dammed at the ends of
   lintels.
6) Mortar droppings fill the air cavity preventing drainage.
7) Weep holes improperly installed.
   - Weep holes covered with mortar and caulking.
   - Weep holes never installed at all.
   - Weep tubes too short to reach back into the air cavity.
8) Use of brick in horizontal applications without provision for
   positive drainage.
   - Copings and tops of garden walls.
   - Brick roofs.
9) Use of non-durable brick in severe weathering climates.
10) Coating of brick with a non-breathable material which traps
    moisture in the wall.

The results of such deficiencies are often severe.

Excessive moisture penetration can lead to deterioration of the brick in the form of cracking, crazing and spalling. Once such deterioration is initiated, it cannot be stopped (or reversed) and delamination continues farther back into the brick exacerbating the problem of moisture penetration. See Figure 5.

Susceptibility of brick to freeze/thaw damage resulting from moisture penetration in freezing climates can be detected by tests.

The current ASTM Standard C216-87 evaluates the durability of face brick based upon its absorption characteristics and the compressive strength of the brick [31]. Unfortunately this test does not always detect non-durable brick. New methods which appear to more reliably predict the durability of face brick have been developed and presented in the literature [32 thru 34].

Excessive moisture penetration may also cause deterioration of the mortar. An example is shown in Figure 6.

Moisture intrusion often results in corrosion of the metal embedments and support system whenever water and oxygen come in contact with the metal. Examples of deterioration the authors have observed include:

1)  Total loss of cross-section of galvanized wall ties.
2)  Total loss of cross-section of painted metal steel and runner support systems.
3)  Corrosion of cadmium plated expansion bolts.
4)  Corrosion of painted relief angles where no flashings were provided.
5)  Total loss of cross-section of metal studs.

Such deterioration has been observed to occur in as little as three years from the time of construction.

Moisture intrusion can also result in excessive and unsightly staining of the masonry. Where adequate barriers are not provided, moisture can also cause damage to interior finishes.

Where moisture control problems exist they are often the result of a combination of several deficiencies in the wall system rather than the result of a single problem. This can make repair efforts more complex in that each deficiency must be evaluated for its contribution to the present problem and to future problems if the deficiency is not corrected.

Since so many factors affect the moisture control of a brick masonry wall, remedial work where this is a problem varies greatly and can range from supplementing the existing maintenance program to complete removal and replacement of the masonry and flashings.

Some of the more common methods for addressing moisture control problems of which the authors are aware include:

FIG. 5 – Spalled masonry resulting from freeze/thaw action.

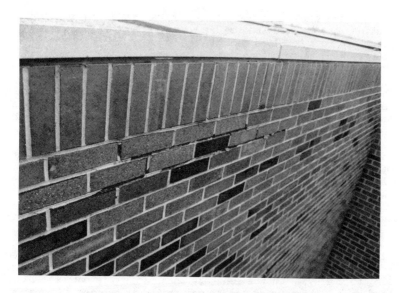

FIG. 6 – Deteriorated mortar joints at a coping.

1) Applying a waterproofing material to the brick.
2) Tuckpointing the mortar joints.
3) Drilling new weep holes or unplugging existing ones.
4) Covering horizontal or nearly horizontal brick surfaces with a metal cap.
5) Replacement of horizontal or inclined brick surfaces with stone supplemented by proper flashings and weeps.
6) Spot removal of only visibly distressed masonry. (This may not address the long-term problem if for example the brick are determined by testing to be inherently non-durable).
7) Total removal and replacement of the masonry, flashings, and in some instances, the support system(s).

Some of these methods may eliminate the problem while others may only mitigate it to some degree. The benefits must be weighed against the costs for each level of repair. Care must be taken such that the repairs themselves do not become problems. Some repairs may actually exacerbate the existing problem or even create new ones. An example of this would be the use of a waterproofing material over a marginal or non-durable brick when the material traps moisture in the brick and thereby accelerates freeze/thaw deterioration. All ramifications of a repair must be analyzed before the repair is attempted.


CLOSURE


Through the years, the use of clay brick masonry in exterior applications has evolved from being a major load bearing portion of the structure to a non-load bearing veneer supported by the structure. With these thinner walls has come greater economy but at the expense of a greater potential for problems.

The performance of the masonry is dependent on design, materials, workmanship and maintenance. This fact has been well documented in the literature. As a result various masonry practices have evolved and have been recommended to all sectors of the industry.

While the practices may have changed somewhat through the years the basic concepts have long been established. Yet it is evident, both in the field and in the literature, that abuse of these basic concepts and practices continues. Such abuse typically results in deterioration of the masonry. Subsequent repairs are costly and may require removal and replacement of the masonry itself. These repair costs far exceed the few dollars saved as a result of such abuses.

In contrast to most other cladding materials when well designed, constructed and maintained masonry mellows with age and becomes more pleasing to the eye. On the contrary when masonry is not well designed and/or constructed and/or maintained it becomes an eyesore and possibly a hazard to public safety.

The masonry industry needs more effective methods of delivering this information to people in decision making positions.

REFERENCES

[1] Watertown Arsenal, "Expansion of Bricks in Water," Report of the Tests of Metals and Other Materials for Industrial Purposes at Watertown, Mass., Fiscal year ending June 30, 1896.

[2] Moore, H. F., Materials of Engineering, McGraw-Hill Book Company, Inc., New York, 1930.

[3] Graham, F. D. and Emery, T. J., Audels Masons and Builders Guide, No. 1, Theo. Audel and Company, New York, 1924.

[4] Plummer, H. C. and Reardon, L. J., Principles of Brick Engineering - Handbook of Design, Structural Clay Products Institute, Washington, D.C., 1939.

[5] Mulligan, J. A., Handbook of Brick Masonry Construction, McGraw-Hill Book Company, Inc., New York, 1942.

[6] McBurney, J. W., "Masonry Cracking and Damage Caused by Moisture Expansion of Structural Clay Tile," Proceedings of the American Society for Testing and Materials, ASTM, Philadelphia, Vol. 54, 1954, pp. 1219-1238.

[7] Myhre, A. M., "Maintenance of Industrial Buildings," Modern Masonry Natural Stone and Clay Products, National Research Council Publication 466, Washington, D.C., 1956.

[8] Plummer, H. C., Brick and Tile Engineering, Structural Clay Products Institute, Washington, D.C., 1962.

[9] "Differential Movement," Technical Notes on Brick and Tile Construction, Nos. 18, 18A & 18B, Structural Clay Products Institute, Washington D.C., 1963.

[10] Principles of Clay Masonry Construction - Student's Manual, Brick Institute of America, McClean, Virginia, 1973.

[11] "Cracking in Buildings," Building Research Digest, No. 75, Building Research Station, Garston, Waterford, England, 1966.

[12] Sorensen, C. P. and Tasker, H. E., Cracking in Brick and Block Masonry, Technical Study No. 43, Experimental Building Station, Australian Government Publishing Service, Canberra, 1976.

[13] Thompson, J. N. and Johnson, F. B., "Design for Crack Prevention," Insulated Masonry Cavity Walls, Publication No. 793, National Academy of Sciences, Washington, D.C., 1960.

[14] Grimm, C. T., "Design for Differential Movement in Brick Walls," Journal of the Structural Division, Proceedings of the American Society of Civil Engineers, Vol. 101, No. ST11, November, 1975.

[15] International Library of Technology - Masonry, Carpentry, Joinery, International Textbook Company, Scranton, Pennsylvania, 1899.

[16] Fishburn, C. C. et al, "Water Permeability of Masonry Walls," Report BMS-7, National Bureau of Standards, Washington, D.C., October 18, 1938.

[17] Fishburn, C. C. , "Water Permeability of Walls Built of Masonry Units," Report BMS-82, National Bureau of Standards, Washington, D.C., April 15, 1942.

[18] Essentials of Good Brick and Tile Construction, Structural Clay Products Institute, Washington, D.C., 1964.

[19] Water Permeance of Masonry, Construction Technology Laboratories, Skokie, Illinois, 1980.

[20] Structural Clay Products Institute and Rau, F. W., Brick Laying II, Delmar Publishers, Albany, 1951.

[21] Recommended Practice for Engineered Brick Masonry, Structural Clay Products Institute, McClean, Virginia, November, 1969.

[22] Amrhein, J. E., Masonry Design Manual, Masonry Industry Advancement Committee, 1972.

[23] Pocket Guide Brick and Tile Construction, Structural Clay Products Institute, Washington, D.C., 1964.

[24] Building Code Requirements for Reinforced Masonry, National Bureau of Standards Handbook 74, Sectional Committee on Building Code Requirements and Good Practice Recommendations for Masonry - A41, Washington, D.C., October 21, 1960.

[25] International Library of Technology - Masonry, Carpentry, Joinery, International Textbook Company, Scranton, Pennsylvania, 1899.

[26] Plummer, H. C. and Reardon, L. J., "Moisture Control," in Principles of Brick Engineering - Handbook of Design, Structural Clay Products Institute, Washington, D.C., 1939, pp. 139-150.

[27] Stoddard, R. P., Brick Structures - How to Build Them, McGraw-Hill Book Company, Inc., New York, 1946.

[28] "Flashing Clay Masonry," Technical Notes on Brick and Tile Construction, No. 7A, Structural Clay Products Institute, Washington, D.C., 1962.

[29] "Moisture Control in Brick and Tile Walls," Technical Notes on Brick and Tile Construction, No. 7B, Structural Clay Products Institute, Washington, D.C., 1965.

[30] "Water Resistance of Brick Masonry," Technical Notes on Brick Construction, Nos. 7, 7A & 7B, Brick Institute of America, Reston, Virginia, 1985.

[31] "ASTM C216-87 Standard Specification for Facing Brick (Solid Clay Units Made From Clay or Shale)," 1988 Annual Book of ASTM Standards, American Society for Testing and Materials, Philadelphia, 1988.

[32] Maage, M., Frost Resistance and Pore Size Distribution in Bricks, Institut fur Bygningsmaterialleare Rapportnummer, BML 80.201, University of Trondheim, Norway, May, 1980.

[33] Crooks, R. W., Kilgour, C. L. and Winslow, D. N., "Pore Structure and Durability of Bricks," Proceedings of the 4th Canadian Masonry Symposium, Department of Civil Engineering, The University of New Brunswick, Fredericton, N.B., Canada, June, 1986.

[34] Winslow, D. N., Kilgour, C. L. and Crooks, R. W., "Predicting the Durability of Bricks," ASTM Journal of Testing and Evaluation, JTEVA, Vol. 16, No. 6, November, 1988, pp. 527-531.

Albert A. Tomassetti

PROBLEMS & CURES IN MASONRY

REFERENCE: Tomassetti, Albert A., Problems & Cures in Masonry - Masonry: Components to Assemblages, ASTM STP 1063. American Society for Testing & Materials, Philadelphia, PA. 1990.

ABSTRACT: Although masonry construction is one of the best types of building methods known to man for cost, durability, beauty, and maintenance, it is subjected to certain problems, on occasion, due to short-comings in design, workmanship, and misuse of materials.

Some of the subjects covered are vari-colored mortar joints, efflorescence, blush (caused by surface coatings), lime deposit (also called lime runs and lime spots), metal wall caps, expansion due to thermal and moisture changes, creep, cleaning, stains, and mortar dry-outs.

All of the problems mentioned here are supported by slides of examples taken from "real" buildings and are coordinated with the material contained in this paper.

KEYWORDS: Vari-colored, Blush, Lime Spots, Creep, White Scum, Dry-Outs, Initial Rate of Absorption (I.R.A.), Striping, Breathing, Cavity, Efflorescence, Spalling.

This paper - titled "Problems & Cures in Masonry" - is the written portion of a collection of slides, taken over the past 25 years, of situations that have caused physical problems to masonry buildings, or caused the appearance to be unacceptable to the design professional and/or owner. It describes many different situations and conditions that commonly occur in masonry construction.

They are: Vari-colored Mortar Joints; Efflorescence; Blush; Lime Spots; Metal Wall Caps; Condensation; Wall Movement - Cracks; Loose Material in Collar Joints (Cavities); Moisture Penetration; Cleaning; Stains - Manganese - Vanadium - White Scum; Dry Outs; Rust from Wire Reinforcement; Mortar Joints in Pavers.

Al Tomassetti is the Executive Director of the Kentuckiana Masonry Institute, Inc., 130 Fairfax Ave., Louisville, KY 40207

1. <u>Vari-colored Mortar Joints</u> may well be the most common problem in masonry, especially in the Eastern half of the U.S. where rain and/or snow occur frequently.

A vari-colored mortar joint condition is one that has mortar joints that vary from light to dark in appearance, yet are on the same wall that was built with the same materials. They are caused by the tooling of joints at varying degrees of moisture - wet mortar will dry light, drier mortar will become darker. This is a result of the variation of the moisture in the units at the time they are laid in the wall, and/or a variation in their absorption rate.

All masonry materials should be protected from the weather from the time they are manufactured until they are placed in the wall. The top of all walls should be covered at all times, other than when working on them, until the coping, or cap, is completed. Of course, this includes all concrete masonry (C/M) units in order to keep them in compliance with the moisture requirements of C90. Clay units must also be covered in order to maintain a constant or uniform moisture condition. If they require wetting because of a high Initial Rate of Absorption (I.R.A.) - typically those that exceed 30 grams per minute - then they should be wet down thoroughly the day before they are to be used and kept covered with a tarpaulin or plastic cover until time to use them.

When an uncovered pallet of brick gets partially wet, those units - when laid in the wall - show small but distinct sections of vari-colored mortar joints. This compounds the problem when we attempt to correct  or at least minimize the problem.

The wetter units provide moisture for a longer period of hydration (curing) which results in those joints attaining higher strength compared to the darker, drier ones. Therefore, the lighter, stronger joints have greater resistance to the cleaning solution when this work is cleaned down. This will result in a difference in the degree of etching, if extreme care is not taken by the person doing the cleaning.

Some commonly used practices to correct or at least minimize this problem, are severe etching with a hydrochloric acid base cleaning solution, "striping", and surface treatments.

Etching with the acid base cleaner helps minimize the problem by removing an excessive amount of the surface paste (matrix) and exposing a lot of the aggregate (sand). The color of the sand then becomes a factor in the finished color through light reflectivity. This is a severe cure because it deteriorates the unit-mortar interface and takes away many years of abrasion resistance from the surface of the mortar, as well as changing the color originally desired by the architect, and/or owner.

"Striping" is a technique of painting the mortar joints with a paste of neat cement (of that used in the mortar) with a toothbrush. This practice has been used successfully in the Southeast states, but is an expensive procedure because of the labor involved.

Surface treatment with stains is also expensive and does not always give the desired results. This system should be investigated and tested prior to its' use. One such practice used in the Southeast is to wash the wall down, then apply a solution made of pigment and mineral spirits. It does help make the wall look more uniform, but it is a "band-aid" solution because in a reasonably short period of time the mineral spirits will evaporate and the pigment will be washed away. Using a proprietary stain - applied with a toothbrush is another alternative, pre-testing is advised.

The best solution is to do the work right the first time and avoid this problem. Uniformity is the key word.

The finished color of a mortar joint is greatly influenced by the absorption of the units, and the color of the units. A sample panel is always recommended, especially with colored masonry mortars.

While we are speaking of mortar joints, this writer would like to make a point for esthetics and recommend any type of mortar joint be used if it will enhance the beauty of the wall. Many sources only recommend the use of a concave or V joint because it helps make the wall a little more water repellent - and that is true. However, "more water repellent" is just that - they may reduce moisture penetration, but do not eliminate it.

Therefore, why not choose the type of joint that increases the beauty of the wall, in combination with the choice of brick and design, and construct the wall using a " breathing cavity system". This type of wall takes into consideration that a 4 inch brick wall will not stop moisture penetration completely, provides ways for it to evaporate from the cavity - as vapor - and prevents any moisture from entering the interior of the building at the same time.

We will discuss the breathing cavity wall in more detail later.

2. Efflorescence is also a very common problem. However, with very few exceptions, it is not destructive and is only present part of the time, usually Spring and Fall. Efflorescence is described by the Brick Institute of America (BIA), as "A powder or stain sometimes found on the surface of masonry, resulting from the deposition of water soluble salts."

There are some folks in our industry, and in academia, who call many types of surface stains "efflorescence". However, for the purpose of this discussion we will refer to it as described in the above paragraph. Lime spots (deposits) and various types of stains will be treated separately later.

In the opinion of the writer, efflorescence, with very few exceptions, does not do any physical harm to a masonry wall unless it is one built with glazed units, has an applied surface coating too high in solids content, or painted with a non-breathing paint that inhibits vapor transmission.

As a general rule efflorescence does not require the use of any special cleaning material to remove it. Left alone it will simply go away if it is exposed to normal wind and rain, as the wall dries out. If a section must be cleaned, a stiff fiber scrub brush and a bucket of water, with a little elbow grease, will remove it very well.

As mentioned earlier, efflorescence generally occurs during the Spring and Fall of the year. If soluble salts are present in any of the products that make up a masonry system they remain there until a sufficient amount of water penetrates the surface to dissolve them. These salts will then migrate toward the surface of the wall, depending on drying conditions.

If drying conditions are slow, as they are in cool weather, (Spring and Fall), the dissolved salts will have time to migrate to the surface and be deposited there, quickly turning to powder. If drying conditions are fast (like Summer), the rate of evaporation of the moisture exceeds the rate of movement of the salts and they are left behind the surface. The point at which the salts remain is called the plane of evaporation. This theory was developed by a Canadian researcher, and is very logical.

The best way to avoid an efflorescence problem is to "Think Water". It is an over simplified statement, but without water present efflorescence does not occur. We encourage the design professionals to keep this statement in mind. To use overhangs where possible, use proper flashing, and flashing details, build breathing cavity exterior walls with a minimum of one inch, (preferably two inch), clear airspace, between the back of the masonry unit and the backup wall. If the architect and/or owner decides to use a water repellent exterior coating, they should be sure it has a vapor transmission ratio that is compatible with the system.

3.   Blush is a whitish film that forms as moisture and salts migrate to the surface of the masonry units and is trapped under the coating. It generally cannot be removed without first removing the coating, which will require an acceptable solvent.

Blush occurs when a high solids water repellent surface coating is applied to a brick wall, or brick pavers, and inhibits the movement of moisture to the surface. There are many water repellent products and sealers on the market and it is important that the architect have the proper information from the manufacturer as to the solids content, and vapor transmission ratio, of a product before he uses it. If this information is not available, a good rule of thumb is to use a product that does not contain more than 6% solids - and test it before final approval.

A very simple test to determine if there is moisture in a wall or paver system, before applying any surface treatment, is to tape a piece of plastic wrap at least 12 inches square to the surface and observe it for a day or so to see if any moisture forms under the plastic film. If it does, a coating of questionable solids content should not be applied at that time.

In many cases a coating for pavers is selected to give the pavers a "wet look", or a sheen. A product that will produce these results is almost certain to contain a high percentage of solids - like 15% or more. Pre-testing and discretion should be used.

4. Lime Spots also known as Lime Runs, or Lime Deposits, are sometimes mistakenly called efflorescence. The difference between them is that efflorescence is a powder that can be brushed away, whereas a lime spot becomes a hard, white formation that must be dissolved by an acid base solution, or physically removed.

Lime spots on a masonry wall or floor come from the mortar system, whether it be a masonry cement, or lime and cement mortar. The lime is formed when the mortar is kept saturated for a prolonged period of time, above a level of 75% Relative Humidity. This condition permits the hydration process to continue as long as it remains at that level or higher, and in so doing produces Calcium Hydroxide as a by product of hydration. Cement chemists tell us that a given amount of portland cement, fully hydrated, will produce between 12% to 22% by weight of the cement as calcium hydroxide. The difference in the range is determined by the $C_3S$ and $C_2S$ content of the cement.

When this lime product, which is in liquid form, reaches the surface of the masonry it combines with the carbon dioxide in the atmosphere and hardens as calcium carbonate. A high lime content in a portland cement lime mortar will increase the potential for lime deposits, if the conditions described above are present.

5. Wall Caps and Masonry Coping can be, and are, common causes of masonry problems. Aluminum caps have two faults, one is that they have an expansion potential about three times that of brick or block, and when they expand they break the joints and/or the sealant at the joints, allowing them to permit water penetration. According to NCMA Tek 53, and BIA Technical Notes 18, a lightweight C/M unit will expand .52 inches per 100 lineal feet per 100 deg. F. temperature change. A clay unit will expand .43 inches, and aluminum will expand 1.54 inches, under the same conditions. There are certainly many joint designs that can, and do, perform satisfactorily and it behooves the design professional to work out those details with the metal fabricator, prior to installation.

Another common fault is when the metal cap is attached to a wooden plate on the top of the wall and only covers an inch or two of the top course of the wall. This condition can permit water to blow up under the cap and enter the wall. A study done by the Portland Cement Association has shown that wind driven rain hitting against a wall will rise one inch for each 10 miles per hour of wind.

Caulking the bottom side of a metal cap, at the masonry face, can help prevent this problem, but it is much safer and longer lasting protection to cover enough of the wall with the cap to avoid this problem.

Masonry caps such as stone or precast concrete coping, brick rolok, or brick soldier caps can also cause problems. Stone or precast concrete coping that are set in mortar, over thru wall flashing, invariably permit moisture penetration through the head and bed joints. The mortar system, whatever it may be, rarely, if ever, attains sufficient bond in the head joints to make them impermeable. The bed joints generally crack, usually at the bottom of the joint, soon after the coping is laid. This is usually due to the small area of bond from the outer edge of the joint to the point where the flashing lays on the top of the wall - this is usually 1/2" to 3/4" - and/or from the thermal and moisture expansion of the coping units.

A safe way to set stone or precast coping is to set them in mortar, then rake all joints (head and bed) a depth of 3/4". When the mortar has dried sufficiently, place a backer rod in each joint and caulk it with a reputable sealant.

If the architect and/or owner insists on using a rolok, or soldier, as a cap, then it is in their best interest to take steps to prevent moisture penetration through them, and into the wall. It is advisable to rake and caulk the bed joint (as above) especially if it is over thru-wall flashing. It is also advisable to rake out, and caulk, a head joint across the width of the wall, every 8 to 10 feet, to act as a control joint.

Most importantly, with a rolok or soldier cap, is to apply a coating that contains enough solids to completely seal the top surface of the wall. If not it will surely permit moisture to penetrate it, for no presently available mortar system will provide a total bond to each unit to make the cap impermeable.

6. Condensation can form on the interior side of an outside wall if the humidity in the room is high, all, or most of the time, and there is no ventilation system to remove it. Condensation that forms against windows travels to the sills and finds its way through the sills and mullion strips to the cavity and outside wall. In case of the inside face of walls, the moisture can cause plaster or drywall to "blister" and pop nails.

Condensation also can form in the cavity of masonry walls and walls built with steel studs and brick veneer. These particular problems have been well documented by others.

I have seen these problems occur in areas with inside swimming pools and in hospital therapy rooms that have hot tubs and baths in them. This problem can be corrected, or avoided, if a competent mechanical engineer will estimate how much air needs to be removed from the room, at constant or given intervals, to take the moisture out of the area and keep the moisture from forming on, and in, the walls and on windows.

7. Wall Movement - Cracking. There are many things that can cause a wall to crack, like foundation or footer settlement, and thermal and moisture changes. This paper will confine its remarks to thermal and moisture changes in concrete, concrete masonry, and clay brick.

We encourage our friends in the design profession to think control joints in concrete masonry (C/M) to accommodate shrinkage after placement, and the subsequent expansion and contraction that will take place due to thermal and moisture changes.  However, these changes do not cause the concrete or C/M system to exceed its original placement size.  With this type of control joint it is safe to use mortar in the joints and rake them out to accommodate the backer rod and caulking material that will be placed in them later.  Special details, based on recommendations published by PCA and NCMA, on how to construct control joints should be followed.

With clay brick, and other fired clay products, we encourage the designer to use expansion joints to accommodate the movement that takes place after the walls, or floors, etc., are in place. Mortar or other materials that can harden and inhibit the expansion of the clay units should not be used in the expansion joints.  It will, of course, be necessary to place a backer rod and caulking compound in the joint to make it resistant to moisture penetration. BIA recommendations should be followed.

8.  Insulation in Masonry Walls.  There is a great awareness today, more than ever, to use insulation in some form in all exterior wall systems.

The two most common types of insulation used in masonry wall systems appear to be rigid and loose fill materials.  Various types of foam systems are also being offered, and we will include them in our comments on loose fill insulation products.

It is the opinion of the writer that loose fill products and foam products that completely fill the cores and/or collar joint space should be used with discretion in a wall system that has any possibility of getting water in it. If water can reach loose fill insulation, or foam, in a collar joint or cores of the masonry unit, it has the potential to saturate the material and keep it wet, or damp, for a long time - if not forever.  Neither product may absorb any water but it can trap it and prevent it from drying out.   In the case of loose fill material it may not only become wet, but the moisture can cause it to compact itself into the lower portion of the wall.

The problem we have observed on numerous occasions, is that the exterior face of the wall, brick or block, remains damp or wet for long periods of time and in areas that are subjected to freeze-thaw conditions this can cause spalling of the surface of the units.  We have no quarrel with these products, only that they can create problems if they are subjected to excessive moisture, in a wall system where the moisture cannot dry out in a reasonable period of time.

Rigid insulation serves well when it is placed in the wall cavity, usually in front of the back up wall with an air space of at least 1 inch to 2 inches between the rigid insulation and the back of the facing units.  Rigid inserts in the cores of C/M units also function satisfactorily for they, as well as the rigid insulation in the wall cavity, have an air space around them that allows any moisture that reaches it to dry out, or evaporate, in some way, even if it has to migrate through the wall - but does not keep the masonry saturated for long periods of time.

9. Creep, as it occurs in concrete frame buildings, has caused damage to masonry by causing the brick to fracture under extreme pressure (compression). Creep has also been called "column shortening" by some engineers. The shortening of the building columns in multi-storied concrete frame building, places extreme vertical presssure on the units at the shelf angles and can cause spalling of the masonry at that point.

This problem is controlled by placing compression joints, usually strips of 1/2" rubber under the shelf angle before it is set in place. In order to avoid the appearance of a thick bed joint at the shelf angle, a lipped brick is used to accommodate the height of the compression joint and the shelf angle.

10. Suggestions to Control Moisture in Masonry Walls. Most masonry walls, especially 4 inch brick and C/M veneer walls are vulnerable to moisture penetration. This statement should not be cause for concern to the designer or the building owner for there are ways to design and build masonry walls to accommodate this moisture without causing any ill effects on the structure, and still maintain the beauty and integrity of the building.

Most masonry walls being built today consist of single wythe (thru wall) C/M, or clay units, or brick veneer over C/M, or steel or wood stud, backup systems. All of these systems can be built to control moisture penetration to the building interior.

With C/M single wythe walls the most common practice is to treat the exterior face of the wall with a water repellent coating like a cement base paint, or similar type product that prevents moisture penetration but still permits the wall to breathe. That is to maintain a vapor transmission low enough to allow any moisture that forms, or penetrates the exterior to migrate to the exterior face without damaging the surface of the units.

With brick or C/M veneer walls it is important to design and build a "breathing cavity wall". This type of system permits any water that penetrates the 4 inch veneer - and it will - to leave the wall without causing any problems. A typical "breathing cavity wall" system with C/M backup is one that has the block backup laid up first, then waterproofing the cavity side with a 3/8" parging of mortar or an application of a waterproof coating. Although the asphaltic or other type coating forms a total vapor barrier, and mortar parging does not, it is the preferred treatment because the architect can visually inspect it before allowing the veneer wall to be built.

After the coating is applied satisfactorily, rigid insulation - either 1 inch or 2 inches thick, depending on the architect or engineer's requirements - should be attached to the outside face of the coated C/M wall. Flashing, which should have been set in place when the block were laid, should be wide enough to accommodate the insulation, a 1 inch or 2 inch air space (2 inches preferred), the width of the brick and extend at least 1/2" beyond the wall face.

It is also advisable to seal the laps of the flashing with a mastic compatible with the flashing material.

The next recommendation is that when the bricklayer is laying out the bond, he place a cotton rope (weep hole) in the bottom of the bed joint, of the first course, at 24 inches center and secure the back of the rope against the rigid insulation.  This rope will be removed when the wall is completed.  On the second course we recommend that a weep hole ventilator be placed in head joints on 24 inch centers.  This will permit air to enter and circulate in the cavity through the ventilators and the hole left by the rope will permit any water, if it accumulates, to leave the wall.

At this point it is recommended that two or three inches of pea gravel be placed behind the brick - in the cavity.  This will prevent any mortar droppings from forming a contiguous barrier at the base of the wall  to block access to the weep hole.

The final step in constructing a "breathing cavity wall" is to place brick size ventilators at six to eight feet intervals within 18 inches of the top of the wall or stopping point - like a shelf angle.  If the designer does not like the appearance of brick size ventilators, weep hole ventilators placed in head joints at two foot intervals will also serve well.  They must also be placed within 18 inches of the top of the cavity wall.

This type of wall system performs exceptionally well.  It permits any water that penetrates the outside wythe to dry out from both sides without causing any interior problems.  It also eliminates, or minimizes, the potential for efflorescence to form.

If the backup system is metal or wood studs the insulation is generally placed between the studs with a sheathing board attached to the exterior side.  This contributes a little rigidity and a little more insulation.  It is highly recommended that 15 or 30 pound felt roofing paper be placed over the sheathing to provide a moisture barrier.  The paper should be lapped a minimum of 4 inches horizontally and where it laps vertically.

The placement of flashing, rope weeps and weep hole ventilators, pea gravel at the bottom  and weep hole ventilators or brick vents at the top  should follow the same steps as described with C/M backup walls and brick veneer.

11.  Cleaning. There are many good papers and tech notes on cleaning masonry.  Two in particular are BIA Technical Notes 20 and one published by the Brick Association of North Carolina.  Here we only want to deal with problems that occur from poor cleaning practices and procedures.

A lot of problems come from cleaning clay masonry with muriatic acid in the hands of unskilled and/or incompetent workmen. The most common problems that result from poor cleaning procedures, and muriatic acid, are acid burn, white scum, manganese stains, and vanadium stains.  Let's take them in that order.

Acid burn generally occurs when the cleaning solution made from muriatic acid is too strong, is applied to a wall that was not prewet sufficiently, and is left on the wall too long. Once this stain occurs it cannot be removed with the same solution that caused it to happen. There are a number of proprietary products available that will successfully remove acid burn. Tech Note 20 also includes recommendations.

White scum is a white film that forms on the surface when a muriatic acid solution has a chance to dry before the wall is rinsed. The scum is the result of the cement paste that was dissolved by the acid solution and allowed to dry. This residue then cannot be removed by a repeat washing of the same acid solution. A special proprietary product is available from a number of sources and Tech Note 20 makes some suggestions also.

We highly recommend that mason contractors and masonry cleaning contractors use the cleaning solution that is recommended by the brick manufacturer. If in doubt, they should use a proprietary cleaning solution that is made especially for the type of unit and the color of the unit they are using. These products are basically a mixture of hydrochloric acid, a refined form of muriatic acid that has a much higher purity level, by comparison, and a blend of detergents and chelating agents. These last two products help the acid dispersion and keep it working on the surface.

Some of these proprietary products also contain metallic inhibitors that prevent the formation of manganese stains and vanadium stains, when used according to the manufacturers suggestions. They are highly recommended when cleaning, brown, tan and light or white colored brick.

Manganese is a product used in brick manufacturing to attain a given range of colors, such as browns and tans. Brick made with manganese are sound and color fast under normal use, however, manganese is soluble in muriatic acid. Muriatic acid solutions improperly used - and many times they are - can cause this brown colored stain to form on the surface of the wall. It usually looks like tobacco juice that forms around the edge of the brick and sometimes runs down over the mortar joints.

Vanadium stain is caused by the dissolving of vanadium salts present within the clay used to manufacture the brick. Like manganese it is not a problem under normal use and normal conditions. It is also highly soluble in muriatic acid and when dissolved leaves a greenish colored stain on the surface as it dries.

Both manganese and vanadium stains can be prevented by using a proprietary cleaning solution that contains inhibitors especially designed for that purpose. Both stains, when they appear, can also be removed by proprietary solutions made for that purpose. BIA Tech Note 20 is, as always, a good source of information.

While we are on the subject of cleaning, I want to point out a few other things that relate to the subject. One is the removal of mud stains that occur at the bottom of nearly every masonry wall - where it rains frequently. We also want to talk about stains from scaffold planks and sand blasting.

Mud stains are not easily cleaned by standard cleaning solutions with a hydrochloric acid base. They are, however, soluble in a solution made of hydrofluoric and phosphoric acid. This type product is available in proprietary form from a number of manufacturers as "restoration cleaner". This product should be handled with care - not only because it is made of acid - but because it will etch polished surfaces like glass, granite, marble, etc. These materials must be protected while this solution is being used.

The scaffold plank nearest the wall can cause a problem if rain strikes the plank and splashes against the wall carrying mortar droppings from the plank's surface with it. This mortar splash has the potential to harden and become extremely difficult to remove. All this can be avoided if the plank nearest the wall is turned back, away from the wall when the scaffold is not in use.

Sand blasting is occasionally used to "clean" new buildings as well as old ones. In truth, this practice is not a cleaning procedure but a destructive method that removes mortar smears (on new buildings), or dirt (old buildings), by removing a portion of the surface of the masonry units at the same time. Whether used in a dry form or with water, sandblasting should rarely, if ever, be used to clean a masonry surface. There are many high quality chemical cleaning materials available to clean masonry surfaces that are very effective. We do not use sand paper to remove dirt from our hands or face - we use soap or a special cleaning solution. Why not apply these same practices to clean our buildings?

12. Dry Outs occur when the water in the mortar system is removed so rapidly that the mortar has little, if any, chance to hydrate and thus gain strength. All mortar systems are made with portland cement as part of the composition, and portland cement is a hydraulic cement product. By way of definition a hydraulic cement product is one that combines with, and hardens in, the presence of water, and the chemical process it follows is called hydration. We think of and take steps to cure portland cement concrete, but we hardly ever think of curing masonry walls, floors, or pavements made with a portland cement base mortar, yet it is equally important.

Cement technicians tells us that a mortar system must have an internal moisture content of 75% Relative Humidity (R.H.) to continue to hydrate. Below that moisture level and below 40 degrees F the hydration process ceases. It is hydration that takes cement paste from a plastic to a crystalline state - and strength gain.

Many of the brick manufactured today have an Initial Rate of Absorption (I.R.A.) of 30 grams per minute, or less, when tested according to the requirements of ASTM C67-87. These units normally do not cause dry outs. Those units that have a higher I.R.A. are usually wetted prior to use in the wall and do not normally cause a problem either.  Those units that have a high I.R.A. and are not wet prior to use in the wall can cause a dry out as well as slow down productivity of the mason.

Dry outs occur more frequently in concrete masonry walls, floors, and pavements because it is not advisable to wet them before laying them in mortar.  C/M units undergo a volume change (increase) due to moisture and thus have limits placed on allowable moisture content in the specifications.  ASTM C90-85 places limits on the moisture content of the units, at the time they are laid in the wall, based on the materials used to manufacture them.

C/M units that are "bone dry" and laid up in hot, windy weather can lose most of the moisture in the mortar in a matter of minutes - and thus experience a dry out.  Experience has shown that if this occurs, it is possible to rejuvenate the hydration process by wetting down the wall, after it is in place - and thus under restraint.  If the dry out is severe it is advisable to wet the wall down thoroughly, then cover it with a plastic cover.  This prevents the moisture from drying out and keeps the temperature at an elevated level (Hot House effect), all of which helps accelerate the hydration of the cement in the mortar, and thus gain strength.

13.  Rust is an occasional problem in masonry walls and is not only unsightly but can become destructive.  There are a number of metal products that are sometimes used in a wall, like bolts, angle irons for lintels, shelf angles, and wire reinforcement. Because of its wide use, wire reinforcement has the greatest potential to cause problems.

Wire reinforcement comes in three types, Brite, Mill Galvanized, and Hot dipped.  With relation to this problem (rust), it is advisable not to use Brite wire in any wall that is exposed to moisture.  It is usually rusting when in the storage yard and has great potential to continue to rust in the wall.  It is an accepted rule of thumb that a metal object can increase its size 2 1/2 times by rusting.  We have seen this condition not only pop out mortar joints but the faces of brick.  We have also seen walls bulge due to the vertical pressure exerted by expansion of the rusting metal.

Mill galvanized wire is much better, and safer, to use than Brite, but it too can cause problems if it is not placed properly in the mortar joint (not too close to the surface), and is not completely enclosed in mortar.  Mill galvanized wire is most likely to start rusting at the welds because the heat of welding removes a lot of the zinc coating during fabrication.

Hot dipped is the best, and of course the most expensive wire reinforcement to use because it is treated after fabrication and completely covered with the zinc coating.

14.  Grouting brick pavers. There are a number of ways to grout pavers, or place mortar joints between them - some are effective and perform satisfactorily, others are not.  Whether or not the pavers are set in mortar over a concrete base, or are set

in sand over a compacted dense grade base makes a difference in whether we should use mortar joints or not.

One of the best systems to use for exterior brick paving is to set them in sand, over a compacted dense grade base, without any mortar in the joints. This method practically eliminates any potential problems, especially water related problems, and is virtually maintenance free except for the occasional need for a weed killer.

Exterior paving over a concrete base, with the pavers set in mortar can cause problems with efflorescence, cracked head joints and freeze-thaw movement. In areas of the country where it rains, we must consider the fact that some water will penetrate the mortar joints and get into and under the pavers. Without special provisions for a drainage system under the pavers, moisture may become trapped and can only leave the masonry by evaporating upward through the mortar joints and the units.

In many cases the designer and/or owner will choose to apply a water repellent coating to the surface - generally for two reasons. One, to help reduce moisture penetration, and the other to give the surface a "wet look". To help reduce moisture penetration, the system should have full mortar joints between pavers and use a surface coating that will help repel water. A water repellent product must also have a vapor transmission ratio that will allow any moisture that does enter the masonry system to migrate - and evaporate - without any adverse effects to the surface. Pre testing such a product is highly recommended.

One must understand the difference between penetrating sealers and surface coatings and their effect on the masonry system - as well as considering the choice of not using any surface treatment.

There are many types of penetrating sealers, such as silicone resins, silanes, siloxanes, and many others that are absorbed into the masonry surface and do not change the appearance of the surface to any appreciable degree. They reduce, but may not totally stop, penetration of moisture. One of the advantages of using such a product is that they have a good vapor transmission ratio and permit moisture that gets into the system to evaporate safely.

Surface treatments are generally products that contain a high solid content - in the range of 15% to 25% or more - and give the surface a "wet look" or "shine". They are effective in reducing moisture penetration to a considerable degree, as well as changing surface appearance. However, there is sure to be some moisture penetration of the surface through minute cracks, etc. When that moisture dries out, it will carry whatever soluble salts are available in the system, as well as calcium hydroxide, to the surface.

When this happens the high solid coating will restrict the moisture from reaching the surface and cause the salts, or lime deposit, to be deposited under the coating. This will produce a whitish film that is commonly called "Blush" which necessitates the removal of the surface coating before one can do anything to remove the salts, or lime deposit that caused the blush.

In many cases it may be advisable to leave the surface "as is" and not do anything to it. The natural look can perform very satisfactorily in most instances.

A mortared paver system can work successfully in interior floors providing the floor is dry before the coating is applied. One method to check a floor, or wall, for moisture is to take a piece of plastic wrap, at least 12 inches square and tape it tightly to the surface. If there is moisture in the system it will show up as condensation under the plastic cover in a day or so. Do not apply a high solid surface coating until you are sure the floor is dry.

There are two mortaring systems that I want to mention that are "No, No's" in my opinion. One is the practice of sweeping a mixture of dry cement and sand over the surface and into the joints between the pavers - then wetting the surface thoroughly. This practice is expected to provide enough moisture to cause the cement to hydrate and harden sufficiently to resist wear. I have never seen it work successfully. It usually hardens to a depth of 1/8 to 1/4 inch and in a matter of time breaks up with normal foot traffic. It makes no difference whether it is used inside or outside; in my opinion it is not a good practice to follow.

Another poor practice is to set pavers with 3/8 or 1/2 inch spacing, then literally dump plastic mortar on the surface and move it around and tuck it into the joints. It is possible to force enough mortar into the joints to fill them, but the problem this system raises is how to remove the mortar from the wearing surface of the pavers. Unless the brick were previously coated with wax or some type of surface treatment that absolutely prevents the penetration of mortar into the pores of the brick, it is literally impossible to remove the mortar smears. Repeated acid washing will destroy interface bond and deeply etch the mortar in the joints. This is not a good practice to follow.

Two good ways to fill mortar joints in pavers is to use a tuck pointers "cookie bag" or a caulking gun - both work well.

Conclusion. The problems listed and explained in this paper are those that the author has investigated over many years. There are, no doubt, many others that my colleagues in the masonry industry have seen that are not included here. Those, plus any disagreements, will be a good basis for future discussion.

It is the intent of the author to identify these problems, and where possible explain how to correct and/or avoid them. In my opinion all of them can be avoided through the cooperation of the designer, mason, and material manufacturers and distributors.

The slide presentation, from which this paper was taken, also contains a number of examples of properly designed and built buildings that are performing exceptionally well - as they should!

Problems occur on a very small percentage of the tremendous numbers of masonry buildings built each year. However, there is no reason why we cannot reduce that number, or eliminate problems entirely, by sharing our knowledge. Together we can share our knowledge on the proper way to build with masonry and enjoy strong, durable, beautiful, fire resistant, and virtually maintenance free buildings.

Walter A. Laska, Charles W. Ostrander,
Robert L. Nelson and Colin C. Munro

RESEARCH OF PHYSICAL PROPERTIES OF MASONRY
ASSEMBLAGES USING REGIONAL MATERIALS

---------------------------------------------------------

**REFERENCE:** Laska, W. A., Ostrander, C. W., Nelson,
R. L., and Munro, C. C., **"Research of Physical
Properties of Masonry Assemblages Using Regional
Materials,"** Masonry: Components to Assemblages,
ASTM, STP 1063, American Society for Testing and
Materials, Philadelphia, 1990.

ABSTRACT: A masonry research program was
conducted in the Metropolitan Chicago market to
determine certain physical properties of wall
assemblages incorporating ASTM C652 [1] brick and
different sizes of ASTM C90 [2] concrete masonry
units. The purpose of the program is to make
design professionals aware of testing masonry
assemblies and the realistic values that can be
obtained by using locally available masonry
materials. This is necessary many times due to a
lack of confidence or the absence of knowledge
on the part of some designers of the adherent
strength that masonry material possess.
This paper consists of the following sections:
Brick and concrete masonry units   ASTM C652 [1],
ASTM C90 [2], C140 [3]
Mortar   ASTM C270 [4]
Concrete masonry unit prisms   ASTM E447 [5]
Brick prisms   ASTM C1072 [6]

   Walter Laska is the Staff Architect and Charles
Ostrander the Executive Director for the Illinois Masonry
Institute, 1550 Northwest Highway, Suite 201, Park Ridge,
Illinois, 60068; Robert L. Nelson is President of Robert
L Nelson & Associates/Masonry Laboratory, 856 Courtbridge
Road, Inverness, Illinois, 60067; Colin C. Munro is a
Chartered Masonry Consultant and principal in The Masonry
Laboratory, 113 South Jefferson Street, Batavia,
Illinois, 60510.

Masonry is designed and constructed to perform as a
homogeneous system, not as individual components. When
designing a structure, an engineer should be aware of the
realistic physical properties of masonry components and
how they perform in conjunction with each other. Using
realistic values will enable the engineer to design more
efficient structures while maintaining a high degree of
confidence.

A research program was conducted in the Chicago
Metropolitan area with the intention of generating
masonry test data which reflected the actual properties
expected under field conditions. This data would then be
made available through publication to design
professionals to assist them in their understanding of
the actual design capacity of masonry.

BRICK AND CONCRETE UNITS ASTM C652 [1], C140 [3]

All tests performed complied with ASTM Standards.
Masonry materials selected (brick units and concrete
masonry units) were subjected to standard ASTM testing
(C140 [3], C652 [1]) to actuate physical properties,
therefore assuring valid test results.

All the brick chosen for the test program were Hollow
Brick units with a nominal width varying from  4in.
(101.60mm) to 6in. (152.40mm). The selection of this unit
type was based upon the recent emergence of single-wythe
design utilized in masonry construction. The results of
all the units tested are listed in Table 1, and conformed
with the physical requirement listed in ASTM C652 [1]
under designation SW Table 1.

TABLE 1.-- Physical properties of
face brick units (average of 5 bricks)

---

| SIZE | | |
|---|---|---|
| Length in (mm) | 11.62(295.15) | 11.62(295.15) |
| Width in (mm) | 3.50 (88.90) | 5.62(142.75) |
| Height in (mm) | 3.62 (91.95) | 3.62 (91.95) |
| WEIGHT - Dry Lbs (kg) | 5.54  (2.51) | 8.25  (3.74) |
| WATER ABSORPTION | | |
| 24 Hour % (g) | 6.0 | 6.3 |
| 5 Hour Boil % (g) | 6.4 | 6.6 |
| SATURATION COEFFICIENT | 0.93 | 0.95 |
| COMPRESSIVE STR.- PSI(MPa) | 9455(65.19) | 9178(63.28) |

---

The specimens conform to ASTM C652 [1] - Grade SW with
exception (see ASTM C652 [1] paragraph 5.1).

The concrete masonry units tested were 6in. (152.40mm),
8in. (203.2mm) and 12in. (304.80mm) in width. These units
were obtained from a randomly selected concrete block
manufacturer and subjected to sampling and testing in
accordance with ASTM designation C140 [3]. All units were
in conformation with the physical property requirements
listed in table 1 and table 2 of ASTM C90 [2], Grade N
Type I. See table 2.

TABLE 2.-- Physical properties of concrete
masonry units (average of 3 units)

-------------------------------------------------------------

```
SIZE
Length in              15.62      15.62      15.62
      (mm)           (396.75)   (396.75)   (396.75)
Width in               5.62       7.62      11.62
      (mm)           (142.75)   (193.55)   (295.15)
Height in              7.62       7.62       7.62
      (mm)           (193.55)   (193.55)   (193.55)
Faceshell in           1.0        1.25       1.25
      (mm)            (25.4)     (31.75)    (31.75)
Web in                 1.25       1.0        1.0
      (mm)            (31.75)    (25.40)    (25.40)
NET AREA
sq in                 40.2       59.0       92.5
(sq cm)             (259.35)   (380.64)   (596.77)
GROSS AREA
sq in                 87.7      119.0      135.0
(sq cm)             (565.80)   (767.74)   (870.97)
MOISTURE CONTENT
% of total Absorption  27.2       27.9       28.3
WATER ABSORPTION
Lbs/Cu Ft             11.2       10.9       11.8
(Kg/cubic meter)    (179.41)   (174.60)   (189.02)
COMPRESSIVE STRENGTH - psi(MPa)
Gross            1305(9.0)   1385(9.55)  1290(8.89)
Net              2846(19.62) 2770(19.10) 2531(17.45)
DENSITY
Lbs/Cu Ft            133.4      135.6      138.4
(Kg/cubic meter)   (2136.86)  (2172.10)  (2216.95)
```

-------------------------------------------------------------

The above units conform to ASTM C90.


MORTARS ASTM C270 [4], ASTM C780 [7]

A partial intent of this program was to simulate a
project engineer and/or architect's interpretation of
ASTM C270 [4] methods for specifying a mortar for his or
her particular project. Two mortar batches were prepared
in a laboratory, meeting the requirements of ASTM C270

[4]. The batches consisted of a Portland Type I cement
(See Table 3), and Type S Hydrated Lime (See Table 4) and
a mason sand conforming to the requirements of ASTM C144
[8].
   The analysis below is typical, as received from the
cement mill. Tested according to ASTM C150 [9].


   TABLE 3.-- Chemical and physical anaylsis of type I
                   Portland Cement

---

                    Chemical Analysis
                    SiO2        20.9%
                    Al2O3        5.3
                    Fe2O3        2.3
                    CaO         64.9
                    MgO          1.3
                    SO3          3.2
                    L.O.I.       1.1
                    Na2O         0.09
                    K2O          0.70
               Total Alkali as
                    Na2O Equiv. 0.60
                    C3S         57.4%
                    C2S         16.6
                    C3A         10.1
                    C4AF         7.1


                    Physical Analysis
           325 Sieve      85.1% pass
           Wagner         1800 cm2/g
           Blaine         3350 cm2/g
           Vicate:     Initial   100min
                       Final     200min
           Gillmore:   Initial   115min
                       Final     220min
           False set (Paste Method)   84%
           Normal Consistency:   25%
           Mortar Air:   10.6%

           Compressive Strength psi (MPa)
               1 day       2200 (15.17)
               3 day       3650 (25.17)
               7 day       4600 (31.72)
              28 day       5800 (39.99)

---

           Autoclave Expansion   0.01%

TABLE 4.-- Results of chemical anaylsis type "S" lime

|                                             | Sample |
|---------------------------------------------|--------|
| Loss on Ignition %                          | 27.3   |
| Carbon Dioxide, $(CO_2)$%                    | <0.5   |
| Free Water $(H_2O)$%                         | 0.0    |
| Sulfur Dioxide $(SO_2)$%                     | <0.1   |
| Combined Iron/Aluminum Oxides $(RO_3)$%      | 0.40   |
| Calcium Oxide $(CaO)$%                       | 42.1   |
| Magnesium Oxide $(MgO)$%                     | 28.3   |
| Calculated Unhydrated Oxides                | 0.0    |
| Calcium Oxide/Magnesium Oxide Ratio         | 1.48   |

The mortar batches were prepared in accordance with the proportion secification requirements, then tested to determine conformance to Table 2 of ASTM C270 [4].

The results of the laboratory mixtures prepared in the laboratory using a Hobart Lab Mixer are listed in Table 5.

TABLE 5.-- Results of mortar test (laboratory mix)

| Mortar Type                        | N            | S             |
|------------------------------------|--------------|---------------|
| Proportion                         | 1:1:6        | 1:1/2:4 1/2   |
| Air Content (%)                    | 3.1          | 3.8           |
| Water Retention (%)                | 92.4         | 91.4          |
| Flow                               | 106          | 106           |
| Compressive Strength psi (MPa)     |              |               |
|    7 days           | 1320(9.10)   | 1745(12.03)   |
|    28 days          | 1790(12.34)  | 2410(16.62)   |

All specimens were stored in accordance to C270, section 5.2.1.

These results indicate that the compressive strengths far exceed the minimum requirements listed in specification C270 [4], table 2.

As the program continued two more batches of mortar utilizing the same materials were prepared in a commercial mortar mixer with a 9 cu. ft. (0.25cubic meters) capacity. One type N and one type S mixture was batched using the proportions established in the laboratory test as outlined in ASTM C270 [4]. The

proportions were 1:1:6 and 1:1/2:4 1/2. The proportions
of these mixes were slightly altered to accurately
simulate field conditions. The hydrated lime content of
these mixtures was 25% higher than the laboratory
controlled mixtures. This is due to the fact that
proportions for field mixing are determined by the volume
of material contained in a bag (a bag of lime is 25%
higher in volume than a bag of portland cement). All
tests performed on these two mortar batches were in
accordance with specification C780 [7]. The results of
the tests are listed on table 6.

TABLE 6.-- Results of mortar test (field mix)

| MIX NO. | 1-N | 3-S |
|---|---|---|
| TYPE | 1:1 1/4:6 (N) | 1:5/8:4 1/2 (S) |
| SAND Cu Ft | 6.0 | 4.5 |
| (cubic meter) | (0.17) | (0.13) |
| BATCH AMOUNTS | | |
| Cement Cu Ft | 1.0 | 1.0 |
| (cubic meter) | (0.028) | (0.028) |
| Lime Cu Ft | 1.25 | 0.62 |
| (cubic meter) | (0.035) | (0.018) |
| WATER Gals | 13.0 | 11.0 |
| (liters) | (49.21) | (41.64) |
| CONE PENETROMETER | 53mm | 47mm |
| AIR CONTENT | 5.3% | 6.9% |
| COMPRESSIVE STRENGTH-28Days | | |
| (3x6) psi (MPa) | 1432(9.87) | 2055(14.17) |
| (2") psi (MPa) | 2046(14.11) | 2711(18.69) |
| MIX TIME | 12 min | 8 1/2 min |

All specimens were stored in accordance to C780, section
A6.6.

Comparison of the mortar properties obtained through
the two different mixing procedure reveals that the field
mixed batches obtained a consistently higher compressive
strength than the laboratory mixed mortar despite having
a higher lime content and a higher air content.

CONCRETE MASONRY PRISMS ASTM E447 [5]

All concrete masonry unit prisms were built and tested
in accordance with specification E447 [5], "Method A".
The mortar used to construct all the prisms was obtained
from the commercial mixer (see table 6). Six groups of
concrete masonry units prisms were built using 6in.
(152.40mm), 8in. (203.20mm) and 12in. (304.80mm) width

units and type N and S mortar. Compressive values were
determined and are shown on table 7.

TABLE 7.-- Results of prism test

| PRISM NO. | MIX NO. | SIZE IN (mm) | f'm psi (MPa) |
|-----------|---------|--------------|---------------|
| CMU-1 | 1-N | 5-5/8×16×15-5/8 (142.88×406.40×396.88) | 2238(1543) |
| CMU-2 | 1-N | " | 1691*(11.66) |
| CMU-3 | 1-N | " | 1455*(10.03) |
| CMU-4 | 1-N | 7-5/8×16×15-5/8 (193.68×406.40×396.88) | 2076*(14.31) |
| CMU-5 | 1-N | " | 2491*(17.18) |
| CMU-6 | 1-N | " | 1813*(12.50) |
| CMU-7 | 1-N | 11-5/8×16×15-5/8 (295.15×406.40×396.88) | 1915(13.20) |
| CMU-8 | 1-N | " | Broken |
| CMU-9 | 3-S | 5-5/8×16×15-5/8 (142.88×406.40×396.88) | 2064(14.23) |
| CMU-10 | 3-S | " | 2450(16.89) |
| CMU-11 | 3-S | 7-5/8×16×15-5/8 (193×68×406.40×396.88) | 2076(14.31) |
| CMU-12 | 3-S | " | 1728(11.91) |
| CMU-13 | 3-S | 11-5/8×16×15-5/8 (295.15×406.40×396.88) | 1702(11.74) |
| CMU-14 | 3-S | " | 1970(13.58) |

* Prisms failed due to the web fracturing.

Two assumptions can be made after reviewing the results
of the prism tests. The allowable values of compressive
strength for concrete masonry that now exist in several
masonry codes are considerably lower than the actual
values obtained from prisms (in this test program) using
mortar made in the commercial mortar mixer (see table 6).
The average compressive value of three 8in. (203.2omm)
wide concrete masonry unit prisms built with type N
mortar is 2127 psi (14.67MPa). According to the ASCE-ACI
530.1 code [10], a compressive value of 1844psi
(12.71Mpa) can be assumed (through interpolation) for
prisms consisting of a type N mortar and a concrete
masonry unit with the same physical properties listed in
table 2. The allowable f'm value determined from the NCMA
code [11] is 1124psi (7.75Mpa) and the allowable f'm
value determined from ACI 531-79 code [12] is 1110psi
(7.65Mpa). These conservatives values can result in the
over design of masonry structures therefore impairing
masonry's cost effectiveness. It should also be noted
that these test results indicate that there is no
significant difference in compressive strength between

prisms built with type N or type S mortar.

Nearly 40 % of the prisms tested experienced failure due to the webs fracturing. This can be attributed to the method in which the prisms were capped. All prisms were built with fully bedded face shells. All webs were void of any mortar joints. The prisms were capped with plaster in accordance with specification C140 [3]. In doing so, the entire net surface area of the prism is capped, not just the bearing area (as call for in the specification). This can not be avoided due to the manner in which this capping procedure is specified and administered. This is also true for capping test specimens with sulfur and granular materials. Now, when the prism is vertically loaded the forces are transferred through the concrete masonry units shells but not the web. This unequal distribution of the loads creates distress or a camber effect on the concrete masonry units webs, causing the webs (in many cases) to rupture, and the prism to fail.


BRICK PRISMS ASTM C1072 [6]

Ten brick prisms were constructed from mortar batches made with the commercial mixers (see table 6). Type S and type N mortars were used. All prisms were built in stack bond consisting of five 3/8" (9.53mm) thick mortar joints and six ASTM C652 [1] brick. The brick were fully bedded with the mortar being tooled, with a concave joint, on one side.

After twenty eight days the prisms were tested in accordance with ASTM C1072 [6]. The prisms were positioned with the tooled joints facing inward. Inspection revealed that approximately 67% of the failures occurred at the top of the mortar joint and 33% of the failures occurred at the bottom of the mortar joint. The flexural bond strength of the mortar joints are indicated in table 8.


Table 8.-- Results of flexural bond test

| PRIMS NO.* | MIX NO.** | SIZE | PSI (MPa) |
| --- | --- | --- | --- |
| FB-1 | 1-N | 4in.(101.6mm) | 56(0.39) |
| FB-2 | 1-N | 4in.(101.6mm) | 61(0.42) |
| FB-5 | 3-S | 4in.(101.6mm) | 68(0.47) |
| FB-6 | 3-S | 4in.(101.6mm) | 53(0.37) |

\* Average Five Mortar Joints
** See table 6

The results of the test indicate an average strength of 60.5psi (042MPa) for type S mortar and a average strength of 58.5psi (0.40MPa) for type N mortar. The building code requirements for engineered brick masonry [4] allows a minimum flexural-tensile strength (normal to bed joints and without inspection) of 24psi (0.17MPa) for type S mortar and 19psi (0.13MPa) for type N mortar. These values are considerably lower than those obtained through the testing program.

CONCLUSION

Due to the results of the test program several assumptions and suggestions can now be made:

The specification for mortar for units masonry (ASTM C270 [4]) as now written is both inaccurate and misleading. New values for compressive strengths must be determined to reflect actual strengths which will be achieved out in the field. These values must be incorporated into a revised specification.

Engineers should conduct prism tests on all loadbearing projects to obtain accurate compressive strengths of masonry (f'm). In doing so more appropriate mortars can be utilized to satisfy a variety of structural and design conditions.

Due to the numerous amount of web failures which occured during the testing of concrete masonry unit prisms, further study should be conducted to determine test procedures which would represent field conditions and which would yield results with a greater degree of confidence. The areas which should be evaluated are (1) capping materials used to level a prism when the weight of the prism is in excess of 75lbs. (2) determining the net area to be used for calculating of f'm when only the face shell is bedded with mortar.

Based on the results of the flexural bond strengths obtained thru the testing program, a minimum ASTM requirement for flexural strength should be established, but only after a complete evaluation of brick and mortar combinations have been made.

REFERENCES

[1]   American Society for Testing and Materials:
C652, "Specifications For Hollow Brick (Hollow
Masonry Units Made from Clay or Shale)", ASTM
Standards, Vol. 04.05.

[2]   American Society for Testing and Materials: C90,
"Hollow Load Bearing Concrete Masonry Units"
ASTM Standards, Vol. 04.05.

[3]   American Society for Testing and Materials:
C140, "Sampling and Testing Concrete Masonry
Units", ASTM Standards, Vol. 04.05.

[4]   American Society for Testing and Materials:
C270, "Mortars for Unit Masonry", ASTM
Standards, Vol. 04.05.

[5]   American Society for Testing and Materials:
E447, "Standard Test Methods for Compressive
Strength of Masonry Prisms", ASTM Standards,
Vol. 04.07.

[6]   American Society for Testing and Materials:
C1072, "Standard Method for Measurement of
Masonry Flexural Bond Strength", ASTM Standards,
Vol. 04.05

[7]   American Society for Testing and Materials:
C780, "Preconstruction and Construction
Evaluation of Mortars for Plain and Reinforced
Unit Masonry, "ASTM Standards, Vol. 04.05.

[8]   American Society for Testing and Materials:
C114, "Standard Methods for Chemical Analysis of
Hydraulil Cement", ASTM Standards, Vol. 04.01

[9]   American Society for Testing and Materials:
C150, "Standard Specification for Portland
Cement", ASTM Standards, Vol. 04.01

[10]   ASCE ACI 530.1, Specifications for Masonry
Structures, Section 1, Table 1.6.2.2, The
American Society of Civil Engineers and
American Concrete Institute, 1988

[11]   Specifications For The Design Of Loadbearing
    Concrete Masonry, Chapter 3, Table 3-1,
National Concrete Masonry Association,
ebruary, 1987

[12]   ACI 531-79, Building Code Requirements for
    Concrete Masonry Structures (Revised 1983) and
Commentary, Part 2, Chapter 4, Table 4.3,
American Concrete Institute, March 1987

John H. Matthys

CONCRETE MASONRY PRISM AND WALL FLEXURAL
BOND STRENGTH USING CONVENTIONAL MASONRY MORTARS

REFERENCE: Matthys, John H., "Concrete
Masonry Prism and Wall Flexural Bond Strength
Using Conventional Masonry Mortars," Masonry:
Components to Assemblages: ASTM STP 1063, J.
H. Matthys, Ed., American Society for Testing
and Materials, Philadelphia, 1990.

ABSTRACT: ASTM C270-86a "Mortar For Unit
Masonry" lists different types of mortar
(i.e., M, S, N, O) that can be produced from
either a proportion or a property
specification. This is the first ASTM
standard that has allowed the use of masonry
cements alone in production of all mortar
types via the proportion specification.
There has been concern by some members of the
building community with regard to masonry
assemblage performance of such mortars. The
Construction Research Center at the
University of Texas at Arlington has
conducted an investigation of the flexural
bond strength of stack bonded prisms using
ASTM C 1072 Bond Wrench Technique and full
scale 4' X 8' walls. This paper describes
the results of prisms constructed using eight
masonry cement mortars and three portland
cement lime mortars. Prisms are evaluated at
28 days and six months age. Both laboratory
curing and outside curing are examined. The
wall tests evaluate two Type S mortars at 28
days.

KEYWORDS: concrete masonry, flexural bond,
bond wrench, conventional mortar, portland
cement lime, masonry cement, transverse
tests, lateral wall load

Dr. John H. Matthys is Professor of Civil
Engineering and Director of the Construction Research
Center at The University of Texas at Arlington, Box
19347, Arlington, Texas 76019-0347

The Construction Research Center at The University of Texas at Arlington under sponsorship from the National Lime Association conducted a comparative performance study of Portland cement lime (Type S lime) mortars to ASTM (American Society of Testing and Materials) equivalent masonry cement mortars [1]. The equivalency used in this investigation was based on the ASTM C-270-86a [2] proportions by volume specification. The reason for this study was simply:

Under the latest revisions of ASTM C-270-86a any listed mortar type for either unreinforced or reinforced masonry is allowed to be produced using masonry cement alone by the proportion specifications. Several groups in the building industry are concerned whether the mortars produced using commercially available masonry cements are equivalent to Portland cement lime mortars with respect to critical masonry assemblage performance characteristics such as masonry bond and masonry water tightness. Many knowledgeable people in the building community feel that such masonry cement mortars are not equivalent to Portland cement lime mortars even though the masonry cements themselves meet the material specifications in ASTM C-91.

The project consisted of evaluating laboratory mortars and field mortars. Field mortars were used with a standard modular 3-cored clay brick with an initial rate of absorption of approximately 25 gms./min./72 cm.$^2$ to construct, test, and evaluate masonry assemblages in compression, shear, flexural bond, and water permeance. Based upon the conditions of construction, materials, testing, and evaluation the data appeared to support the conclusion that (1) with regard to only mortar property criteria requirements of ASTM, like mortars produced using either MC or PCL according to ASTM proportion criteria are equivalent; (2) with regard to masonry assemblage property performance of like proportioned mortars, the MC and PCL are not equivalent particularly with respect to shear strength, flexural bond strength, and water penetration resistance. Results of the 28 day prism flexural bond strengths were published in reference 3; results of the six months prism flexural bond strengths were published in reference 4; results of the 28 day shear strengths (diagonal tension) were published in reference 5. Additional information is needed on the comparative performance of concrete block if one is to address all of the most common units used in masonry construction.

SCOPE OF INVESTIGATION

The program reported here was developed based on the following concerns:

1.  ASTM C-270 currently allows any type of mortar
    to be produced by using MC alone.  Are these
    mortars equivalent to PCL mortars with respect
    to flexural bond strength and water permeance
    for concrete block assemblages?

2.  Some current building codes give the same or
    higher allowable bond strength for concrete
    block as compared to clay brick for a given
    mortar binder; i.e., either PCL or MC.

3.  Some current codes provide for only a 25%
    reduction in allowable flexural bond for
    concrete block masonry when MC is used instead
    of PCL as compared to a 50% reduction for the
    case of clay brick.

4.  Most flexural bond strength data is generated
    from small scale specimens via ASTM E-518
    and/or ASTM C-1072.  Very little data exists on
    a direct comparison of prism test to full scale
    panels as defined in ASTM E-72 [6].

To address the above concerns the Construction
Research Center at UTA under sponsorship by the
National Lime Association has conducted two separate,
yet related, projects involving MC, PCL, and concrete
block.

(1) The first project "Conventional Masonry Mortar
    Investigation of Concrete Block" addressed the
    concern of mortar equivalency for concrete
    block with respect to flexural bond strength
    and water penetration.  The portion of this
    project reported in this paper deals only with
    the "prism study" of flexural bond strength at
    28 days age.

(2) The second project "Flexural Bond Strength of
    Concrete Block Assemblages" specifically
    addressed the performance of concrete block
    prism flexural bond strength versus full scale
    concrete "block wall" flexural bond strength at
    28 day age.

A survey of concrete block used throughout the
U.S.A established selection of 20.3 cm. X 20.3 cm. X
40.6 cm. medium weight concrete block as manufactured
under ASTM C-90 [7] for construction of all
assemblages.  The mortars to be examined for the prism
study basically consisted of the eight MC (four Type S
and four Type N) and three PCL (one Type S, one Type N,
and one Type O) used in the previous clay brick study
[1].  The mortars to be examined for the wall study
would be the Type S PCL and one of the Type S MC used
in the prism study.

The flexural bond strength prism performance would be evaluated based on single wythe stack bonded prisms tested according to ASTM C-1072 [8] at 28 days and six months. Only the 28 days results are presented in this paper. The flexural bond strength of 1.2 m. X 2.4 m. walls (full scale walls) would be evaluated based on the transverse uniform loading system in ASTM E-72 conducted at test age of 28 days.

SPECIMEN CONSTRUCTION AND TEST PROCEDURES

The mortars used in both the prism and wall study were produced from the proportion specification of ASTM C-270 and mixed to field flow of 120 to 130 using a 0.34 cubic meter commercial paddle type mixer. Half the sand and half the estimated water were gradually charged into the mixer. Total elapsed time equalled four minutes. The cement and the lime (or masonry cement) were gradually added during a four minute interval. During the next four minutes the remainder of the sand with additional water was added. A flow test was run. Additional water was added if needed with a final one minute mixing. Cone penetrometer readings were taken. Air contents were measured using a pressure meter.

For the prism study twelve 5.1 cm. compression cubes were made for each mix. Six cubes were cured in an environmentally controlled laboratory, three to be tested at 28 days, three to be tested at six months. Six cubes were cured outside and covered with polyethylene for 28 days, three to be tested at 28 days and three to be tested at six months. For five months they were exposed directly to weather. For each MC mortar 15 three-unit high stack bonded block prisms were constructed using face shell bedding. Nine MC prisms were built in the environmentally controlled laboratory, six to be tested at 28 days and three to be tested at six months. For each PCL mortar 18 three-unit high stack bonded block prisms were constructed using face shell bedding. Twelve PCL prisms were built in the environmentally controlled laboratory, nine to be tested at 28 days and three to be tested at six months. Six MC and PCL prisms were built outside and wrapped with polyethylene for 28 days, three to be tested at 28 days and three to be tested at six months. All prisms were built by the same two masons. For prism construction retempering of the mortar was allowed. All prisms were tested for flexural bond strength according to ASTM C-1072. See Figure 1 for sketch of bond wrench machine.

Two Type S mortars, one a PCL and one a MC used in the prism study, were produced by the proportion specification of ASTM    C-270 for use in the wall

study.  Six cubes for each mortar type were cast and
cured in lab with assemblage specimens until testing at
28 days.  Air content, flow, and penetrometer readings
were taken on each mortar.  All assemblages were
constructed by the same two masons using face shell
bedding and 0.95 cm. concave tooled joints.  Nine one-
unit wide stack bonded prisms and three 1.2 m. wide by
2.4 m. high single wythe walls in one half running bond
were constructed.  All mortar in these assemblages was
from the same batch with retempering being allowed as
required by the masons.  All specimens were built by
the same two masons in the environmentally controlled
laboratory air (72 degrees F., 55% RH).  Specimens were
cured in the laboratory 28 days prior to testing.  The
wall and prism specimens for each Type S mortar were
constructed in less than a two hour period.  The prisms
were tested in the bond wrench machine.  The full scale
walls were tested in a portable reaction frame using an
air bag for generating uniform loading (See Figure 2).
The vertical wall span was 2.3 m.  Air pressure was
measured using a water manometer.  This test
arrangement results in a loading that produces tension
perpendicular to the bed joints.


TEST RESULTS

    For the prism study Table 1 lists the plastic
mortar properties.  Table 2 lists the compressive
strength of 5.1 cm. X 5.1 cm. X 5.1 cm. cubes cured
outside and in the laboratory.  In general the
coefficients of variation were reasonably good.  The
outside cured cube specimens typically exhibited a
larger strength than the laboratory air cured cubes.
Since these are field mortars, they are not required to
meet any specification strength requirements.  The 28
day average flexural bond strengths are given in Table
3 for both the lab cured prisms and the outside cured
prisms.  The coefficients of variation are high.  The
PCL Type N mortar exhibited significantly higher bond
than the Type N MC regardless of cure conditions.  In
general the PCL Type S mortar showed higher flexural
bond strength than any of the Type S MC regardless of
cure conditions.  As shown the difference between the
mortar types was not as pronounced for Type S mortars
as for Type N mortars.  It should be noted that for the
lab cured specimens only 10% of the tested mortars
exhibited an ultimate flexural bond strength equal to
or larger than the allowable flexural bond strength
given in several masonry codes.  For the outside cured
specimens only 30% of the mortars tested exhibited an
ultimate flexural bond strength equal to or larger than
the allowable flexural bond strength.

    For the wall study the characteristics of the
mortar (mortar mix I and H in prism study) are given in

Table 4. The flows and cone penetrometer values for both mortars were approximately the same. As expected the PCL exhibited low air content, the MC high air content. The PCL showed a substantially higher compressive cube strength as compared to the MC mortar. The coefficient of variation of both mortars was quite good. The three unit high stack bonded prisms exhibited a large COV as has been previously found for these materials using this test method [1]. Comparison of Table 4 mortar properties to the properties of the same mortars used in the prism study, Table 1 and 2, indicate generally good agreement. The MC prism wall study bond strength was 98 kPa (Table 5) compared to 79.2 kPa (Table 3) in the prism study. The PCL prism wall study bond strength was 166.6 kPa (Table 5) compared to 137.2 kPa (Table 3) in the prism study. The MC wall study prisms flexural bond strength was approximately 59% (58% in the prism study) of the PCL's prism flexural bond strength. The MC walls flexural bond strength was approximately 56% of the PCL walls flexural bond strength. For this concrete block data the prisms exhibited a lower flexural bond strength than the walls. The ratio of wall to prism data was approximately 1.8 for the PCL mortar and 1.7 for the MC mortar.

CONCLUSIONS

Prism Study

Based upon the conditions of materials, construction, testing and evaluation used in the studies, the data suggests:

1.  For Type N mortars the PCL was superior to the MC mortars particularly with respect to the outside cured prisms.

2.  For Type S mortars the PCL performed significantly better than the average performance of the MC mortars. On a one-to-one comparison one of the MC lab prisms and one of the MC outside prisms performed as well as or better than the PCL prism.

3.  The small percentage of ultimate flexural bond values of prisms that equal or exceed current allowable values in many masonry codes suggest a reexamination of such allowable values are in order.

Wall Study

1.  The PCL mortar developed a significantly higher flexural bond strength than the MC for concrete

block prisms and concrete block walls;
typically 70% to 80% higher.

2.  The coefficient of variation on concrete block
    prism bond strength are high regardless of the
    mortar type.  The coefficients of variation on
    full scale walls were reasonably good.

3.  Flexural bond strength of walls built in one
    half running bond were higher than the flexural
    bond strength of corresponding stackbonded
    prisms for concrete block assemblages.

ACKNOWLEDGEMENTS

The projects mentioned in this paper were
financially supported by the National Lime Association
of Arlington, Virginia.

REFERENCES

[1]  Matthys, John H., "Conventional Masonry Mortar
     Investigation," National Lime Association,
     Arlington, Virginia, August 1988.
[2]  "Standard Specification For Mortar Unit Masonry,"
     ASTM C-270-86a, American Society For Testing and
     Materials, Philadelphia, PA 19103.
[3]  Matthys, John H., "Flexural Bond Strengths of
     Portland Cement Lime and Masonry Cement Mortars,"
     8th International Brick/Block Masonry Conference
     Proceedings, Dublin, Ireland, 1988.
[4]  Matthys, John H., "Brick Masonry Flexural Bond
     Strengths Using Conventional Masonry Mortars," 5th
     Canadian Masonry Symposium, Vancouver, B.C.,
     Canada, 1989.
[5]  Matthys, John H., "Brick Masonry Diagonal Tension
     (Shear) Tests," 5th Canadian Masonry Symposium,
     Vancouver, B.C., Canada, 1989.
[6]  "Standard Methods of Conducting Strength Tests of
     Panels For Building Construction," ASTM E-72-86,
     American Society For Testing and Materials,
     Philadelphia, PA 19103.
[7]  "Standard Specification For Hollow Load-Bearing
     Concrete Masonry Units," ASTM C-90-85, American
     Society For Testing and Materials, Philadelphia,
     PA 19103.
[8]  "Standard Method For Measurement of Flexural Bond
     Strength," ASTM C-1072-86, American Society For
     Testing and Materials, Philadelphia, PA 19103.

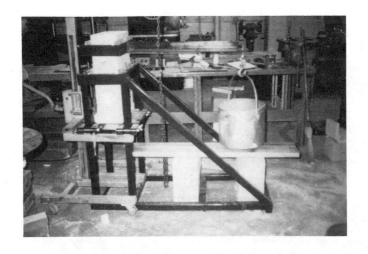

Figure 1a -- Bond Wrench Equipment

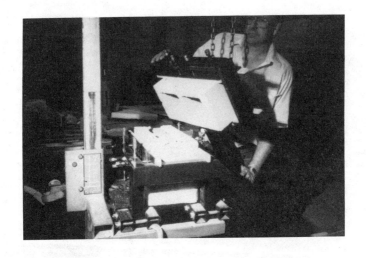

Figure 1b -- Joint Bond Failure

Figure 2b -- Full Scale Wall Test

Figure 2a -- Lateral Load Test Apparatus

TABLE 1 -- Prism Study
Mortar Properties

| Mix No. | Flow Table % | Cone Penetrometer mm | Air Content % |
|---------|--------------|----------------------|---------------|
| **Type N Mortars** | | | |
| A-MC | 121 | 35 | 12.0 |
| C-MC | 130 | 58 | 13.5 |
| E-MC | 129 | 52 | 13.0 |
| G-MC | 129 | 55 | 12.5 |
| J-PCL | 130 | 58 | 2.1 |
| **Type S Mortars** | | | |
| B-MC | 130 | 57 | 14.0 |
| D-MC | 130 | 52 | 11.3 |
| F-MC | 129 | 58 | 10.7 |
| H-MC | 130 | 55 | 15.2 |
| I-PCL | 125 | 69 | 1.7 |
| **Type O Mortars** | | | |
| K-PCL | 125 | 66 | 1.1 |

TABLE 2 -- Prism Study
Mortar Compression Strengths
kPa - 5.1 cm. X 5.1 cm. X 5.1 cm. CUBES - 28 DAYS

| Mix | Outside* Cured kPa | V% | Lab Air** Cured kPa | V% |
|-----|--------------------|-----|---------------------|-----|
| **Type N Mortars** | | | | |
| A-MC | 10969 | 5.7 | 10143 | 2.0 |
| C-MC | 5250 | 8.2 | 5502 | 2.7 |
| E-MC | 9821 | 12.0 | 8519 | 7.5 |
| G-MC | 8162 | 6.2 | 5264 | 3.3 |
| J-PCL | 11921 | 3.8 | 8505 | 11.4 |
| **Type S Mortars** | | | | |
| B-MC | 15743 | 2.2 | 13713 | 3.0 |
| D-MC | 23044 | 11.6 | 17444 | 5.2 |
| F-MC | 17941 | 2.3 | 14861 | 2.6 |
| H-MC | 14476 | 14.9 | 9030 | 7.1 |
| I-PCL | 19075 | 4.2 | 16989 | 4.4 |
| **Type O Mortars** | | | | |
| K-PCL | 4585 | 7.0 | 3843 | 2.1 |

*Covered with polyethylene until test date.
**Exposed to lab air (72 degrees F., 55 RH) until test
  date.
7kPa = 1 psi

TABLE 3 -- Prism Study
Flexural Bond Strengths
28 Day Average Values

| Mix | Outside* Cured | | Lab Air** Cured | |
|---|---|---|---|---|
| | kPa | V% | kPa | V% |
| Type N Mortars | | | | |
| A-MC | 40.3 | 59.0 | 97.2 | 51.3 |
| C-MC | 19.2 | 85.8 | 79.6 | 63.3 |
| E-MC | 23.4 | 57.2 | 78.7 | 40.5 |
| G-MC | 19.6 | 72.9 | 17.1 | 102.0 |
| J-PCL | 244.5 | 54.5 | 128.8 | 53.8 |
| Type S Mortars | | | | |
| B-MC | 52.4 | 26.6 | 53.0 | 62.0 |
| D-MC | 171.2 | 76.1 | 74.2 | 74.5 |
| F-MC | 124.6 | 68.8 | 138.9 | 69.6 |
| H-MC | 124.8 | 107.1 | 79.2 | 68.6 |
| I-PCL | 181.4 | 26.2 | 137.2 | 44.9 |
| Type O Mortars | | | | |
| K-PCL | 104.9 | 53.5 | 115.5 | 52.2 |

*Covered with polyethylene until test date (3 prisms).
**Exposed to lab air (72 degrees F., RH) until test
  date (6 prisms).
7 kPa = 1 psi

TABLE 4 -- Wall Study Data
Mortar Properties

| Mix Type | Flow % | Cone mm | Air Content % | Compressive Strength kPa | V% |
|----------|--------|---------|---------------|--------------------------|-----|
| PCL (I)  | 122.5  | 55      | 3.2           | 16401                    | 7.0 |
| MC (H)   | 129.4  | 57      | 14.5          | 10969                    | 3.3 |

7 kPa = 1 psi

TABLE 5 -- Wall Study Data
Assemblage Properties

| Mix Type | Flexural Bond Strength Prism kPa | V% | Walls kPa | V% |
|----------|----------------------------------|------|-----------|------|
| PCL (I)  | 166.6                            | 61.6 | 297.5     | 26.7 |
| MC (H)   | 98.0                             | 35.7 | 165.9     | 10.3 |

7 kPa = 1 psi

DISCUSSION

"Concrete Masonry Prism and Wall Flexural Bond Strength Using
Conventional Masonry Mortars" - J. H. Matthys

Question (S. K. Ghosh, Portland Cement Association):
  Coefficients of variation are very high for joint tests of
  stack-bonded prisms.  Values ranged from 40% to over 100%.
  The results of tests of larger components, such as walls and
  wallettes tested by various investigators including the
  author, indicate that coefficients of variation on the order
  of 15-25% are representative of the inherent variability of
  concrete block masonry.  Thus, the results of prism tests
  using the bond wrench do not appear to be a reasonable
  indicator of the flexural strength of concrete block walls.
  Would the author comment on this?

Answer (J. H. Matthys, University of Texas at Arlington):

In an unpublished study conducted by the author on flexural bond strength of portland
cement lime (PCL) mortars using normal weight concrete block in stack bonded prisms
tested by the ASTM C1072 "bond wrench" apparatus, COV ranged from a low of 29.5% to a
high of 73.1%. In an unpublished study conducted by Dr. Russell Brown on flexural bond
strength of PCL and masonry cement (MC) mortars using normal weight concrete block in
stack bonded prisms tested by ASTM C952 (eccentric load), the COV's for PCL mortars
ranged from a low of 27.8% to a high of 41.5% while the COV's for MC mortars ranged from
a low of 15.2% to a high of 60.6%. The current paper under discussion addresses the
flexural bond strength of medium weight concrete block units used in stack bonded prisms
and tested by the ASTM C1072 "bond wrench". COV's for PCL ranged from 45 to 62%;
COV's for MC ranged from 36 to 102%. For all the above studies, the prisms were built by a
mason laying to a line. No jig or other alignment device was used. All prisms for the above
data were air cured in the laboratory without covering or wetting specimens. Whether the
high COV's are related to materials, construction, curing, size of specimen, shape of units,
or test methods needs to be examined.

A project conducted at the University of Texas at Austin by Dr. Klingner for the National
Lime Association addressed the flexural bond strength of stack bonded prisms for various
PCL mortars using standardized concrete brick, ottawa sand, prism construction jig,
specimen cured in plastic bags, and tested by ASTM C1072. The materials, procedure, and
test used are that of the new Uniform Building Code Standard #2430 "Standard Test
Method For Flexural Bond Strength of Mortar Cement". COV's for these tests ranged
typically from 15% to 25%. Additional testing using the identical materials and procedures
except with the replacement of the various PCL with various masonry cements was
conducted by the National Concrete Masonry Association Research and Development
Laboratory and presented to the International Conference of Building Officials in the form of

a report titled "Summary of Flexural Bond Strength Research on Concrete Masonry Prisms Using Masonry Cement Mortars" and dated December 1989. COV's presented ranged from 16 to 25%. One should note that although rigid standardized procedures were followed for both studies above, the units used were concrete brick and not concrete block.

Using rigid standardized procedures, obviously reasonably good COV can be obtained for concrete brick prism tested by the bond wrench. Whether the same applies to concrete block needs to be determined. Also it is known that using a prism construction jig and curing specimens in plastic bags increases bond strength. Unpublished data shows using a jig to build prisms can result in a 50% increase in bond. How such tests are related to typical field masonry is yet to be addressed.

With respect to the discusser's interest of potentially using the results of stack bonded prisms tested by the bond wrench as an indicator of the flexural strength of concrete block walls, the following comments are appropriate:

1.   The only published data to the author's knowledge that directly compares stack bonded concrete block prism tests to 4' x 8' uniform flexural wall tests is presented in this paper; i.e., nine prisms and three walls for PCL mortar and nine prisms and three walls for MC mortar. Although the published data is obviously small, it appears some general trends exist.

2.   The average flexural bond strength of stack bonded concrete block prisms for masonry cement was ≈ 58% of the corresponding portland cement lime prisms in the prism study.

3.   The average flexural bond strength of stack bonded concrete block prisms for masonry cement was ≈ 59% of the corresponding portland cement lime prism in the wall study.

4.   The average flexural bond strength of the 4' x 8' concrete block walls for masonry cement was ≈ 56% of the corresponding portland cement lime walls in the wall study.

5.   The ratio of average 4' x 8' wall flexural bond strength to average prism flexural bond strength (bond wrench method) is 1.8 for PCL lime mortars and 1.7 for MC mortars.

In summary although the COV's of reported stack bonded concrete block prisms bond tests are high while that for corresponding reported wall tests are low to moderate, it appears that the relationship of flexural MC mortar bond to flexural PCL mortar bond is consistent regardless of whether prisms or walls are used. Also it appears that the relationship of average wall strength to average prism strength is quite constant for medium weight concrete block regardless of mortar used. Thus it appears that one may reasonably use properly factored concrete block prism bond results as an indicator of concrete block wall strength.

DISCUSSION

"Concrete Masonry Prism and Wall Flexural Bond Strength Using
Conventional Masonry Mortars" - J. H. Matthys

Question (S. K. Ghosh, Portland Cement Association):
  Flow was apparently used in this study to control the
  consistency of mortar.  This is not the way mortar
  consistency is normally controlled either in the laboratory
  or in the field.  In the field it is the mason who
  determines the consistency of mortar since he must work with
  it and since mortar consistency affects the quality of his
  work.  In the laboratory, the cone penetrometer method
  specified in ASTM C780 is preferred.  Would the author
  elaborate on why the mortars used in both the prism and wall
  study were produced and mixed "to field flow of 120 to 130"?

Answer (J. H. Matthys, University of Texas at Arlington):

In the initial mixing of masonry cement mortars and portland cement lime mortars the
quantity of water used was determined by the mason laying the prism/wall in light of the
unit used.  Flow determination of these mixes established the "benchmark" of flow in the
range of 120 to 130.  For both the prism and wall study both flow and ASTM C780 cone
penetrometer measurements were made and reported.  In general the flows and
penetrometer values shown are consistent for both masonry cement and portland cement
lime mortars.  For both studies the mason was allowed to retemper the mortar as needed.

DISCUSSION

"Concrete Masonry Prism and Wall Flexural Bond Strength Using
Conventional Masonry Mortars" - J. H. Matthys

<u>Question</u> (S. K. Ghosh, Portland Cement Association):
  The author used only one portland cement and one Type S lime
  for producing PCL mortars.  Is it implied that these
  materials produce mortars representative of all PCL
  mortars?  To the best of this writer's knowledge, both
  portland cement and lime vary from manufacturer to
  manufacturer as well as from a single manufacturer.  Does
  not this bring into question the validity of comparisons
  made between the performance of PCL mortars and MC mortars?

<u>Answer</u> (J. H. Matthys, University of Texas at Arlington):

  The portland cement used for the PCL mortars was a single brand Type I cement
  meeting ASTM C150.  The lime used for the PCL mortars was a single brand Type
  S lime meeting C207 and possessing average plasticity characteristics.  The
  resulting PCL mortars using these two products in combination were felt to result in
  typical PCL mortar performance.  The concern by the discusser over the variation
  of cement and lime and the potential resulting mortar property variations might be
  significantly lessened by referring to the paper presented in this same symposium
  by Dubovoy and Ribar of Construction Technology Laboratories.  In Phase II of
  Dubovoy's study selected PCL mortar mixes were tested to provide data on flexural
  bond strength and water penetration.  Three different limes were used in
  combination with portland cements from the same geographical areas as the limes.
  Dubovoy's and Ribar's conclusion in this phase was "flexural bond values
  developed by portland cement and lime mortars were essentially unaffected by
  variability in lime characteristics".

Madan L. Mehta

THE POTENTIAL FOR TRAFFIC NOISE REDUCTION BY THIN MASONRY PANELS

REFERENCE: Mehta, Madan., "The Potential for Traffic Noise Reduction by Thin Masonry Panels", Masonry: Components to Assemblages, ASTM STP 1063, John H. Matthys, Editor, American Society for Testing and Materials, Philadelphia, 1990.

ABSTRACT: Heavy reinforced concrete and masonry walls are commonly used as protective barriers against highway noise. Recently, prefabricated thin brick wall panels have been proposed to isolate traffic noise from city streets and highways. This paper reviews the theoretical background on noise level reduction achieved by barriers erected on one side of the road and examines its application to thin masonry panels.

KEYWORDS: barrier attenuation, barrier insertion loss, barrier transmission loss, infinitely long barrier, finite length barrier, ground attenuation.

Noise barriers have been used extensively in recent years to protect against traffic noise from highways and city streets. The types of barriers currently in use include earth berms, vertical walls and combinations of the berm and the wall. In wall type barriers, various shapes, profiles and surface treatments have been investigated in scale model tests for their acoustical performance [1]. Several of these different shapes and profiles have been used in practice and subsequently investigated for their performance through field measurements [2].

The wall type barrier most commonly employed is a vertical plane made of a thick concrete or masonry wall. However, in view of the escalating costs of these barriers, a prefabricated system consisting of thin (nearly 60 mm thick) brick panels with intermediate vertical supports to withstand lateral loads, has been recently proposed as an alternative [3]. This paper examines the acoustical performance of such barriers erected on one side of the road, that is, the effect of reflections from the opposite side of the road has been neglected.

The acoustical performance of a barrier is primarily a function of: (i) sound diffracted over the barrier, (ii) sound absorbed by the ground, and (iii) sound transmitted through the barrier. These factors are discussed below in the above order.

Dr. Madan Mehta is Professor of Architecture at the University of Texas at Arlington, Box 19108, Arlington, Texas 76019.

## DIFFRACTION

As far as the diffracted sound is concerned, a barrier's performance is measured by the reduction of sound pressure level that it provides at a specific point, as compared with the sound pressure level in absence of the barrier. The terms commonly used to define this reduction are: *barrier attenuation* and *barrier insertion loss*.

Barrier attenuation is defined as the reduction in sound pressure level in free-field conditions. Here the effect of the ground on which the barrier is erected is omitted. Barrier insertion loss takes into account the modification introduced by the absorption of the ground on both sides of the barrier and is defined as the difference in sound pressure levels with and without the barrier but in the presence of the ground in both cases. Both the barrier attenuation and insertion loss refer to noise reduction in excess of that occurring due to the divergence of sound waves with the distance of travel.

### Point Source Attenuation

An analytical solution of barrier attenuation was first provided by Sommerfield [4] for the simplest configuration: a semi-infinite barrier (a barrier of finite height but infinite length) whose diffracting edge is wedge shaped and the incident sound is a plane wave. Since Sommerfield's exact solution was mathematically complex, simpler but approximate solutions were sought by several investigators.

The literature on the subject refers to Redfearn's work [5] as one the earliest attempts in this direction which assumes that for a receiver in the shadow zone, the sound field may be considered to radiate from a virtual line source - the top edge of the barrier. From his analysis, Redfearn produced easy-to-use graphs which gave the values of attenuation in the shadow zone of the barrier, for a point source and a semi-infinite plane barrier as a function of two variables: the diffraction angle, $\gamma$, and (h/$\lambda$), where h is the effective barrier height, Fig. 1, and $\lambda$, the wavelength of sound. Both, the sound source and the receiver were assumed to lie in a plane perpendicular to the barrier.

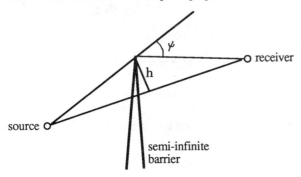

Fig. 1. Diffraction angle, $\gamma$, and effective barrier height, h.

Two limitations handicapped Redfearn's analysis. (i) It could not be applied in situations where the source and receiver lay in a plane inclined to the plane of the barrier. (ii) The barrier attenuation was expressed as a function of two variables, (h/$\lambda$) and $\gamma$, which meant that a large number of curves was required to cover all possible cases of interest [6].

The limitations in Redfearn's work were overcome by Maekawa [7] who, from an extensive experimental data collected from scale modeling of barriers and field measurements, introduced a design graph, Fig. 2, showing a relationship between $A_p$ and Fresnel number, N, where,

$A_p$ = the attenuation by a rigid, thin semi-infinite plane barrier, lit by a point source,

$N = 2\delta/\lambda$ ,

$\delta$ = path length difference = (a + b - d) ,

d = length of the straight line drawn between the source and the receiver, called the *line of sight* , and

(a+ b) = the shortest path length between the source and receiver over barrier edge, Fig. 3.

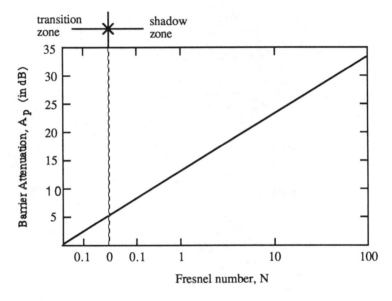

Fig. 2. Maekawa's design graph for attenuation by a thin semi-infinite plane barrier

In Fig. 2, the value of $A_p$ is nearly 5 dB when N = 0. This is the case in which the line of sight just clears the edge of the barrier. The reason for this non-zero value of $A_p$ is that the sound radiated from the edge of the barrier affects not only the shadow zone but also a small illuminated zone, due to its interference with the sound coming directly from the source. Thus, the value of $A_p$ does not fall to zero abruptly as the receiver moves from the shadow zone to the illuminated zone but does so gradually.

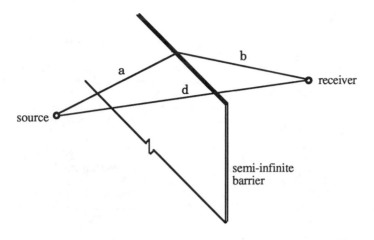

Fig. 3. Barrier and source-receiver geometry

In the part of the illuminated zone where $A_p$ lies between 0 and 5 dB, called the *transition zone* , the value of $A_p$ , as obtained from Maekawa's data, varies from 5 dB at $N= 0$, to 0 dB at $N = 0.2$. From this data, the following empirical relationships, for the shadow zone and transition zone were obtained [8].

$$A_p \text{ (shadow zone)} = 20 \log \frac{\sqrt{2\pi N}}{\tanh \sqrt{2\pi N}} + 5 \tag{1}$$

$$A_p \text{ (transition zone)} = 20 \log \frac{\sqrt{2\pi N}}{\tan \sqrt{2\pi N}} + 5 \tag{2}$$

Equation 1 does not hold in situations where the diffraction angle is large. This occurs when both the source and the receiver are close to the barrier, i.e. $(a + b) \gg d$. Here the value of $A_p$ is greater than that predicted by the equation due to a large spherical sound divergence. This can be accounted for [8] by adding the term, $20 \log [(a + b)/d]$ , to Eq 1. However, since this factor is of little or no significance in traffic noise problems, it has been excluded in subsequent discussion.

It has been shown [9,10] that, in place of Eq 1, Maekawa's data can be approximated to the following simpler relationship, for barrier attenuation in shadow zone, and it is this expression that will be developed further.

$$A_p = 10 \log (20N + 3) \tag{3}$$

## Line Source Attenuation

**Semi-infinite barrier** : Under otherwise identical conditions, the attenuation of a barrier due to a line source is smaller than that obtained from a point source. Thus, if two sources are considered, a point source and a line source, both of which produce the same sound pressure level at a receiver in the absence of a barrier, and if a semi-infinite barrier is interposed between them at the same relative locations, it will be seen that the barrier is more effective in the case of the point source than in line source. This is due, mainly, to the more nearly cylindrical divergence of sound rays from the edge of the barrier lit by a line source than a point source.

For an incoherent line source (such as free-flowing traffic), barrier attenuation can be obtained by assuming that the line source consists of numerous small segments, each of which behaves as a point source, and adding the effects of all such sources at the receiver. If it is further assumed that these sources have equal acoustic power then using the laws of decibel addition, barrier attenuation due to a line source can be expressed as:

$$A_n = 10 \log \frac{\displaystyle\sum_{i=1}^{n} 10^{(U_i/10)}}{\displaystyle\sum_{i=1}^{n} 10^{(U_i - A_i)/10}} \tag{4}$$

where,

$A_n$ = total attenuation due to a line source consisting of n segments,

$U_i = 10 \log (1/d_i)^2$ = sound intensity level produced by the i-th source (of unit acoustic power) which is at a distance $d_i$ from the receiver, under free-field conditions, and in absence of the barrier, and

$A_i$ = barrier attenuation for the i-th source, as obtained from Eq. 3.

Attenuation values obtained from the analysis of Eq 4 show close agreement with those obtained experimentally [11]. In the particular case of an infinitely long incoherent line source, Beranek [8] reports that barrier attenuation is approximately 1 to 5 dB lower than that obtained from a point source. The 5 dB difference is obtained at N = 10 and nearly 1 dB difference is obtained at N = 0.1. If this is incorporated in Eq 3, the attenuation due to an infinitely long incoherent line source parallel to the barrier can be expressed as:

$$A = 10 \log (20N + 3) - (20N)^{0.3} \tag{5}$$

where N is to be measured in a plane perpendicular to the barrier and passing through the receiver.

Since traffic noise levels are usually expressed in dBA, it is convenient to express barrier attenuation also in dBA. The dBA values of barrier attenuation can be obtained by applying A-weighting function to attenuation values (as obtained from Eq 5) and including the effect of the spectral distribution of traffic noise.

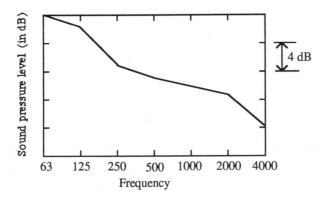

Fig. 4. Typical highway noise spectrum [Ref. 12]

Highway noise spectrum is a complicated function of various parameters such as the mean traffic speed, characteristics of road surface, road gradient, percentage of heavy vehicles, etc. However, if a typical highway noise spectrum is assumed as that given in Fig. 4, then the attenuation of an infinitely long barrier for highway noise can be calculated using the laws of decibel addition, and expressed as a function of path length difference, $\delta$. The result is shown in Fig. 5.

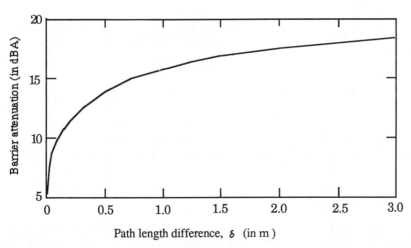

Fig. 5. Barrier attenuation for traffic noise as a function of path length difference

The curve of Fig. 5 tends to flatten with increasing values of $\delta$ and it has been reported [8,15] that 20 dBA is virtually the maximum achievable attenuation by a semi-infinite plane barrier. In actual practice, the attenuation may be less than the above value when the effects of the ground and local meteorological factors are taken into account.

A mathematical approximation to the curve of Fig. 5 is given by the following expression. This expression is similar to that given by Fisk [13] for the attenuation of $L_{eq}$ values as a function of $\delta$.

$$A = 15.5 + 6.0 \log \delta \qquad (6)$$
$$\text{for,} \quad \delta \geq 0.03 \text{ m}$$

Finite length barrier : In practice, barriers must be of finite length. Thus, some sound will leak round the ends of the barriers. The attenuation of a barrier of finite length is roughly proportional to the angle that it subtends at the receiver as measured in the horizontal plane [14, 15]. If this angle is denoted by $\alpha$, then for an infinitely long barrier, $\alpha = 180^\circ$.

The attenuation of a given infinitely long barrier decreases sharply as $\alpha$ deviates from its maximum value ($180^\circ$). The decrease is more severe for a barrier which provides a high value of attenuation when it is infinitely long. Thus, if the maximum attenuation of a barrier (obtained when $\alpha = 180^\circ$), called the *attenuation potential* of the barrier, is 20 dB, its attenuation will fall to 8 dB if $\alpha = 140^\circ$, giving a decrease of 12 dB. On the other hand, if barrier potential is 5 dB, its attenuation at $\alpha = 140^\circ$ will be nearly 4 dB, giving a decrease of only 1 dB. The attenuation of a barrier with $\alpha$ less than $80^\circ$ is almost negligible, regardless of its potential. For example, in case the barrier potential is 20 dB, its attenuation at $\alpha = 80^\circ$ is only 3 dB.

The values of attenuation for finite length barriers have been developed in the form of graphs [14]. The following expression may be used as first approximation for the attenuation achieved by a finite length barrier.

$$A_\alpha = A^{\left(\frac{\alpha}{\pi}\right)} \qquad (7)$$

where, $A_\alpha$ is the attenuation of a barrier subtending an angle $\alpha$ at the receiver, and A, its attenuation potential. For more accurate values, the graphs are recommended.

## GROUND ATTENUATION

The presence of the ground can greatly affect the barrier's performance. Consider first the case without the barrier. In this case, the total sound pressure at the receiver is due to the combined effect of two sound rays: (i) the direct ray, ray 1 and (ii) the reflected ray, ray 2, Fig. 6(a). When a barrier is interposed between the source and the receiver, two additional rays will affect the sound pressure at the receiver. These are produced by reflections from the ground on the receiver side of the barrier, Fig. 6(b). Thus, in the presence of a barrier, the total sound pressure at the receiver, $p_t$, is the sum of pressures due to four rays, i.e.,

$$p_t = p_1 + p_2 + p_3 + p_4 \tag{8}$$

where,

$p_1$ = sound pressure due to the ray from the source itself

$p_2$ = sound pressure due to the ray from the image of the source

$p_3$ = sound pressure due to the ray from the source after reflection from the ground on the receiver side

$p_4$ = sound pressure due to the ray from the image of the source after reflection from the ground on the receiver side.

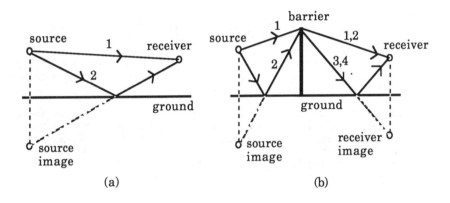

(a)                                    (b)

Fig. 6  Sound rays between the source and the receiver.
(a) in the absence of a barrier   (b) in the presence of a barrier

If the ground on the source side of the barrier is acoustically hard (perfectly reflecting), and if a point source is assumed to be located extremely close to (almost in contact with) the ground, then with reference to Figs. 6 (a) and (b), the sound pressures at the receiver due to rays 1 and 2 are equal i.e. in Eq 8, $p_1 = p_2$. Since both rays are in phase, a 6 dB increase in sound pressure level at the receiver is obtained in comparison with the free-field sound pressure level. This increase occurs regardless of whether or not the barrier is present. If the ground on the receiver side is also hard and the receiver is also assumed to be located close to the ground, then $p_3 = p_4$, giving an additional 6 dB increase in sound pressure level at the receiver.

Thus, if the ground on both sides of the barrier is hard and both the source and the receiver are close to the ground, the barrier attenuation is degraded by 6 dB. Now, if everything remains as before but the receiver is moved to a position above the ground, the reflected rays (rays 3 and 4) and those coming from the source and its image (rays 1 and 2), in Fig. 6(b), will have different path lengths. This will produce an interference pattern between the two sets of rays which is a function of their path length differences. If the source emits a broad band noise (e.g., traffic noise) and if the difference in attenuation between the two sets of rays, caused by their different path lengths, is ignored, there will be a 3 dB degradation in the attenuation provided by the barrier in place of the 6 dB obtained earlier, due to incoherent addition of the rays. In fact, the actual degradation will be much smaller than 3 dB due to greater attenuation suffered by the reflected rays.

The ground effect becomes complicated when (i) the ground is acoustically soft (absorbing) which produces a complex reflection (phase shift) of sound waves and (ii) the point source is replaced by a line source. In general, it can be said that the insertion loss of a barrier on a soft ground ( a ground covered with grass or vegetation) is smaller than that on a hard ground i.e., the barrier is less effective on an acoustically soft ground than on a hard one. The reason is that without a barrier, the sound travels over ground with a smaller angle of incidence. Since the absorption of a surface is greater at smaller angles of incidence, the attenuation of soft ground in the absence of a barrier is high. The interposition of a barrier increases the effective height of the source, increasing the angle of incidence of sound with the ground, and hence reduces the attenuation provided by the ground.

Precise values of ground attenuation cannot be obtained because of the non-uniformity in attenuation properties of the ground and the enormous variation in ground types. However, experimental evidence [13, 18] suggests that, for a receiver position of 1.5 m or more above the ground, a hard ground has virtually no effect on the insertion loss of a barrier, i.e., the attenuation of noise from free-flowing traffic over hard ground is the same as for the free field - at the rate of 3 dB per doubling of distance.

The values of attenuation by soft grounds have been reported by several researchers [e.g. 16, 17]. The general conclusion is that 1.5 dB per doubling of distance represents a good measure of the excess attenuation, EA, of a soft ground in the absence of the barrier. Note that EA is the attenuation in excess of that obtained under free-field conditions. More precisely, EA may be expressed as [18]:

$$\text{EA (soft ground)} = 5.2 \log(D/3h) \quad \text{for, } 1 \le h \le D/3 \tag{9}$$

$$= 0 \quad \text{for, } h > D/3$$

where, h = height of source above ground, and D is the horizontal distance of receiver from the source

How is the value of EA modified in the presence of a barrier? One approach would be to obtain an empirical expression similar to Eq 9 , where h would represent the height of the barrier edge in place of the height of the source. However, to the author's knowledge, no such relationship has been reported. The best estimate currently available [15] is that EA varies between 0 to 4 dB, depending on the ratio (D/R), where D and R are the horizontal distances of the receiver and the source respectively from the barrier. This may be approximately represented by the following expressions.

$$
\begin{array}{ll}
\text{EA} = 0, & \text{for } D/R < 1.3 \\
\text{EA} = 6.3\log (D/R) - 0.7 & \text{for } 1.3 \le D/R \le 5.1 \\
\text{EA} = 4 & \text{for } D/R > 5.1
\end{array} \tag{10}
$$

## TRANSMISSION LOSS

The overall effect of a barrier at the receiver is the combined effect of its insertion loss, IL, and transmission loss, TL. This can be stated simply as:

$$R_o = -10 \log \left[ 10^{-(IL/10)} + 10^{-(TL/10)} \right] \tag{11}$$

where $R_O$ is the overall sound pressure level reduction provided by the panel.

Equation 11 shows that the barrier's TL should be much greater than than its IL if the overall effect of the barrier is not influenced by its transmission loss. This situation is obtained, if TL is nearly 10 dB (or more) higher than IL. Thus, since the practical upper limit of IL is 20 dB, it implies that there is no acoustical benefit in using barriers with transmission loss higher than 30 dB. Note however, that in case TL = IL, the overall sound pressure level reduction provided by a barrier is 3 dB less than its insertion loss.

The approximate transmission loss of a single layer homogenous panel can be determined by the *plateau method*. In this method, the TL-frequency relationship of a panel consists of three regions: (i) region 1 in which TL increases by 6 dB per octave, (ii) region 2 in which TL is constant, the plateau region, and (iii) region 3 in which TL increases initially by 10 dB, gradually falling to a 6 dB increase per octave [19].

In region 1, the transmission loss follows the (limp) mass law, which for normal sound incidence is given by [Ref. 19, p. 297]:

$$TL = 10 \log \left[ 1 + \left( \frac{\omega M_s}{2 \rho c} \right)^2 \right] \tag{12}$$

where, $\omega$ = angular frequency = $2\pi f$, $\rho$ = density of air, c = speed of sound in air, and $M_s$ = the surface density of barrier's material.

Substituting $\rho = 1.2$ kg m$^{-3}$, c = 343 m sec$^{-1}$, $M_s = 115$ kgm$^{-2}$ (surface density of 60 mm thick brick masonry panel), and recognizing that within the frequencies of our interest, the term $(\omega M_s / 2 \rho c)^2 \gg 1$, the above expression reduces to:

$$TL = 20 \log (f) - 1 \tag{13}$$

In practice, the traffic noise does not have a normal incidence on the barrier but does so linearly. This reduces the barrier's TL below that given by Eq 13. For field incidence (which occurs in building enclosures), the reduction is nearly 5 dB[19] and for linear incidence, the reduction is approximately 8 dB[20]. Thus, the approximate transmission loss of a 60 mm thick brick panel may be expressed as:

$$TL = 20 \log (f) - 9 \tag{14}$$

From Eq 14, the transmission loss values at 63 Hz, 125 Hz, and 250 Hz are: 27 dB, 33 dB, and 39 dB respectively.

The plateau region for a brick masonry panel is 37 dB high with a frequency ratio of 4.5 [19]. With the above information, the TL-frequency relationship for a 60 mm thick brick panel can been drawn, Fig. 7, which shows that its transmission loss is greater than 30 dB except for a small region near 63 Hz. The 60 mm thick brick panel is, therefore, acoustically adequate as a highway noise barrier.

Note that a masonry panel thinner than 60 mm will provide TL values lower than those given by Fig. 7. Since the 60 mm thick panel is just adequate in the low frequency range, this thickness may be regarded as the minimum thickness for traffic noise barriers.

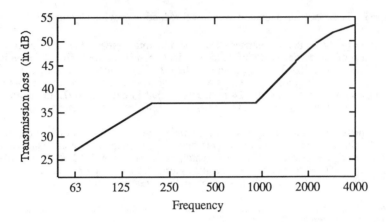

Fig. 7.  Approximate transmission loss of a 60 mm thick brick panel

The graph of Fig. 7 may only be considered as a first approximation. The plateau method assumes a diffuse sound field on both sides of a panel which is obviously not present in a highway situation. Additionally, the height of plateau and frequency ratio are based on measurements on panels of thicknesses that are commonly used. Since a 60 mm thick brick panel is not commonly used, the plateau height and frequency ratio values may differ (albeit by a small amount) from those assumed in Fig. 7. Besides, the mass law itself is based on several simplifying assumptions. Thus, the precise values of transmission loss of a 60 mm thick brick wall panel must be obtained from measurements.

Since the transmission loss of a panel is greatly affected by openings, holes and cracks, it is important to pay attention to mortar joints in a brick wall as well as the joints between the panel and the supporting vertical elements. Partially filled mortar joints due to poor workmanship can severely reduce the transmission loss of a barrier and hence the overall sound pressure level reduction provided by it.

The experimental verification of the analysis presented in this paper is the subject of future work by the author and is proposed to be done in two parts: (i) laboratory measurement of transmission losses of thin masonry panels along with their supporting elements, and (ii) field verification of the insertion loss provided by these barriers.

ACKNOWLEDGEMENTS

The author acknowledges with appreciation the encouragement provided by Mr. William Bailey, Manager, Technical Services, Acme Brick Plant, Fort Worth, Texas for his encouragement in the preparation of this paper.

REFERENCES

[1]    May, D.N. and Osman, M.M. " Highway Noise Barriers: New Shapes" Journal of Sound and Vibration 71(1), 1980, pp 73-101.

[2]     May, D.N. and Osman, M.M. " The Performance of Sound Absorption, Reflection and T-Profile Noise Barriers in Toronto" Journal of Sound and Vibration 71(1), 1980, pp 65-71.

[3]     Bailey, W. " Personal Communication with the Author", 1988

[4]     Sommerfield A. "Mathematische Theories de Diffraktion" Mathematische Annalen, 47, 1896, pp 317-74.

[5]     Redfearn, S.W. "Some Acoustical Source-Observer Problems" Phil. Mag. Ser. 7 No. 30, 1940, pp 223-236.

[6]     Kurze, U.J. "Noise Reduction by Barriers", Journal of the Acoustical Society of America, Vol. 55(3), 1974, pp 504-518.

[7]     Maekàwa, Z. "Noise Reduction by Screens", Mem. Faculty of Engineering, Kobe University Vol. 11, 1965, pp 29-53.

[8]     Beranek, L.L., Noise and Vibration Control , McGraw Hill Book Company , 1971, pp 175.

[9]     Moreland, J.B. and Musa R.S. " The Performance of Acoustic Barriers" International Conference on Noise Control Engineering, Washington D.C., 1972, pp 95-104.

[10]    Tatge, R.B. " Noise Reduction of Barrier Walls", Arden House Conference, 1972.

[11]    Isei, T., Embleton T.F. and Piercy J.E. "Noise Reduction by Barriers on Finite Impedance Ground", Journal of the Acoustical Society of America, 67(1), 1980, pp 46-58.

[12]    Rosenberg, Carl. "Blocking the Bombardment of Noise", Architecture Incorporating Architectural Technology, March 1989, p 70.

[13]    Fisk, D.J. "Attenuation of $L_{10}$ by Long Barriers", Journal of Sound and Vibration, 38(3), 1975, pp 305-316.

[14]    Scholes, W.E., Salvidge, A.C. and and Sargent J.W. "Motorway Noise Propagation and Screening", Journal of Sound and Vibration, 38(3), 1975, pp 281-303.

[15]    U.S. Department of Housing and Urban Development., Noise Assessment Guidelines, HUD-PDR-735(1), 1984, p 25.

[16]    Attenborough, K. "Predicted Ground Effect for Highway Noise" Journal of Sound and Vibration, 81(3), 1982, pp 413-424.

[17]    United Stated Department of Transportation, Federal Highway Administration, Noise Barrier Design Handbook , FHWA-RD-76-58, 1976, p 2-14.

[18]    Hothersall, D.C., and Chandler-Wilde, S.N. "Road Traffic Noise Attenuation", Journal of Sound and Vibration, 115(3), 1987, pp 459-472.

[19]    Beranek, L.L., Noise Reduction, Robert E. Krieger Publishing Company, 1980, p 30.

[20]    May, D.N., "The Optimum Weight of Highway Noise Barrers", Journal of Sound and Vibration, 68(1), 1980, pp 1-13.

Mark Merryman, Gilberto Leiva, Nicholas Antrobus,
and Richard E. Klingner

IN-PLANE SEISMIC RESISTANCE OF
TWO-STORY COUPLED CONCRETE MASONRY WALLS

---

REFERENCE: Merryman, M., Leiva, G., Antrobus, N. and Klingner,
R. E., "In-Plane Seismic Resistance of Two-Story Coupled
Concrete Masonry Walls," Masonry:  Components to Assemblages,
ASTM STP 1063, John H. Matthys, Editor, American Society for
Testing and Materials, Philadelphia, 1990.

ABSTRACT:  To study the in-plane resistance of multi-story
masonry coupled walls, 6 full-scale reinforced masonry
specimens, each two stories high, are being constructed and
tested in the laboratory as part of the U.S.-Japan masonry
program (TCCMAR).  All specimens are of fully grouted hollow
concrete masonry.  Four specimens are pairs of walls, coupled
by floor slabs with and without lintels.  Two other specimens
are single walls with door and window openings.  In this paper,
the design, construction, and testing of two of the coupled
wall specimens are described.  The specimens were loaded
vertically by constant loads representing gravity loads on the
shear wall's tributary area, and horizontally by quasi-static,
reversed cyclic shear loads applied in the plane of the walls
at the two floor levels.  Test results are presented,
discussed, and compared with analytical predictions of specimen
performance.  Implications of the results for design and test
procedures are discussed.

KEYWORDS:  analysis; design; earthquake; masonry; walls

Messrs. Merryman and Leiva are Graduate Research Assistants at the
University of Texas at Austin.  Mr. Antrobus is a Project Engineer at
Law Engineering in Atlanta, GA, and a former Graduate Research Assistant
at the University of Texas at Austin.  Dr. Klingner is the Phil M.
Ferguson Professor in Civil Engineering at the University of Texas,
Austin, TX  78712.

INTRODUCTION, OBJECTIVES AND SCOPE

## Introduction

The U.S. Coordinated Program for Masonry Building Research, funded by the National Science Foundation, consists of a set of separate but coordinated tasks, intended to address the basic issues of masonry material and structural response to gravity and seismically induced loads [1]. The program is divided into 10 tasks: 1) materials; 2) mathematical models; 3) walls; 4) intersections; 5) floors; 6) construction; 7) small-scale models; 8) design methods; 9) full-scale building; and 10) design recommendations and criteria development.

In Task 3.1(c) of the TCCMAR Program, 6 full-scale reinforced masonry specimens, each two stories high, are being constructed and tested in the laboratory. All specimens are of fully grouted hollow concrete masonry. Two types of specimens will be constructed, and are designated as Type 1 and Type 2. Two Type 1 specimens will be constructed, each representing a shear wall of a two-story building. Four Type 2 specimens are being constructed, each representing a pair of coupled shear walls in a two-story building. Each specimen will be loaded vertically by constant loads representing gravity loads on the shear wall's tributary area, and horizontally by varying loads at the two floor levels.

## Objectives and Scope

The objectives of Task 3.1(c) are to examine how the in-plane seismic resistance of multistory concrete masonry walls is affected by floor-wall joints, wall openings, and floor elements. The specific objectives of the Type 2 specimen tests are:

a)  to examine the cyclic shear resistance of the coupled wall system which each specimen represents

b)  to examine the shear strength and in-plane response of the floor-wall joints

c)  to examine the coupling effectiveness (under reversed cyclic loads) of plank floor systems, with and without masonry lintels

d)  to verify the behavior of the lateral loading system

e)  to test the analytical models being developed in other TCCMAR tasks

The emphasis of this paper is the design, construction, and testing of the first two specimens (Specimens 2a and 2b), shown in Fig. 1. Preliminary test results are presented, discussed, and compared with previous analytical predictions of wall performance. Further details are given in Refs. 2 and 3.

## SPECIMEN DESCRIPTION AND MATERIALS

### Description of Specimen 2a

As shown in Fig. 1, the Type 2 specimens are coupled walls. The first Type 2 specimen, denoted here as Specimen 2a, has one central door opening 3.33 ft (1015 mm) wide and 8.0 ft (2440 mm) high without a lintel over the opening. For the original prototype from which Specimen 2a was derived, the floors were assumed to span perpendicular to the plane of the coupled walls, and were constructed of cast-in-place reinforced concrete 8 inches (203 mm) thick. The floors had a width of 36 inches on each side of the walls, giving a total specimen width of 77-5/8 inches (1972 mm).

### Description of Specimen 2b

The second Type 2 specimen, denoted here as Specimen 2b, was almost identical in appearance to Specimen 2a. Instead of cast-in-place floors, however, Specimen 2b has floors constructed of 6-inch precast, prestressed concrete planks, spanning parallel to the plane of the specimen and covered with a 2-inch (51-mm) cast-in-place topping. The two floor systems are shown in Fig. 2.

### Materials

Concrete units conformed to Type I of ASTM C90. Portland cement-lime mortar conformed to proportion specifications for Type S, ASTM C270. Grout conformed to the coarse grout specification of ASTM C476, included Grout-Aid (a proprietary admixture), and was placed with a slump of about 11 inches. Reinforcement conformed to Gr. 60 of ASTM A615. Material properties for Specimens 2a and 2b are summarized in Tables 1 and 2 respectively.

### Structural Details of Specimen 2a

The walls were constructed of hollow lightweight bond beam units 6 inches thick, laid in running bond. Vertical reinforcement consisted of 5 #4 bars (13 mm) placed at 16 in. centers (406 mm) in each section of wall. Vertical reinforcement was lap spliced to dowels in the base, using a 40d lap. Horizontal reinforcement in the first story was #4 bars (13 mm) every course. Horizontal reinforcement in the second story was #4 bars in every other course. All horizontal bars were anchored to the end vertical bars with 180-degree hooks.

As shown in Fig. 2, floors were of cast-in-place concrete, 8 in. (203 mm) thick. Transverse reinforcement in the top of the slab consisted of #5 bars (16 mm) spaced at 10 in. (254 mm), and in the bottom of the slab, of #4 bars (13 mm) spaced at 10 in. (254 mm). Longitudinal reinforcement consisted of #3 bars (10 mm) spaced at 12 in. (305 mm). Additional longitudinal reinforcement, consisting of four #4 bars (13 mm), was placed in the slab directly over the shear walls. These #4 bars were enclosed by #3 ties (10 mm) spaced at 3 inches (76 mm) in the coupling slab, and at 8 inches (203 mm) in the walls.

## Structural Details of Specimen 2b

Walls of Specimen 2b were identical to those of Specimen 2a. As shown in Fig. 2, floors were of precast, prestressed concrete planks, 6 in. (152 mm) thick, reinforced with 4 strands, 3/8 inch (9.5 mm) in diameter, Gr. 270 (186 MPa). Reinforcement in the topping consisted of 6-gage welded wire fabric, spaced at 6 inches (152 mm) in each direction. Additional longitudinal reinforcement, consisting of four #4 bars (13 mm), was placed in the slab directly in the plane of the coupled walls. These #4 bars were enclosed by #3 ties (10 mm) placed at 3-in. centers (76 mm).

## TEST SETUP

### General

The overall test setup consisted of the following elements: 1) reaction system; 2) precast base beams; 3) vertical loading frames; 4) lateral loading frame; and 5) sway bracing.

### Test Setup for Specimen 2a

The walls of the prototype structure used to develop Specimen 2a were assumed to be spaced at 20-foot intervals, and were subjected to gravity loads from the floor slabs which spanned perpendicular to them. Because the slabs of Specimen 2a were only about 6 feet wide rather than 20 feet, additional vertical forces representing these gravity loads had to be applied to the walls of Specimen 2a. These loads were applied to the top of the second story of each coupled wall by hydraulic actuators attached to a steel frame mounted on the reaction floor. Lateral loads, representing seismic loads, were applied to the outer edges of the floor and roof at the midpoints of each coupled wall by hydraulic actuators attached to steel frames mounted on a massive reaction wall. Simple steel link sway bracing was attached to the outer edges of the second floor and the roof and anchored to the reaction wall running parallel to the specimen's in-plane centerline.

### Test Setup for Specimen 2b

The test setup for Specimen 2b was identical to that for Specimen 2a, except that no additional vertical loads were applied. This is because the floors of the prototype structure for Specimen 2b were assumed to span parallel to the coupled walls, and did not transmit gravity loads to the walls.

INSTRUMENTATION

Specimen 2a was provided with 137 channels of instrumentation to monitor the following aspects of specimen behavior:  1) applied loads (measured using pressure transducers attached to the hydraulic actuators in the case of vertical loads, and load cells attached to the hydraulic actuators in the case of horizontal loads); 2) overall lateral displacements (measured using linear potentiometers attached between the specimen and a fixed reference frame; 3) flexural deformations of each wall (measured using series of linear potentiometers placed vertically along both edges of each wall); 4) shearing deformations of each wall (measured using pairs of linear potentiometers, oriented diagonally in each wall panel); 5) slip (measured using linear potentiometers placed between the wall and the base, and between the base and the laboratory floor); 6) end rotations of coupling beams (measured using pairs of linear potentiometers placed above and below both ends of each coupling beam); and 7) strains and stresses in reinforcement and concrete (measured using electrical resistance strain gages).  Specimen 2b was instrumented similarly.

OVERALL PERFORMANCE OF TEST SPECIMENS

Loading History for Specimen 2a

Specimen 2a was subjected to the Sequential Phased Displacement (SPD) loading history shown in Fig. 3 and developed earlier in the TCCMAR Program [4].  The test was begun under load control, and changed to displacement control after the start of yielding of the longitudinal reinforcement of the walls.  Readings were taken at the displacement peaks, and at frequent intervals between those peaks.

Test Results for Specimen 2a

The observed history of base shear versus overall drift ratio (top lateral displacement divided by the specimen heigth of 204 inches) for the entire test is shown in Fig. 4.  In Fig. 5, the envelope of base shear versus overall drift ratio is compared with analytical predictions made before the test using a sequential collapse analysis.

As shown in Figs. 4 and 5, Specimen 2a reached a maximum base shear of 95.9 kips (427 kN), at a overall drift ratio of 0.81% and a top displacement of 1.69 in. (42.9 mm).  Specimen behavior was basically flexural (governed by bending rather than shear), and was limited by in-plane slip of the walls with respect to the base, and by buckling of the walls' longitudinal bars.  Predicted base shears corresponding to significant events agreed well with observed values.  Significant events for northward loading are summarized in Table 3.

## Observations Regarding Behavior of Specimen 2a

Specimen 2a behaved as intended, in that load capacity was limited by formation of a flexural mechanism--that is, by formation of sufficient flexural hinges to produce a mechanism.  Shearing cracks formed near the bases of both walls, but did not widen.  Displacement capacity was limited by buckling of the longitudinal bars at the base of both walls, and by the subsequent lateral (out-of-plane) slip of the bases of both walls with respect to the base beam.  As shown by the load-overall drift ratio curves of Fig. 4, Specimen 2a showed satisfactory maintenance of strength and stiffness, and satisfactory energy dissipation up to overall drift ratios in excess of 1%.  Flexural cracks developed on the top and bottom faces of both slabs, near the inside edges of the coupled walls.  The cracks extended the entire width of the slabs.  There was no evidence of cracking between the walls and the slabs.

## Loading History for Specimen 2b

As shown in Fig. 6, Specimen 2a was subjected to a modified version of the Sequential Phased Displacement (SPD) loading history. Fewer cycles were used, and larger displacement increments were used from one stage of the testing to the next.  As before, the test was begun under load control, and changed to displacement control after the start of yielding of the longitudinal reinforcement of the walls.

## Test Results for Specimen 2b

The observed history of base shear versus overall drift ratio for the entire test is shown in Fig. 7.  In Fig. 8, the envelope of base shear versus overall drift ratio is compared with analytical predictions made before the test using a sequential collapse analysis.

As shown in Figs. 7 and 8, Specimen 2b reached a maximum base shear of 88.6 kips (394 kN), at a overall drift ratio of 1.66% and a top displacement of 3.46 in. (87.9 mm).  Specimen behavior was basically flexural, and was marked by yielding of the flexural reinforcement in both walls, and by the eventual loss of the compression toes in both walls, with some buckling of the walls' longitudinal bars.  Predicted base shears corresponding to significant events agreed well with observed values.  Significant events for northward loading are summarized in Table 4.

## Observations Regarding Behavior of Specimen 2b

Specimen 2b behaved as intended.  Load capacity was limited by formation of a flexural mechanism.  Shearing cracks formed near the bases of both walls, but did not widen.  Displacement capacity was limited by strain hardening of the longitudinal bars at the base of both walls, and by splitting of the compression toes in both walls.  As shown by the load-overall drift ratio curves of Fig. 7, Specimen 2b showed satisfactory maintenance of strength and stiffness, and satisfactory energy dissipation up to overall drift ratios in excess of 1.5%.  As previously, flexural cracks developed at the ends of the coupling slabs. Unlike Specimen 2a, however, the flexural cracks in the planks were completely closed at the end of the test, indicating that the planks had behaved in an essentially elastic manner.  Shearing cracks developed between the walls and the planks, increasing the lateral flexibility of the coupled wall system.


GENERAL CONCLUSIONS REGARDING SPECIMEN BEHAVIOR


In terms of its specific objectives, the tests of Specimens 2a and 2b can be considered a success:  1) the specimen showed satisfactory cyclic shear resistance; 2) the specimen showed satisfactory floor-wall joint behavior; and 3) the specimen behaved as intended and as anticipated in design.


ACKNOWLEDGEMENTS

        The work described here is part of the U.S. Coordinated Program for Masonry Building Research (TCCMAR), and is supported by the National Science Foundation (Grant No. ECE-8611860).  The NSF Program Manager is Dr. A. J. Eggenberger.  The views expressed in this paper are those of the authors alone, and are not necessarily those of other TCCMAR participants nor of the National Science Foundation.

REFERENCES

[1]   Technical  Coordinating  Committee  for  Masonry  Research,  U.S.
      Coordinated Program for Masonry Building Research:    Summary
      Report, August 1986.

[2]   Antrobus,  N.,  Leiva,  G.,  Merryman,  M.  and  Klingner,  R.  E.,
      "Preliminary  Report  on  Testing  of  Specimen  2a,  TCCMAR  Task
      3.1(c):    In-Plane  Seismic  Resistance  of  Two-Story  Concrete
      Masonry Walls with Openings," Proceedings, 4th JTCCMAR Meeting,
      October 16-19, 1988, Rancho Bernardo, California.

[3]   Merryman, Mark, "In-Plane Seismic Resistance of Two-Story Concrete
      Masonry Coupled Walls," M.S. Thesis, the University of Texas at
      Austin, August 1989.

[4]   Porter, M. L., "Sequential Phased Displacement (SPD) Loading for
      TCCMAR Testing," presented at Keystone, Colorado, September 1986.

METRIC (SI) CONVERSION FACTORS

| | | | | |
|---|---|---|---|---|
| 1 ft. | = | 0.305 m | 1 kip | = | 4.448 kN |
| 1 in. | = | 25.4 mm | 1 ksi | = | 6.895 MPa |
| 1 psf | = | 4.882 kg/m$^2$ | 1 psi | = | 0.006895 MPa |

TABLE 1 -- Results of standard material tests, Specimen 2a.

| Material | Number, Type and Age of Specimens | Average Compressive Strength psi    (MP ) | | Coeff. Var. |
|---|---|---|---|---|
| Concrete Units | 3 units, gross area<br>3 units, net area | 830<br>1750 | ( 5.72)<br>(12.1 ) | 11%<br>11% |
| Lab. Mortar | 3 2-in. cubes, 14 days<br>3 2-in. cubes, 31 days | 3260<br>3380 | (22.5 )<br>(23.3 ) | 4%<br>3% |
| Field Mortar | | | | |
| Story 1 | 3 2-in. cubes, 16 days<br>3 2-in. cubes, 28 days<br>2 3-in. cyl.,   16 days<br>5 3-in. cyl., test day | 800<br>690<br>600<br>1150 | ( 5.52)<br>( 4.76)<br>( 4.14)<br>( 7.93) | 3%<br>10%<br>6%<br>5% |
| Story 2 | 3 2-in. cubes, test | 1230 | ( 8.48) | 2% |
| Grout | | | | |
| Story 1 | 3 3-in. prism, 28 days<br>1 3-in. prism, test<br>5 3x6 yin. core, test | 5320<br>5410<br>4040 | (36.7 )<br>(37.3 )<br>(27.9 ) | 4%<br>-<br>19% |
| Story 2 | 3 3-in. prism, test<br>4 3x6 in. core, test | 4690<br>4420 | (32.3 )<br>(30.5 ) | 8%<br>10% |
| Masonry Prisms | | | | |
| Story 1 | 3 prisms, test day | 2010 | (13.9) | 19% |
| Story 2 | 4 prisms, test day | 2340 | (16.1) | 8% |
| Slab Concrete | | | | |
| Floor 1 | 3 6-in. cyl., 7 days<br>3 6-in. cyl., test | 4250<br>5280 | (29.3)<br>(36.4) | 12%<br>1% |
| Floor 2 | 5 6-in. cyl., test | 3660 | (25.2) | 9% |

TABLE 2 -- Results of standard material tests, Specimen 2b.

| Material | Number, Type and Age of Specimens | Average Compressive Strength psi    (MP ) | | Coeff. Var. |
|---|---|---|---|---|
| Concrete Units | 3 units, gross area<br>3 units, net area | 830<br>1750 | ( 5.72)<br>(12.1 ) | 11%<br>11% |
| Lab. Mortar | 3 2-in. cubes, 14 days<br>3 2-in. cubes, 31 days | 3260<br>3380 | (22.5 )<br>(23.3 ) | 4%<br>3% |
| Field Mortar<br>    Story 1<br><br>    Story 2 | 9 2-in. cubes,156 days<br>4 3-in. cyl., 156 days<br>9 2-in. cubes,100 days<br>7 3-in. cyl., 100 days | 1640<br>1330<br>1770<br>1650 | (11.31)<br>( 9.17)<br>(12.20)<br>(11.38) | 13%<br>4%<br>9%<br>10% |
| Grout<br>    Story 1<br><br>    Story 2 | 4 3-in. prism,154 days<br>3 3x6 in. core, 154<br>4 3-in. prism, 88<br>4 3x6 in. core, 88 | 5480<br>3250<br>4930<br>2470 | (37.78)<br>(22.41)<br>(33.99)<br>(17.03) | 6%<br>9%<br>10%<br>27% |
| Masonry Prisms<br>    Story 1<br>    Story 2 | 4 prisms, test day<br>4 prisms, test day | 3090<br>2510 | (21.31)<br>(17.31) | 10%<br>14% |
| Slab Concrete<br>    Floor 1<br>    Floor 2 | 6 6-in. cyl., test<br>5 6-in. cyl., test | 5220<br>3670 | (35.99)<br>(25.30) | 2%<br>6% |

TABLE 3 -- Observed behavior of Specimen 2a, northward loading.

| Load Point | Specimen Behavior | Base Shear kips | (kN) | Top Displ. in. | (mm) |
|---|---|---|---|---|---|
| 39 | Flexural cracking of tension wall | 24.2 | (107.6) | 0.036 | (0.3) |
| 58 | Flexural cracking of compression wall | 24.2 | (107.6) | 0.045 | (0.4) |
| 95 | Yield of longitudinal steel in tension wall | 48.3 | (226.4) | 0.11 | (1.8) |
| 131 | Cracking and yield of bottom slab; yield of compression wall of bottom slab | 58.5 | (260.2) | 0.16 | (4.1) |
| 169 | Cracking and yield of top slab | 62.4 | (277.6) | 0.24 | (6.1) |
| 170 | Flexural cracking above lap splices | 66.8 | (297.1) | 0.28 | (7.1) |
| 207 | Diagonal cracks in tension wall | 73.2 | (325.6) | 0.41 | (10.4) |
| 279 | Diagonal cracks in compression wall | 86.7 | (385.6) | 0.86 | (21.8) |
| 317 | Toes of both walls start to crush; wide flexural cracks at wall bases and splices | 89.4 | (397.7) | 1.13 | (28.7) |
| 357 | Maximum load | 95.9 | (426.6) | 1.69 | (42.9) |
| 384 | Face shells spall at toe of compression wall | 77.9 | (346.5) | 1.67 | (42.4) |
| 414 | Extreme compression bar buckles in compression wall; walls slide on base | 80.5 | (358.1) | 2.23 | (56.6) |

TABLE 4 -- Observed behavior of Specimen 2b, Northward Loading.

| Load Point | Specimen Behavior | Base Shear | | Top Displ. | |
|---|---|---|---|---|---|
| | | kips | (kN) | in. | (mm) |
| 52 | Flexural cracking of 2nd story tens. wall; yield of 1st story tens. wall | 34.9 | (155.2) | 0.112 | (2.8) |
| 63 | Flexural cracking of 2nd story comp. wall; yield of 1st story comp. wall; cracking of 1st flr slab top face | 43.9 | (195.3) | 0.202 | (5.1) |
| 135 | Diagonal shear cracking of both 1st story walls; cracking of 1st story slab bottom face | 65.5 | (291.3) | 0.56 | (14.2) |
| 171 | Yield of 2nd story tens. wall base | 80.8 | (359.4) | 1.25 | (31.8) |
| 172 | Cracking of comp. toe both walls; face shell spall at toe of comp. wall; | 88.0 | (391.4) | 1.69 | (42.9) |
| 208 | Maximum load and top displacement; longitudinal shear cracking of bottom face of both slabs | 88.1 | (391.9) | 3.46 | (87.9) |
| 217 | Fracture of extreme tens. bar of comp. wall | 46.0 | (204.6) | 2.51 | (63.8) |
| 219 | Loss of comp. toe, comp. wall | -55.4 | (-246.4) | -2.34 | (-59.4) |
| 220 | Extreme comp. bar of comp. wall buckles | 6.9 | (30.7) | 0.00 | ( 0.0) |

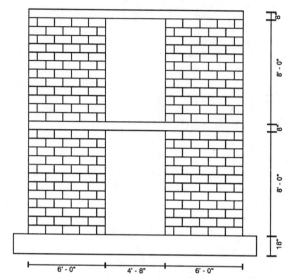

**Fig. 1:**    Type 2 specimens

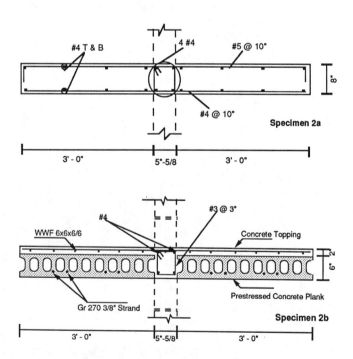

**Fig. 2:**    Floor system details, Specimens 2a and 2b

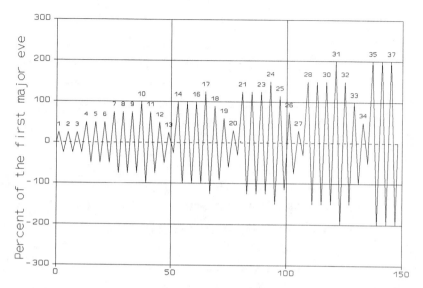

**Fig. 3:**    Sequential Phased Displacement (SPD) loading history used
for Specimen 2a

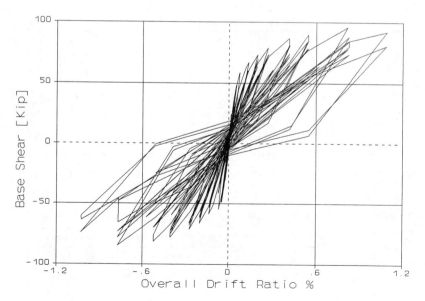

**Fig. 4:**    Observed load versus overall drift ratio, Specimen 2a
(September 10-14, 1988)

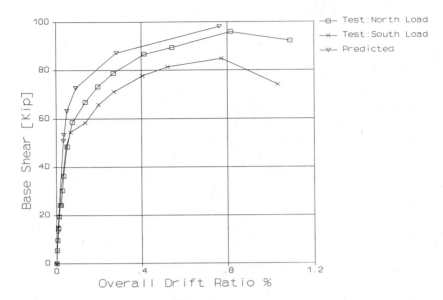

**Fig. 5:**    Comparison between predicted and observed load-overall drift ratio envelopes, Specimen 2a

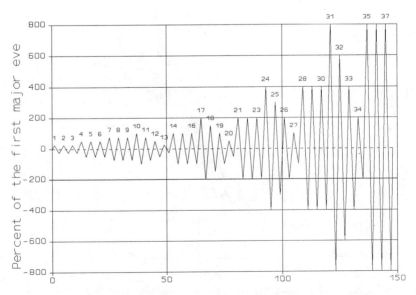

**Fig. 6:**    Modified Sequential Phased Displacement (SPD) loading history used for Specimen 2b

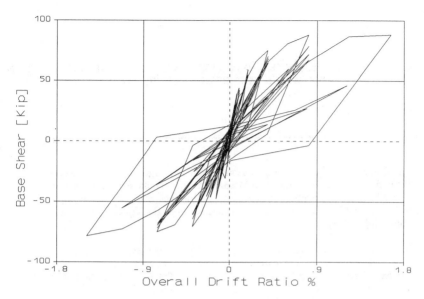

**Fig. 7:**    Observed load versus overall drift ratio, Specimen 2b
(March 30 - April 3, 1989)

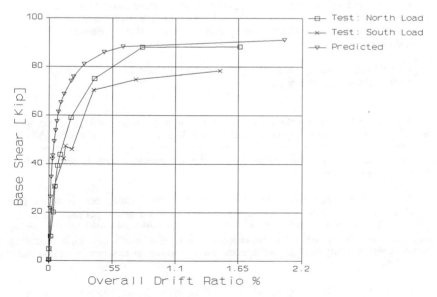

**Fig. 8:**    Comparison between predicted and observed load-overall
drift ratio envelopes, Specimen 2b

Norbert V. Krogstad

MASONRY WALL DRAINAGE TEST - A PROPOSED METHOD FOR FIELD EVALUATION
OF MASONRY CAVITY WALLS FOR RESISTANCE TO WATER LEAKAGE

REFERENCE: Krogstad, N.V., "Masonry Wall Drainage Test - A Proposed
Method for Field Evaluation of Masonry Cavity Walls for Resistance to
Water Leakage," Masonry: Components to Assemblages, ASTM STP 1063,
John H. Matthys, Editor, American Society for Testing and Materials,
Philadelphia, 1990.

ABSTRACT: A procedure for the field testing of wall drainage systems is
proposed. The procedure combines a new test method called the "Wall
Drainage Test" with field adapted versions of ASTM E514 to evaluate new
and existing masonry drainage cavities and flashing systems. The wall
drainage test introduces water directly into masonry cavities at rates
based upon values of water penetration measured in field versions of
ASTM E514. To assure that water is introduced uniformly along the back
face of the outer wythe during such testing, realistic limits are
recommended for several parameters including the velocity and volume of
water at points of entry into the cavity as well as limits on the minimum
number and maximum spacing of the entry points. Procedures for using the
test method as a quality control check for new masonry walls and as an
investigative tool to evaluate existing masonry wall leakage problems are
discussed.

KEYWORDS: Water penetration, masonry, wall drainage test, masonry
cavity, field testing, interior leakage, flashing

Most people familiar with masonry know that some water will penetrate into and
through the outer masonry wythe of traditionally designed veneer and cavity walls during
wind-driven rains. This water, however, does not usually present a problem to the users
of such buildings. Water penetration into or through masonry walls is only damaging to
the buildings or its tenants:

1. If the drainage cavity behind the veneer or the flashing system at the base of
the veneer section is unable to handle the volume of water which penetrates through the
veneer. If this is the case, either water enters directly into the interior spaces or
enters into the backup walls. Water entering into interior spaces can result in damage to
interior finishes, equipment or can become an annoyance to the tenants of the building.

Norbert V. Krogstad, Senior Architect/Engineer, Wiss, Janney, Elstner Associates, Inc.,
Consulting and Research Engineers, 330 Pfingsten Rd., Northbrook, Illinois 60062

Water entering into the backup wall can lead to deterioration of the backup or can adversely increase the humidity of interior spaces.

2.    If large volumes of water pass through the veneer on a regular basis, such large volumes of water may accelerate corrosion and deterioration of wall ties, and shelf angles due to the numerous wetting and drying cycles.

3.    If large volumes of water penetrate into and collect within the masonry veneer itself either by being absorbed by the materials or by collecting in the cores of the masonry units.    This may accelerate deterioration of the veneer especially in severe freeze-thaw environments or the deterioration of metal elements in the cavity due to the resulting high humidity levels in the cavity space.

CURRENT ASTM TESTS

At this time, no ASTM tests are applicable to masonry veneer and cavity walls which directly address these issues.    ASTM E514 titled "Standard Test Method for Water Penetration and Leakage Through Masonry" is only to be used as a laboratory test. Currently there is no field version of this test accepted by ASTM.    Even if a field version of E514 is adopted, this test by itself would not adequately evaluate the wall system.    This is due to the limitations of the current lab and field versions of ASTM E514 which only measure the quantity of water that is penetrating into or through a masonry wall.    The field versions of ASTM E514 do introduce some water into the cavity but because the area tested is usually only 12 sq ft (1.1 sq meters) they test far too small a percentage of the wall surface to adequately evaluate the flashing system, as shown in Fig. 1.

Fig. 1 - Field adapted ASTM E514 test

The usage of ASTM E1105 titled "Standard Test Method for Field Determination of Water Penetration of Installed Exterior Windows, Curtain Walls, and Doors by Uniform of

Cyclic Static Air Pressure Difference" may be applicable to masonry veneer walls as well. This test method, however, is not particularly suited for masonry walls. In order to adequately test the cavity drainage and flashing system of a masonry wall it is generally necessary to test a large area of a wall. Testing a relatively small wall area may not generate leakage that would develop in an actual rain storm. This is because water which may collect on flashings during rain storms which effectively test the entire wall may simply run horizontally on flashings during a small scale test. The problem could be solved in part by creating water dams along the flashing at the ends of the test region. This procedure, however, can be cumbersome and would also require performing a large number of tests to adequately test the drainage system, since the length of sections tested would be relatively small.

A second problem with using ASTM E1105 for masonry walls relates to the difficulty of creating a realistic pressure differential across the face of the masonry. Because masonry veneer and cavity walls usually have an open cavity space behind the outer wythe, it is usually not practical to establish a pressure drop across the face of the masonry with an interior chamber. Instead, a chamber would have to be built on the outside of the wall and enclose the spray grid. Whereas this is certainly possible to do, such a test may not be economically feasible for many projects especially if a large percentage of the wall is to be tested.

PROPOSED TEST PROCEDURE

An alternate method of testing masonry walls would be to combine a field modified version of ASTM E514, such as that proposed by Clar Monk in ASTM STP 778, titled "Adaptations and Additions to ASTM Test Method E514 (Water Permeance of Masonry) for field conditions" (1) with a new test method introduced in this paper called the "Wall Drainage Test." The wall drainage test is a very simple test which would introduce water directly into the masonry cavity space along the back face of the outer wythe to simulate water penetration through the masonry which would occur during a rain storm. The test can be performed by drilling a series of small holes through the outer wythe of masonry. Water can be directed through these holes in a controlled fashion, based upon the results of Field ASTM E514 tests. The flow of water could be controlled by using a small hose with an in-line flow regulator and flow meter for each water entry point, as shown in Figs. 2 and 3. Using this method, large areas of wall could easily be tested. To date this author has performed such tests on lengths of wall up to 36 ft, (11 meters) but considerably longer lengths could reasonably be tested. Failure would occur if leakage was observed into the interior or backup.

A test procedure combining a field modified version of ASTM E514 with the wall drainage test could be implemented as follows:

1. Representative areas of the wall would be selected for testing by a field modified version of ASTM E514. It is suggested that a minimum of five such areas be selected. Test areas could include slightly cracked regions provided such cracks are representative and provided any results from cracked areas are combined using a weighted average based on actual occurrence of these cracks on the wall with the results of at least five tests on uncracked areas.

2. Field modified ASTM E514 tests would be performed on the selected areas using parameters based upon actual recorded worst case occurrences of wind and rain for the region. If it is not practical to determine the actual worst case conditions, a conservative approach for many regions of the country would be to use parameters established in the laboratory version of ASTM E514. When using a field modified version of ASTM E514, some attempt should be made to record both the amount of water which

penetrates into the masonry and the amount of water which penetrates through the masonry. One method of doing this would be to run such tests over an extended period of time. The beginning portion of such tests would include both absorption and permeance. After the

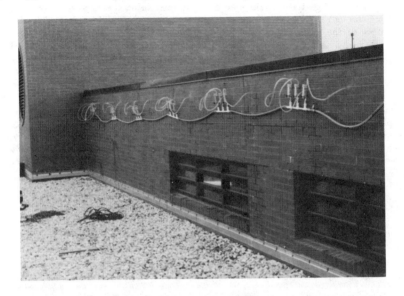

Fig. 2 - Typical wall drainage test set-up.

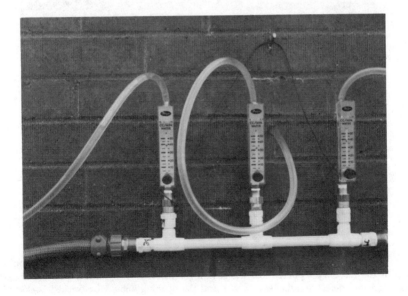

Fig. 3 - Water flow rate meter and flow regulator at water entry points.

readings have stabilized, it is presumed that the area tested is saturated and that the rate after this point would represent only penetration through the veneer thickness.

3.    Wall drainage tests would then be performed using the average values of water penetrating through the masonry from the field modified ASTM E514 tests to determine the effectiveness of the wall cavity and flashing system.    Test areas should be chosen to include as many features of this wall as possible such as vertical expansion joints, corners, etc.    Continuous observations should be made from the interior during the test to check for signs of interior leakage.    Openings should be made so that it is possible to determine if water is entering the backup system as well.    An alternative approach to making openings in the backup wall would be to measure both the total volume of water introduced into the wall and that coming out at the flashing and weepholes.    This can be accomplished by attaching a gutter system directly below the flashing and weepholes to collect the water, as shown in Fig. 4.    The approach has the advantage of identifying a leakage problem that may otherwise go unnoticed but has the disadvantage in that the test is more    difficult to run and takes considerably more time since the test must be run until readings stabilize.

Fig. 4 - Gutter system attached below flashing and weepholes to measure water leaving the wall during testing.

4.    If interior leakage occurs during the wall drainage test, openings should be made in the masonry to determine the reason for the failure.    When a potential defect is discovered, the defect should be repaired and the opening should be bricked back.    When the masonry is sufficiently set, a second test should be performed to confirm the effectiveness of the repair.    Retesting in this way would permit exploring several different repair options.

LIMITS OF PARAMETERS FOR WALL DRAINAGE TEST

The intent of the wall drainage test is to simulate in some reasonable fashion, the path that would be followed by water which penetrates through masonry walls. In order to effectively do this, it is very important to establish some limits on the test parameters to assure a reasonably uniform distribution of water along the back face of the masonry veneer wall. The following limits are suggested:

1.   Water entry points should be spaced no farther apart than one-half the vertical height above the element in the wall cavity and flashing system to be tested or a maximum of 24 in. (610. mm), as shown in Fig. 5.   The element in the wall cavity and flashing system which is tested is usually the flashing at the base of the wall, however, it may also be window head flashings or suspected locations of mortar bridges.

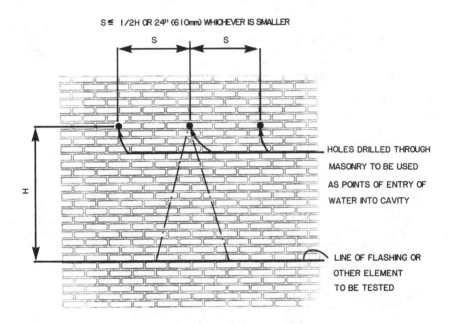

Fig. 5 - Spacing of points of water entry

2.   The quantity of water introduced at each entry point should equal the tributary area represented by the entry point times the average leakage rate per sq ft recorded in the field modified E514 tests but not greater than 300 cc/min.   (The spacing of entry points should be reduced if necessary to assure this rate.)   The tributary areas should be the gross area represented by each entry point minus any nonmasonry areas, such as, portions of vents or windows.   The gross area is the actual height of masonry above the element to be tested and below upper levels of flashing times the spacing between entry points.

3.    To assure that the velocity of water leaving the end of the tube is very small, the outside face of the water stream should extend no farther than .5 centimeters in front of the end of the tube at 50 centimeters below the tube when the tube is oriented horizontally, as shown in Fig. 6. This provision can be easily met at 300 cc/min or less by using 1/4 in. (6.35 mm) inside diameter plastic tubing.

4.    Holes at entry points should be drilled at the top of a head joint at a 30 degree angle downward and must extend completely through the outer wythe. This provision is intended to prevent water from flowing back out of the holes during the test.

5.    Tubes which are inserted into the holes should be held at least 1/2 in. (12.7 mm) back from the end of the hole, as shown in Fig. 7. This provision assures that water flows along the back face of the masonry.

6.    If the entire length of flashing is not tested, additional water entry points should be added beyond the length of flashing or other element to be tested amounting to 50 percent on each side, as shown in Fig. 8. If this is not possible, water dams should be constructed on the flashing at both the ends of the test area provided these dams do not interfere with the testing. The purpose of these water dams would be to prevent lateral flow of water which would not happen in actual rain storms that test the whole wall simultaneously. Water dams could be constructed by making openings in the masonry at the end of the test area to expose flashing. The dams could be sealed to the flashing and the cut edge of the masonry with sealant or mastic for the full height of the flashing.

7.    The minimum length of test area should be 10 ft (3 meters) or the actual length of the masonry wall.

8.    If it is desirable to measure both the volume of water entering the wall as well as the volume of water leaving the wall at flashing and weepholes, water dams should be installed on the flashing at the end of the test area and a gutter system should be installed beneath the flashing and weepholes capable of collecting water leaving the wall at this level. Tests of this kind should be performed until the rate of water loss has stabilized for a period of not less than 1 hour.

USE OF WALL DRAINAGE TEST FOR NEW AND EXISTING MASONRY

The wall drainage test in combination with the field modified ASTM E514 tests could be used both as a quality control procedure for new construction and as an investigative tool for existing buildings which have leakage problems possibly associated with masonry walls.

In new construction, a mock-up wall should be constructed with the materials to be used in the building. After the wall has cured for 28 days, three or more tests should be performed in the wall using E514. Measurements should be made of both the water penetrating the outside face and water penetrating through the walls. If the results of the water penetration tests are not acceptable, a new mock-up wall could be constructed with different materials and retested. Once construction begins, wall drainage tests could be performed on the new masonry walls at a rate equal to twice the average through-wall penetration rates measured in the mockup test as soon as 3 days after the wall is completed. The factor of two would serve as a performance factor to account for potential increased water penetration rates due to aging. Problems with the cavity drainage and flashing system could be identified early on in the construction and costly future repairs could be avoided. After wall areas have cured for 28 days, field modified

ASTM E514 tests should be performed on representative areas.  If the penetration rates are equal to or less than those in the mockup wall, no further wall drainage tests would be required.  If the measured penetration rates are greater than those on the mockup, additional wall drainage tests should be conducted along with a careful evaluation of other possible problems which may arise out of high penetration rates.

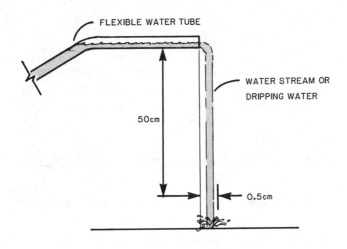

Fig. 6 - Limits of discharge of water from hose

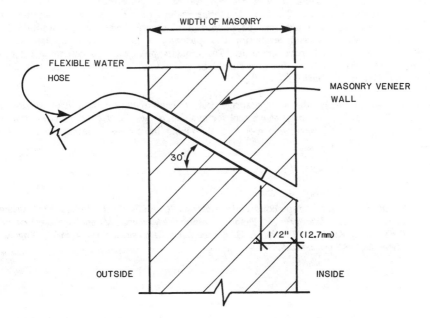

Fig. 7 - Position of hose in masonry wall

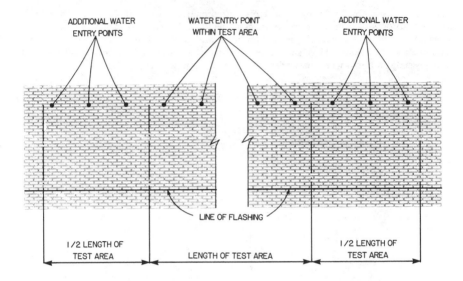

Fig. 8 - Requirement for addition water entry points if entire flashing length is not tested

This test procedure can also be used very effectively in the determination of potential leakage problems in existing construction. First, several field modified ASTM E514 tests should be conducted on representative wall areas. If mortar bridging is not suspected to be a problem, the wall drainage test can be performed at a level two-thirds the wall height above the flashing. If mortar bridges or other cavity irregularities are suspected, wall drainage tests can be performed at several levels on the wall. When performing tests at several levels for this purpose, it is important to start at the bottom of the wall and move upwards. The tributary areas used to determine the flow rates at the entry points in this case should be based upon only the area of wall above the level of the entry points.

If used rationally, the wall drainage test in combination with a field version of ASTM E514 can become an effective tool for evaluating the effectiveness of new masonry walls as well as an investigative tool for determining the causes of leakage problems in existing masonry walls.

REFERENCES

(1)  Monk, C. B., Jr., "Adaptations and Additions to ASTM Test Method E514 (Water Permeance of Masonry) for Field Conditions," *Masonry: Materials, Properties, and Performance, ASTM STP 778,*  J. G. Borchelt, Ed., American Society for Testing and Materials, 1982, pp. 237-244.

*Ahmad A. Hamid[1], George F. Assis[2], and Harry G. Harris[3]*

# TOWARDS DEVELOPING A FLEXURAL STRENGTH DESIGN METHODOLOGY FOR CONCRETE MASONRY

**REFERENCE** : Hamid, A. A., Assis, G. F., and Harris, H. G., "Towards Developing a Flexural Strength Design Methodology for Concrete Masonry, <u>Masonry: Components to Assemblages ASTM STP 1063</u>, John Matthys, Editor, American Society for Testing and Materials, Philadelphia, 1990.

**ABSTRACT:** In order to develop an appropriate flexural strength design methodology for concrete masonry, information about the nonlinear stress distribution in the compression zone is needed. This paper presents experimental results from grouted concrete masonry prisms tested under eccentric compression loading in the out-of-plane direction which is part of Task 1.2(a) of the U.S.-Japan Coordinated Program on Masonry Building Research. The stress parameters $k_1k_3$ and $k_2$ are presented for 150 mm (6 in.) concrete masonry and are used to predict the ultimate moment capacity of full-scale reinforced masonry walls. Good agreement was obtained between the predicted and the experimental moment capacities. An equivalent rectangular stress block is presented. It is concluded that a design methodology similar to that for reinforced concrete is feasible for grouted concrete masonry.

**KEYWORDS:** Concrete masonry units (CMU), compressive strength, flexural design, masonry, material properties, moment capacity, stress block, stress parameters, ultimate strength

1,2,3 Professor, Graduate student, Professor, respectively, Department of Civil and Architectural Engineering, Drexel University, Philadelphia, PA., 19104.

In order to develop an appropriate design methodology for concrete masonry based on a strength limit state, information about the stress distribution in the compression zone is needed. The stress parameter $k_2$ defines the location of the stress resultant (Fig. 1) while the stress parameter $k_1k_3$ is the ratio between eccentric and concentric compressive strengths. Task 1.2(a) of the U.S.-Japan Coordinated Program on Masonry Building Research (TCCMAR) [1,2] is aimed at determining the stress-strain relationships under concentric and eccentric loading and the stress parameters for the compression stress block. The program covers different parameters such as unit size and strength, geometry, and direction of loading.

In this paper, the stress parameters $k_1k_3$ and $k_2$ for 150 mm (6 in.) concrete masonry under out-of-plane loading are presented. These values are used to predict the ultimate moment capacity of reinforced concrete masonry walls tested under out-of-plane loading as part of Task 3.2(b) of the TCCMAR program [3]. A correlation between predicted and experimental results is presented. An equivalent stress block, similar to the "Whitney" stress block for reinforced concrete is suggested.

## STRESS PARAMETERS

Three-course, half CMU prisms were tested under eccentric compression loading using the test set-up shown in Fig. 2. The loading system consists of a two-channel MTS loading system which operates on displacement control. The concentric load ($P_A$) is applied with a 328-kip actuator. A 55-kip actuator, placed at an eccentricity of 0.76m (30 in.), is used to apply the eccentric loading ($P_E$) via a feedback control system. An 8" diameter spherical head is placed between the concentric actuator and a stiff beam which is used to transfer the eccentric loading to the specimen (Fig. 2). Four LVDT's, mounted between the bearing plates at the four corners, are used to measure the maximum fiber compressive strain and to monitor any accidental eccentricity. Two control LVDT's, placed at the location of the neutral axis, control the displacement of the 55-kip actuator($P_E$) to maintain a triangular strain distribution with zero strain at the extreme fibers as shown in Fig. 1.

Applying a concentric loading $P_A$ and an eccentric loading $P_E$, the stress parameter governing the position of the stress resultant ($k_2$) can be derived from equilibrium considerations:

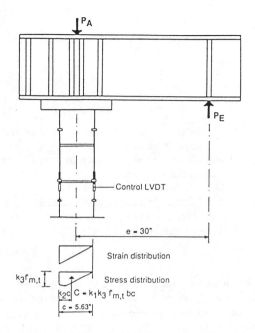

**Fig. 1 -    Stress    Parameters    for    the    Compression Stress    Block**

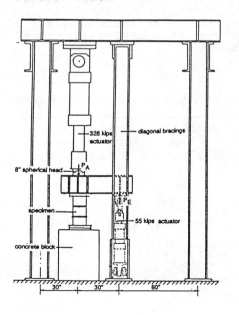

**Fig. 2 -    Test    Set-Up    for    Eccentric    Compression Loading**

$$k_1 k_3 = \frac{P_A - P_E}{f'_{m,t}\, bc} \qquad (1)$$

$$k_2 = \frac{1}{2} - \frac{P_E\ (e)}{(P_A - P_E)c} \qquad (2)$$

where

$P_A, P_E$ = concentric and eccentric loads, respectively,

$f'_{m,t}$ = maximum prism compressive strength under concentric load,

b    = thickness of the prism,

c    = distance from neutral axis to extreme compressive fiber, and

e    = eccentricity of load $P_E$

Figures 3 and 4 show the variation of the stress parameters $k_1 k_3$ and $k_2$ versus maximum compressive strain, respectively. The stress parameters corresponding to peak $P_A - P_E$ are listed in Table 1 for the three test prisms. It is apparent that the obtained stress parameters are similar to those for reinforced concrete [4] which suggests that a strength design methodology similar to that for reinforced concrete is feasible. The maximum strain of 0.0025 is, however, lower than the 0.003 value commonly used for unconfined concrete.

**Table 1- Test Results from Masonry Prisms**

| Specimen | $k_1 k_3$ Indiv. | Mean | $k_2$ Indiv. | Mean | $\varepsilon^*$ Indiv. | Mean |
|---|---|---|---|---|---|---|
| 1 | 0.766 |       | 0.420 |       | 0.0026 |        |
| 2 | 0.792 | 0.793 | 0.451 | 0.433 | 0.0027 | 0.0025 |
| 3 | 0.821 |       | 0.433 |       | 0.0022 |        |

* Compressive strain at the extreme fibers at peak load

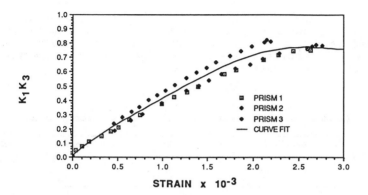

Fig.3  -    Stress   Parameter   $k_1 k_3$   vs.   Extreme   Fiber
Compressive   Strain

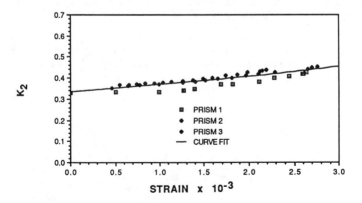

Fig. 4  -    Stress   Parameter   $k_2$   vs.   Extreme   Fiber
Compressive   Strain

## APPLICATION TO FULL-SCALE WALLS

In Task 3.2(b) of the U.S.-Japan Coordinated Program for Masonry Building Research, full-scale reinforced CMU masonry walls were tested under out-of-plane loading [3]. Details of the test specimens and loading are shown in Fig.5. The experimental values for the ultimate moment capacities for three walls having different percentage of steel are listed in Table 2. The location and properties of reinforcing steel are also listed.

### Table 2- Results of Reinforced Concrete Masonry Walls

| Wall | Area of Steel, mm$^2$ (in$^2$) | $\rho$bal.[1] | b cm (in) | d cm (in) | Yield Str.[2], MPa (ksi) | Ult. Mom.[3], kN·m (k-in.) Exper. | Predict. | Ratio Exper. Predict. |
|------|------|------|------|------|------|------|------|------|
| W1 | 258 (0.40) | 0.016 | 120.5 (47.44) | 7.44 (2.93) | 380 (55.0) | 8.23 (72.25) | 6.78 (59.55) | 1.21 |
| W2 | 400 (0.62) | 0.012 | 121.3 (47.75) | 7.19 (2.83) | 461 (66.8) | 14.70 (129.10) | 12.26 (107.60) | 1.20 |
| W3 | 774 (1.20) | 0.013 | 121.0 (47.63) | 7.21 (2.84) | 436 (63.2) | 21.28 (186.80) | 20.78 (182.40) | 1.02 |

1. Based on actual $f_y$ and $E_s$ from axial tension test, prism compressive strength of 14.15 MPa (2050 psi) and maximum compressive strain of 0.0025.
2. From axial tension test.
3. Based on the stress parameters listed in Table 1 and prism compressive strength of 14.15 MPa (2050 psi).

The moment capacity for an under-reinforced rectangular section can be predicted from static equilibrium.

$$C = T \quad \text{or} \quad k_1 k_3 f'_{m,t} bc = A_s f_y \tag{3}$$

$$M_u = A_s f_y (d - k_2 c) \tag{4}$$

$$= A_s f_y \left( d - k_2 \frac{A_s f_y}{k_1 k_3 f'_{m,t} b} \right) \tag{5}$$

where

    b  = width of wall section
    d  = actual depth of reinforcing bars
    $A_s$ = area of reinforcing steel bars
    $f_y$ = actual yield strength

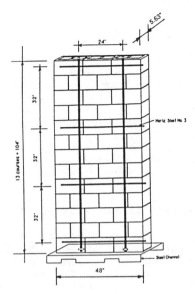

a) Isometric View of the Wall Panel

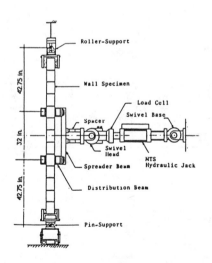

b) Test Set-Up

Fig.5 -  Full-Scale  Tests  of  Reinforced  Block
Masonry  Walls

The above expressions are similar to those used for reinforced concrete[1]. Using Eqn. (5), one can estimate the ultimate moment capacity of the wall provided that $k_1k_3$ and $k_2$ values are known. The compressive strength under concentric loading, $f'_{m,t}$, was obtained by testing three running bond prisms. The prisms were 16"x6"x24" high. The average compressive strength at 28 days was 2050 psi. Using the average values of the stress parameters listed in Table 1, the predicted values of the ultimate moment of the three walls can be calculated using Eqn.(5) and are listed in Table 2. A good correlation between the experimental and predicted values is obtained which clearly indicates that an ultimate strength design method is justifiable.

**EQUIVALENT RECTANGULAR STRESS DISTRIBUTION**

The complex stress distribution can be replaced by a fictitious simple rectangular shape provided that this distribution results in the same total compression force C applied at the same location as in the actual distribution at the point of failure. An equivalent rectangular stress block for concrete masonry is shown in Fig.6 which is similar to the "Whitney" stress block for reinforced concrete [5] currently used in ACI-318 "Building Code Requirements for Reinforced Concrete" [6].

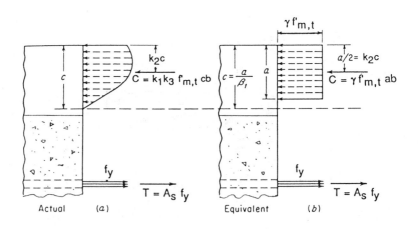

**Fig. 6 - Equivalent Rectangular Stress Block**

From the first condition that the total compression force be the same,

$$C = k_1 k_3 f'_{m,t} cb = \gamma f'_{m,t} ab \qquad (6)$$

From which $\gamma = \dfrac{k_1 k_3 c}{a}$ \qquad (7)

From the second condition that the location of the compression force C be located at the same distance,

$$\frac{a}{2} = k_2 c \qquad (8)$$

From Eqs. (7) and (8)

$$\gamma = \frac{k_1 k_3}{2 k_2} \qquad (9)$$

Substituting the values of $k_1 k_3$ and $k_2$ in the above equation yields $\gamma = 0.91$ compared to the 0.85 value for concrete with $f'_c < 4000$ psi. The height of the rectangular stress block, a, can be calculated from Eqn. (6) and Eqn. (9). The experimental value of $k_2$ result in a value of $\beta_1$ equals 0.87 which is very close to the 0.85 value for concrete.

## CONCLUSION

From the limited test data on grouted concrete masonry presented in this paper it is apparent that a flexural strength design methodology for concrete masonry similar to that for reinforced concrete is feasible. More test data which cover different geometric and material parameters are needed to propose generic stress parameters for concrete masonry.

## ACKNOWLEDGEMENTS

The support of this research by the National Science Foundation through Grant No. ECE-8517019 to Drexel University is gratefully acknowledged. Tests were conducted using a MTS loading system which was partially funded by the National Science Foundation through Grant No. 8412474. Mason's time was made available through the Delaware Valley Masonry Institute and D. M. Sabia and Company and is gratefully acknowledged.

## REFERENCES

[1]    Hamid, A.A., Assis, G.F. and Harris, H.G.,"Material Properties for Grouted Block Masonry," Report No. 1.2(a)-1, U.S.-Japan Coordinated Program for Masonry Building Research, Department of Civil and Architectural Engineering, Drexel University, Philadelphia, PA, August 1988.

[2]    Hamid, A.A., Assis, G.F. and Harris, H.G.,"Compression Behavior of Grouted Block Masonry - Some Preliminary Results," Proceedings of the Fourth North American Masonry Conference, Los Angeles, CA, August 1987.

[3]    Hamid, A.A., Abboud, B.E., Farah, M.W. and Harris, H.G., "Flexural Behavior of Vertically Spanned Reinforced Concrete Block Masonry Walls," Proceedings of the Fifth Canadian Masonry Symposium, Vancouver, Canada, June 1989.

[4]    Hognestad, E., Hanson, N. W., and McHenry, D., "Concrete Stress Distribution in Ultimate Strength Design,"Journal of the American Concrete Institute, V27, No. 4, December 1955, pp. 455-475.

[5]    Whitney, C., "Plastic Theory of Reinforced Concrete Design," Proceedings ASCE, December 1980; Transactions ASCE, Vol. 107,1942, pp. 251-326.

[6]    American Concrete Institute, "Building Code Requirements for Reinforced Concrete", Report of ACI Committee 318, Standard Building Code, Detroit, MI, 1983.

Douglas E. Volkman, P.E.

ASPECTS OF BLAST RESISTANT MASONRY DESIGN

---

REFERENCE: Volkman, D. E., "Aspects of Blast Resistant Masonry Design," Masonry: Components to Assemblages, ASTM STP 1063, John H. Mattys, Editor, American Society for Testing and Materials, Philadelphia, 1990.

ABSTRACT: Blast resistant design should be examined for building code incorporation, due to the potential of explosions occurring in an industrial society. Specifically, public and commercial structures of concrete masonry construction need additional building code criteria, since these buildings have high density populations to protect. Presently, blast resistant design is accomplished by using government published manuals, but these do not address industry standard construction. A design procedure is presented to illustrate and regulate the methodology. Using this procedure, a common wall section is shown capable of withstanding an air blast load of 4.54 kg (10 lbs) of TNT, located 0.91 m (3 ft) above ground surface and 30.48 m (100 ft) from a structure. Building code criteria in this order of magnitude is sufficient to protect against blast, resist progressive failure, and yet not be an economic impediment. Design details and adequate inspection must be observed to ensure blast resistant integrity.

KEYWORDS: blast resistant design, building code, detonation, reflected pressure, modelling, ultimate strength

Blast resistant design is a customary method of structural design for certain experimental and storage facilities at Department of Defense and Department of Energy installations throughout the United States. Comprehensive design and analysis manuals are available through the Government which allow engineering personnel, with a reasonable knowledge of structural mechanics, to determine the structural effects resulting from the dynamics of high explosive detonations. These manuals have been developed through extensive military testing to provide good correlation of yield in steel and reinforced concrete structures from blast loads. One of the most thorough and recent publications available is the six-volume set of engineering manuals, Structures to Resist the Effects of Accidental Explosives[1]. Through well referenced examples,

Mr. Volkman is a staff member at Los Alamos National Laboratory, P.O. Box 1663, Engineering 3, Mail Stop M984, Los Alamos, NM 87545.

the information contained in the separate volumes is woven together to serve as a guide in solving problems. Properties of blast loading from either internal or external explosions, structural dynamics, reinforced concrete design, and structural steel design are the four main topics that comprise the essence of this publication. Although the last volume in this series of manuals introduces blast resistant design of masonry and prestressed concrete, the design emphasis as a whole is placed on reinforced concrete and structural steel.

Most buildings in the United States are designed in accordance with the specifications of one of the major building codes. Due to the importance placed on seismic loading, the Uniform Building Code is probably the most conservative building code addressing lateral loads in masonry structures. However, only a small section of one chapter in this code mentions the structural requirements necessary to sustain blast loading, other than venting a structure to relieve internal pressures. This provision specifies that building components subject to blast exposure be designed to resist a minimum internal pressure of 4.79 kPa (100 psf)[2]. Curiously though, no mention is made of the exposure effects from external explosions.

Industrial complexes, mining operations, construction and demolition activities all employ explosives to varying degrees. Furthermore, many of these operations have close proximity to communities. An accidental explosion occurring within the confines of an industrial complex has the potential to exert significant shock loads through the ground or air to affect the environment beyond the plant perimeter. Development of new sites requires mining companies to handle and use explosive material. Over estimating of the charge necessary to loosen rock obstructions at the mine site can have the same effect as an industrial accident. In supplying the needs of these legitimate consumers with explosive materials, shipment either by truck or train necessitates trafficking hazardous materials through rural and urban areas. Any collision involving one of these vehicles containing a cargo of explosives can present serious destructive consequences, especially if the accident occurs within an urban setting. The blast load possibilities presented so far are derived from unplanned occurrences. Deliberate, planned bombings are the forte of terrorists. Fortunately in the United States, the probability of terrorist activity is remote, but cannot be categorically ignored.

BUILDING CODE CONSIDERATIONS

A building code standard for strengthening concrete masonry structures to withstand blast loads needs to focus on applicability. Public and commercial buildings of any shape, footprint size, and height are of prime concern because of the concentration of people in these facilities. An additional benefit derived from a blast load standard is increased resistance of total structural failure. By sustaining localized failures, prevention of major tragedies, resulting from progressive type yielding, can be averted.

Acoustics need to be addressed for architectural treatment of blast resistant building designs. Walls and roofs should be designed to lower sound transmissions. Exterior finish should be a dense, close grain type surface to enhance reflectance of sound[3].

Fragmentation loads are capable of causing considerable damage. These result from unsecured objects within the vicinity of an explosion, which become launched projectiles by the sudden overpressures of the explosion.   In depth examination of blast resistant design must ascertain the effect of fragmentation loading on a structure.

Successful construction of concrete masonry depends on attention to details.   Explicit design in conjunction with proper inspection is necessary for a good end product.   The following criteria should be considered for inclusion in a building code standard to ensure a better blast resistant structure:

1.   Reinforcing steel bars need to have sufficient development length to deliver full dynamic tensile strength.   For end anchorage, use an embedment length of at least 40 bar diameters.   Reinforcing steel bar splices must have lap length based on dynamic strength characteristics of the material and splices need to be located in low stress regions.

2.   Wall anchorage to both the foundation and roof should be adequate to transfer loads.

3.   Interior masonry walls should be connected to intersecting walls with S-type steel hooks wrapped around vertical reinforcement. Hooks should be located on alternate joints.

4.   Joint mortar should cover the full horizontal projected surface and not just the exposed face of the masonry.

5.   Horizontal wire reinforcement on alternate joints is required to provide a mechanism for load distribution.

6.   Vertical steel reinforcement should be located at all building corners and wall openings.


BLAST LOAD CHARACTERISTICS

Explosive materials are classified as solid, liquid, or gas. Solid explosives are primarily high explosives and are the subject of this article.

Data on blast effects are available for many high explosives.   For simplification, all explosive materials are related to the measured effects of a bare, spherical trinitrotoluene (TNT) charge.   This is accomplished by determining the equivalent weight of TNT for the explosive in question.   The equivalent charge weight is achieved by proportioning the heat of detonation of the explosive to the heat of detonation of TNT and multiplying the proportion by the weight of the charge.

Detonation of an explosive material is what causes a blast to occur.   The result of this process is a very rapid and stable chemical reaction in the substance, which reduces the explosive to a hot, dense, high pressure gas.   The volume expansion resulting from the detonation is immense and results in source pressures somewhere between 53,090 kPa (7,700 ksi) and 33,784 kPa (4,900 ksi)[4].

An atmospheric explosion creates a very intense spherical shock wave expanding equilaterally from the origin of the explosion. Reflection from ground surface or building walls occurs as the wave moves along its propagation path. The effect of the interaction of this rebounding wave with the initial blast air current causes a strengthening of the shock. The shock wave completely engulfs any obstruction encountered and exerts pressure on all exposed surfaces throughout the duration of the blast effect.

Confined explosions occur within structures. Examples of facilities where confined explosions can occur are magazines storing explosive materials, chemical laboratories, grain elevators, any building in which natural gas leaks, gasoline tanks, as well as many other situations where the potential exists for an explosion to occur from within. A confined explosion is synonymous with an internal explosion. This type loading is presently acknowledged in building codes, though coverage is incomplete.

Unconfined explosions occur externally to structures. The three types of unconfined explosions are free air burst, air burst, and surface burst. Of these three, the air burst represents the external explosive loading most applicable to building code incorporation as structural design criteria. An air burst explosion is characterized by an explosive detonated slightly above the ground surface (Fig. 1). The consequence of ground separation decouples the explosive from the ground, eliminating immediate reflection which reduces total pressure magnification. After reflection does occur from the ground, the rebounding pressure wave combines with the air propagated wave producing an increase in pressure at the front. However, the pressure magnification is still less than would be observed from a surface blast, in which the blast occurs at ground surface.

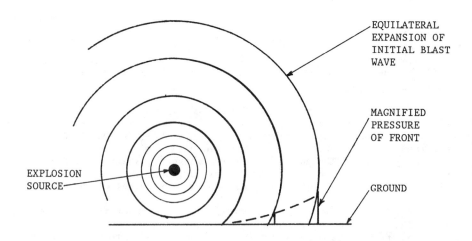

FIG. 1 -- Air burst explosion.

Components of the pressure wave consist of two distinct impulses (Fig. 2). The initial impulse consists of high positive pressure, often referred to as overpressure or pressure exceeding normal atmospheric pressure. The positive impulse decays very rapidly and is followed by a negative pressure impulse, or vacuum, of significantly lower magnitude. The wall of a structure, undergoing such an explosion load, would sustain peak, side-on, positive pressure, which decays quickly over time followed subsequently by a stress reversal resulting from the suction of the negative pressure, but lasting over a longer period of time. Structural members are designed to withstand the stress and deflection incurred for the positive phase of blast loads, as well as the consequential response from the negative phase. A building may sustain both positive and negative pressures simultaneously, depending on the duration of the load and the size of the building.

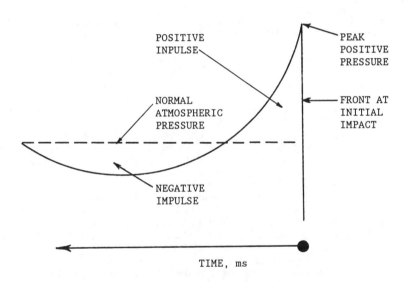

FIG. 2 -- Pressure progression at fixed location.

The magnitude and distribution of a blast load on a structure is a function of three factors. First is the explosive properties, primarily consisting of material composition, energy output, and charge weight. Next is the location of the source detonation relative to the structure. Last is the magnitude and reinforcement of the pressure wave by its reflective interaction with the ground or structure itself.

Some of the measured characteristics of blast phenomenon are accomplished through scaled properties. Distance, wave length, time, and impulse are represented as scaled quantities, that is, the unit of measure divided by the cubed root of the charge weight.

LOAD AND STRUCTURE MODELLING

The primary function of current blast resistant design manuals is
to predict structural adequacy in the advent of an accidental explosion. Much of the structural analysis and design techniques deal with
the elasto-plastic and plastic range of response. The presumption
behind this approach is the need for precision in producing yield of
structural materials for a certain quantity and arrangement of an
explosive charge. Conversely, building codes establish design criteria
and specify the use of standard methods of design to provide a level of
structural quality for a material to function in the elastic range.
Therefore, to manipulate the complexities of blast resistant design
into a usable building code standard necessitates idealizing loads and
structures, which can be readily adapted for the construction industry.

The time endured concept of dynamic load factor (DLF) equates the
maximum deflection of a dynamic load to that deflection resulting from
the static loading condition of an equal magnitude force[5]. Under
idealized conditions of triangular or rectangular impulse load, the DLF
correlates well within the elastic range of structures. Blast impulse
load for the positive phase, which produces maximum structural response, can be idealized as a triangular dynamic load. The peak positive pressure makes initial contact at time equal to zero. This peak
pressure reduces to zero through the time of duration of the loading in
an approximately linear function, thus creating a triangular impulse.

Modelling of a structure is always the most difficult task associated with design because of the numerous degrees of freedom for movement. For simplification, a single degree of freedom system, consisting of a spring and mass is the usual model for structural components
(Fig. 3). The single degree of freedom system is arranged to emulate
the motion of significant response for the structural component. Since
structural systems are more complicated than single degree of freedom
models, adjustments must be made to the model parameters for true correlation with the actual structure. Three dynamic load factors are
used to modify load, mass and structural resistance. Damping is considered negligible, since it has little effect on the maximum response
of the structure which is the first cycle of motion. Traditional
concrete masonry wall design can be made equivalent to the elastic
response of a one degree of freedom idealized system by utilizing the
appropriate dynamic load factors[6].

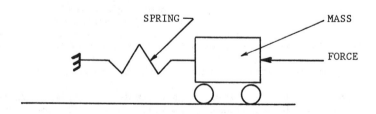

FIG. 3 -- Model of structural response.

MASONRY CONSIDERATIONS

The American Society of Civil Engineers published research stating that unreinforced masonry walls display inherent strength in resisting blast loads. The strength was greater than could be accounted for by flexural considerations alone. By relating to other research, a theory was proposed to advance the understanding of this inherent strength. While undergoing the rapid straining associated with blast loads, the mortar failed in tension, thus causing a yielding from flexure. The extra strength of concrete masonry appeared after this mode of failure. Segmental in construction, the masonry wall produced a load induced deflection in the shape of an arch. Exhibiting strong compressive strength, the masonry wall transformed its bending response into arch action, which produced greater than anticipated resistance to blast loading before ultimate failure[7]. However, due to the inability of unreinforced masonry to elastically sustain blast load, and in the attempt to prevent progressive type failure, use of steel reinforced concrete masonry is the only acceptable means for providing a blast resistant masonry structure.

Depending on geographical location, the Uniform Building Code requires the use of vertical and horizontal reinforcing steel in concrete masonry unit wall construction to resist seismic lateral loads. Utilizing standard practice for seismic construction while incorporating the previously introduced blast analysis techniques, design criteria for blast loads can be established. Pursuant to these goals, single wythe, running bond, concrete masonry unit construction with both joint and vertical cell reinforcement is studied for suitability as a blast resistance standard.

For consistency with the blast analysis techniques, all design is based on ultimate strength to determine the yield of the structural member. Due to the impulse nature of blast loading, the concrete masonry undergoes rapid rates of straining, not unlike the behavior of other construction materials. Incorporating this response for the design allows the use of greater material strengths[8] than would be obtainable under static loading conditions (Table 1).

TABLE 1 -- Blast load material strength.

| Material | Stress Component | Blast Increase Factor |
| --- | --- | --- |
| Concrete Masonry Unit | Flexure | 1.19 |
| | Shear | 1.00 |
| | Compression | 1.12 |
| Steel Reinforcement | Flexure | 1.17 |

DESIGN PROCEDURE

The cohesiveness of a methodical procedure is an attractive expedient to tie the concepts of blast design with commercial masonry construction. While using a building code as a forum, a blast design manual should be used for blast loading synthesis and dynamic analysis, with the entire process fused by the design procedure. This will enable engineering personnel, unfamiliar with this type of design to maintain confidence in the calculations they generate. Therefore, the following design procedure, conforming to these guidelines, is recommended for flexural members:

1.  Determine all known parameters of the design. This is accomplished by sketching the structure and component to be designed. Show supports, spans, known dimensions, openings, charge location, charge size, other loads, and so forth. Next, list the properties of the concrete masonry and reinforcing steel materials. This includes the thickness of the masonry block, assumed joint and vertical cell reinforcement size and spacing, mortar type, specified compressive strength of the masonry, the modulus of elasticity of the masonry, the tensile yield stress of the steel reinforcement, and the modulus of elasticity of the steel.

2.  Calculate the ultimate strength design properties of the member. For flexural design, the equivalent rectangular compressive stress block needs to be determined, from which the ultimate moment capacity can be computed. The material strengths in these calculations need to be corrected for the blast increase factors (see Table 1), as well as adjusted for the standard reliability factors associated with ultimate strength design. In addition, the average of the gross and cracked section, moment of inertia needs to be established.

3.  Determine the dynamic transformation equivalents of the structure. For elastic response of a simple span, the equivalent elastic stiffness, the load/mass equivalent factor, the equivalent mass, and the natural period of vibration has to be calculated[9].

4.  Based on charge size and location, develop the blast load properties and scaled characteristics. For an air burst explosion, the scaled height of the charge, the scaled horizontal distance of the charge, the angle of incidence of the charge to the structure are primary computations. Assuming full ground reflection of the blast before striking the structure, both the ground reflected pressure and the scaled ground reflected impulse must be calculated to determine the reflected pressure acting on the structure and the accompanying scaled time of duration[10].

5.  With the information generated, determine the ratio of the time of duration for the positive phase impulse to the natural period of vibration of the structure. From this ratio, the dynamic load factor is selected[5].

6.  Multiply the dynamic load factor times the calculated reflected pressure on the structure to determine the equivalent static pressure. This result needs to be increased by the standard ultimate strength design load factors for infrequent live load in order to establish a design pressure.

7.  Using the design pressure, calculate the design moment, and compare this result with the ultimate moment capacity of the structural member from step two. Sufficiency is attained if the latter moment exceeds the design moment.

BUILDING CODE PROVISIONS

For an appraisal of the order of magnitude involved in blast resistant masonry design, consider using the design procedure for a wall
with a 3.05 m (10 ft) vertical span.  The wall consists of 20.32 cm
(8 in.) thick concrete masonry, reinforced with 1.27 cm (0.5 in.)
diameter steel bars at 1.22 m (4 ft) centers and wire reinforcing on
alternate horizontal joints.  This represents steel reinforcement consistent with building code requirements in seismically active geo-
graphic areas.  The blast resistance of a wall of this dimension and
strength is approximately a 4.54 kg (10 lb) charge of TNT located
0.91 m (3 ft) above ground and 30.48 m (100 ft) from the structure.
Explosive loading criteria in this range needs to be included in the
building codes for public and commercial structures that are located in
geographically low risk seismic zones.  Essentially, these provisions
would elevate the minimum lateral loading requirements to enhance
public protection in densely populated structures.

Additional work is necessary to firmly establish appropriate blast
load requirements.  However, the information and techniques are available for consolidation into a useful code.  Flexural aspects of blast
resistant masonry design is a major constituent for inclusion in a
building code.  Surface spalling of the masonry requires additional
analysis for closer, higher yield explosions.  Satisfactory results
should be obtained if loading criteria deemed adequate for flexural
resistance is used.  Other issues requiring further investigation are
deflection, shear requirements, anchorage, and support rotation.  Expansion of the design procedure to incorporate these components is
necessary for a comprehensive design, and the eventual formation of
building code provisions.

ACKNOWLEDGEMENTS

The author acknowledges with appreciation the effort and assistance of Debra Archuleta-Lugo and M. Dean Keller, P.E., as well as the
support extended by Rita C. Volkman, Wanda S. Kuzma, Edward Jennings
Volkman, and the author's wife.

REFERENCES

[1]    Department of Defense Explosives Safety Board, Sponsor, Struc-
       tures to Resist the Effects of Accidental Explosions Volumes I-
       VI, Special Publication ARLCD-SP-84001, U.S. Army Armament
       Research and Development Center, Dover, New Jersey, 1984-1987.

[2]    International Conference of Building Officials, Ed., "Require-
       ments for Group H Occupancies," in Uniform Building Code,
       International Conference of Building Officials, Whittier,
       California, 1988, pp. 74.

[3]    Beall, C., "Cementitious Masonry Units," in Masonry Design and
       Detailing, Prentice-Hall, Inc., Englewood Cliffs, New Jersey,
       1984, pp. 75.

[4]     Ayvazyan, H., Dede, M., Dobbs, N., Whitney, M., Bowles, P., Baker, W., and Caltagirone, J. P., "Blast Loads," in Structures to Resist the Effects of Accidental Explosions Volume II, Special Publication ARLCD-SP-84001, U.S. Army Armament Research and Development Center, Dover, New Jersey, 1986, pp. 5.

[5]     Biggs, J. M., "Rigorous Analysis of One-degree Systems," Introduction to Structural Dynamics, McGraw-Hill Book Co., New York City, 1964, pp. 34-81.

[6]     Dede, M., Sock, F., Schramm, S. L., and Dobbs, N., "Dynamically Equivalent Systems," in Structures to Resist the Effects of Accidental Explosions Volume III, Special Publication ARLCD-SP-84001, U.S. Army Armament Research and Development Center, Dover, New Jersey, 1984, pp. 67-73.

[7]     McKee, K. E., and Sevin, E., "Design of Masonry Walls for Blast Loading," Journal of the Structural Division Proceedings of the American Society of Civil Engineers, Volume 84, No. ST1, January 1958, pp. 1512-1 to 1512-18.

[8]     Dede, M., Schramm, S. L., Dobbs, N., and Caltagirone, J. P., "Dynamic Strength of Material," in Structures to Resist the Effects of Accidental Explosions Volume VI, Special Publication ARLCD-SP-84001, U.S. Army Armament Research and Development Center, Dover, New Jersey, 1985, pp. 13.

[9]     Dede, M., Sock, F., Schramm, S. L., Dobbs, N., "Dynamic Design Factors," in Structures to Resist the Effects of Accidental Explosions Volume III, Special Publication ARLCD-SP-84001, U.S. Army Armament Research and Development Center, Dover, New Jersey, 1984, pp. 74-82.

[10]    Ayvazyan, H., Dede, M., Dobbs, N., Whitney, M., Bowles, P., Baker, W., and Caltagirone, J. P., "Unconfined Explosions," in Structures to Resist the Effects of Accidental Explosions Volume II, Special Publication ARLCD-SP-84001, U.S. Army Armament Research and Development Center, Dover, New Jersey, 1986, pp. 15-28.

# SUMMARY

The papers in this ASTM STP 1063 are arranged in two broad categories, namely Components and Assemblages. Components are important from a quality control and production point of view. However the critical test is the performance of the assemblage particularly in light of real world situations.

The papers presented address performance of conventional and new materials under ASTM specifications, substantiation and/or deficiencies of existing ASTM specification criteria, critical examination of application of ASTM test methods, problem areas in masonry and proposed solutions, and structural design and analysis methodologies.

This publication thus contains information of interest to persons producing masonry materials and/or ingredients, masonry components, and masonry assemblages; to persons specifying masonry products; and to persons investigating and/or designing masonry systems. Potential modification to existing specifications and test methods and development of new standards may result from this symposium.

## COMPONENTS

**Bailey, Matthys,** and **Edwards** have investigated the IRA of both bed surfaces of individually cored, flashed, extruded, and molded brick. Current ASTM C 67-87 allows either bed surface to be used in conducting the IRA test. The study indicated that flashed brick and molded brick may give significantly different IRA values depending on the bed surface tested.

**Dunstan, Keck,** and **Hays** present basic properties of ASTM C 90 medium and light weight block utilizing a new lightweight aggregate produced from a mixture of fly ash and hydrated lime. This product, Aardelite, met the requirements of ASTM C33 while producing an economical and technically sound concrete masonry unit.

The current ASTM standard for paving brick, C 902, while acknowledging the need for skid/slip resistance, fails to identify a recommended test procedure. **Trimble** examines existing slip resistance test methods for application to brick pavers and recommends action for implementation of a method into brick standards.

**Hedstrom** and **Hogan's** report addresses many of the concerns of the relationship between grout performance in concrete masonry to grout proportions and to current ASTM C 1019 evaluation procedure. This information will impact provisions of ASTM C 476 and ASTM C 1019 and also grout criteria in bulding codes.

The variation of aggregate gradation on properties of mortar has been of concern for some time since most masonry sands currently available do not meet the limit requirements of ASTM C 144. **Buchanan** and **Call** report the results of an ASTM C 12.04 round robin test to evaluate the effects of broadening C 144 limits. Testing indicates that acceptable mortars can be produced from aggregates slightly exceeding C 144 values.

**Ribar** and **Dubovoy** examine the bond strength and water tightness of masonry assemblages with various combinations of masonry cement mortars and clay bricks. Brick properties of IRA and surface texture in conjunction with mortar properties were the primary factors in bond development.

**Dickelman** traces the development of clay plasticizers for use in masonry mortars. Physical properties of bentonite clay plasticized masonry mortars are compared to conventional mortars when tested in accordance to ASTM C 270. This information will be of direct use to ASTM Committee C 12.09 on Modified Mortars. This committee is currently addressing development of a specification for such mortars.

ASTM Committee C 12 on Mortar For Unit Masonry initiated the development of a specification for Ready Mix Mortar in the early 1980's. **Gates, Nelson,** and **Pistilli** present background information and generated test methods and data in comparison to conventional mortars used for development of the specifications. In summary this report concludes that ready mix mortars are at least equivalent to and in most instances superior to conventionally produced mortar.

During development of the ready mix mortar specifications, a provision allowing use of cylinder specimens in place of mortar cubes was inserted based on a similar provision for conventional mortars found in ASTM C 780-87. To substantiate this action for ready mix mortars, **Schmidt, Brown,** and **Tate** investigated the relationship of compressive strength for cubes and cylinders for ready mix mortars.

ASTM C 270-86a was the first ASTM Standard that allowed production of any mortar type by proportion specification using masonry cement alone as the binder. **Matthys** produced several Type N and Type S Mortars (masonry cement and portland cement lime) using the proportion specification of ASTM C 270 and compared their resulting properties to the property specification requirements of C 270.

**Haver, Keeling, Somayaji, Jones,** and **Heidersbach's** paper reviews the available literature on corrosion in masonry. Based on experimental comparisons, some commonly used methods of corrosion analysis in concrete are not accurate for masonry. The paper points out the need for research in this area.

**Matthys** and **Singh's** experimental study on compressive strength comparison of 2" cubes versus 3" x 6" cylinders for (1) lab mortars cured in a moist room (2) jobsite mortar cured in lab air and (3) jobsite mortar cured outside should help to establish conversion factors for existing mortar specifications and those currently under developments.

**Sriboonlue** and **Wallo's** experimental study on constituent proportions of portland cement lime mortar and grout defines trends in stiffness characteristics.

## ASSEMBLAGES

The laboratory flexural bond strengths of one Type N and one Type S prepackaged masonry cement mortar was investigated by **McGinley** for ten different clay bricks. Bond strengths of 2 psi to 55 psi with coefficients of variation from 30% to 150% were observed. It appears that the IRA of clay units can have a greater influence on the flexural bond strength than is generally accepted by the masonry industry.

**Gabby** presents a compilation of clay masonry flexural bond stresses for portland cement lime mortars as determined from ASTM E 72, ASTM E 518, and ASTM C 1072. Comparison between test methods shows that there is insignificant differences between results from ASTM E 72 and ASTM C 1072. This information should be of use to ASTM Task Group C 12.03.03/C.15.05.02 which is currently addressing flexural bond performance criteria for masonry.

Nondestructive techniques are becoming more predominant in the overall consideration of evaluating, strengthening, and retrofitting masonry structures. ASTM Task Group C 15.04.06 is currently developing strandards in this area. **Noland's** update reviews existing NDE technologies and points out those that should be considered as possibilities for masonry evaluation.

Brown and Borchelt investigate the relationships between ASTM C 652 hollow clay brick units and portland cement lime mortar type to assemblage strength and stiffness. This information will be of value in addressing performance specification and building code criteria.

A common system of construction in commercial application is attachment of stone panels to buildings. Cement criteria of anchoring procedures are generally empirical. Amrhein, Hatch, and Merrigan investigate the strength of common anchor systems and show large factors of safety to existing code criteria.

A field adapted ASTM E 514 water permeability test is often used by masonry consultants to investigate moisture penetration problems in brick masonry. Such results have influenced decisions on existing structures when debated both inside and outside of courtrooms. Brown addresses the significant differences between the standard ASTM E 514-74 test method and the current field adapted ASTM E 514 method.

Crooks and Herget review the development of masonry construction systems in the U.S. Theory and practical experience have produced recommended masonry practices. Such practices are often abused, particularly in light of differential movement, workmanship, and moisture control. The authors give examples of observed abuse and subsequent repair solutions.

Based on 25 years of experience, Tomasetti reveals situations that have caused physical problems to masonry structures or unacceptable appearance to the design professional and/or owner. Reasons for these occurances and solutions to the problems are addressed.

Laska, Ostrander, Nelson and Munro report results of a study to generate masonry test data which reflect the actual properties of masonry expected under field conditions. This information will be of use in correlating component properties to assemblage properties and lab properties to field properties. Also an assessment of actual strength versus published code values is made. The authors suggest changes to the ASTM C 270 mortar specifications.

In recent years, with the adoption of ASTM C 270-86a and development of ASCE/ACI 530 Masonry Code, there has been significant interest in the flexural bond capacity of masonry assemblages. Matthys presents bond results of ASTM C 1072 stackbonded concrete block prisms and airbag tests of 4' x 8' concrete block walls using both masonry cement and portland cement lime mortars. This data will be of use to Task Group C 12.03.03/C 15.05.02 in addressing flexural bond performance criteria of masonry.

**Mehta** reviews the theoretical background on noise level reduction by barriers and its application to thin masonry panels.

In existence is a U.S. - Japan Coordinated Program for Masonry Building Research funded by the National Science Foundation. This program consisted of coordinated research projects that address basic issues of masonry material and structural response to gravity and seismically induced loads. **Antrobus, Leiva, Merryman,** and **Klingner** present results of their experimental evaluation of fullscale reinforced masonry two story structures. This work along with others from the TCCMAR program will provide the technical basis for improved building code provisions and appropriate design and analysis methodologies.

**Krogstad** examines the current ASTM tests for water penetration in light of masonry veneers and cavity walls. He then proposes a new test method for field testing which can be used as a quality control check for new masonry walls or as an investigative tool in evaluating existing masonry leakage problems.

The major structural materials in the U.S.A. have developed an ultimate strength design methodology for structural design. To keep pace clay and concrete masonry need to accurately define the nonlinear stress distribution in compression at ultimate condition. **Hamid, Assis,** and **Harris** present such experimental data for grouted concrete masonry.

The potential of explosions occuring in an industrial society is severe enough that a blast resistant design should be considered for public and commercial structures. **Volkman** examines blastload characteristics, current building code criteria, and a proposed design procedure.

## CLOSING REMARKS

The information presented in these papers covers a wide breadth of masonry activity currently in existence from testing laboratories, trade associations, universities, research centers, material manufacturers, ASTM committees, and consulting firms. As usual, reported results often provide potential solutions to problems while also creating other areas

needing attention.    It is significant that this most
traditional universal building material of components
and assemblages is being technically addressed.    Let
us hope this situation continues and expands.  The end
result can be no less than proper application and more
confidence in the use of these enduring materials.


John H. Matthys
Professor, Civil Engineering
Director, Construction Research Center
University of Texas at Arlington
Symposium Chairman

# Author Index

## A

Amrhein, J. E., 279
Antrobus, N., 378
Assis, G. F., 403
Atkinson, R., 248

## B

Bailey, W., 5
Borchelt, J. G., 263
Brown, M. L., 147
Brown, M. T., 299
Brown, R. H., 263
Buchanan, Jr., C. E., 63

## C

Call, B. M., 63
Crooks, R. W., 309

## D

Dickelman, B., 108
Dubovoy, V. S., 85
Dunstan, Jr., E. R., 27

## E

Edwards, J., 5

## G

Gabby, B., 235
Gates, R. E., 123

## H

Hamid, A. A., 403
Harris, H. G., 403
Hatch, R. H., 279
Haver, C., 173
Hay, P., 27
Hedstrom, E. G., 47
Heidersbach, R., 173
Hergot, F. A., 309
Hogan, M. B., 47

## J

Jones, D., 173

## K

Keeling, D., 173
Kingsley, G., 248
Klingner, R. E., 378
Krogstad, N. V., 394

## L

Laska, W. A., 339
Leiva, G., 378

## M

Matthys, J. H., 5, 164, 194, 350
McGinley, W. M., 217
Mehta, M., 366
Merrigan, M., 279
Merryman, M., 378
Munro, C. C., 339

## N

Nelson, R. L., 123, 339
Noland, J., 248

## O

Ostrander, O. W., 339

## P

Pistilli, M. F., 123

## R

Ribar, J. W., 85

## S

Schmidt, S., 147
Singh, R., 194
Somayaji, S., 173
Sriboonlue, W., 206

**T**

Tate, R., 147
Tomassetti, A., 324
Trimble, B. E., 38

**V**

Volkman, D. E., 413

**W**

Wallo, E. M., 206

# Subject Index

## A

Aardelite aggregate, 27
Absorption, initial rate
    of, 5, 85, 217, 324
Aggregate
    C 144: 63
    gradation, 63
    lightweight, 27
Air content, 85, 164
    ready-mixed mortar, 123
Anchor systems, stone, 279
ASTM standards
    C 67: 5
    C 90: 27, 339
    C 140: 339
    C 144: 63
    C 270: 108, 123, 164,
            194, 206, 339, 350
    C 476: 47, 206
    C 514: 85
    C 618: 27
    C 652: 263, 339
    C 780: 123, 147, 194
    C 902: 38
    C 1019: 47
    C 1072: 85, 235, 339, 350
    E 72: 235
    E 447: 339
    E 514: 299, 394
    E 518: 235
Attenuation, 366

## B

Barriers, sound
    attenuation, 366
    insertion loss, 366
    transmission loss, 366
Bentonite, 108
Blast resistant masonry, 413
Bond
    C 1072: 85, 235, 339, 350
    strength, 85, 123, 217, 235
    wrench, 85, 217, 350
Breathing, 324
Brick
    C 652: 263, 339
    clay, 5, 85, 217
    hollow, 263
    paving, 38
    slip resistance, 38
    wall panels, 366

    water penetration, 5, 85, 299
Building code, blast resistant
    design incorporation, 413

## C

Carbonation, 173
Cavity walls, 173, 299, 394
Cement (See Mortar)
Chlorides, 173
Coefficient of friction, 38
Compressive strength, 47, 85,
    164, 206, 263, 403
    ready-mixed mortar, 123, 147
    shape effects on, 147, 194
Concrete masonry units, 235,
    350, 403
    C 90: 27, 339
    masonry grout, 47
Corrosion, reinforcing
    metal, 173, 324
Cube compressive strength,
    147, 194, 206
Cylinder compressive strength,
    147, 194, 206

## D

Detonation, 413
Differential movement, 309
Drainage wall, 299
Durability, 324

## E

Earthquake resistance, masonry
    walls, 378
Efflorescence, 324
Explosion resistant design, 413
Extended-life mortar, 123

## F

Flashing, 394
Flexural bond (See Bond)
Flexural design, 403
Fly ash
    C 618: 27
Friction, coefficient of, 38

**G**

Grout, 403
   C 476: 47, 206
   C 1019: 47
   masonry, 47
   portland cement lime, 206
   physical properties, 47

**I**

Industry, masonry, practices,
   309, 324, 413
Initial rate of absorption, 5,
   85, 217, 324

**J**

Japan and United States masonry
   programs, 378, 403

**K**

Kerfs, 279

**L**

Leakage, 299, 394
Lime deposits, 324

**M**

Masonry (See also Mortar)
   brick
      C 652: 263, 339
      clay, 5, 85, 217
      hollow, 263
      paving, 38
      slip resistance, 38
      wall panels, 366
      water penetration, 5, 85,
         299
   cavity, 394
   clay, 235
   concrete units, 27, 47, 235,
      339, 350, 403
   construction, problems and
      cures, 309, 324
   design, blast-resistant, 413
   industry practices, 309, 324,
      413
   leakage, 299, 394
   nondestructive evaluation, 248
   mechanical properties, 248
   panels, 366
   prisms, 108, 123

   brick, 217, 263, 339
   C 1072: 85, 339
   concrete masonry, 47, 217,
      339, 350, 403
   E 447: 339
   research programs, 339, 378,
      403
   ultimate strength, 403, 413
   wall assemblies, 85, 108,
      173, 309, 339, 378
   water penetration, 5, 85,
      123, 299
   control, 309
   E 514: 299, 394
Metal corrosion, in masonry, 173
Modelling, 413
Modulus of elasticity, 47, 206, 263
Moisture control (See Water
   penetration and Water control)
Moment capacity, 403
Montmorillonite, 108
Mortar, 47, 173
   C 270: 108, 123, 164, 194,
      206, 339, 350
   C 780: 123, 147, 194
   conventional, 147, 164, 350
   extended-life, 123
   masonry cement, 63, 85,
      164, 194, 217, 350
   plasticizers, 108
   portland cement lime, 63, 164,
      194, 206, 235, 263, 350
   ready-mixed, 123, 147, 194

**N**

Noise reduction, 366
Nondestructive evaluation, 248

**P**

Particle size distribution, 85
Paving brick
   C 902: 38
Plasticizers, masonry mortar, 108
Portland cement (See Mortar)
Prisms, 108, 123
   brick, 217, 263, 339
   C 1072: 85, 339
   concrete masonry, 47, 217,
      339, 350, 403
   E 447: 339
Pull-outs, 279

## R

Ready-mixed mortar, 123, 147, 194

## S

Seismic resistance, concrete
    masonry walls, 378
Slip resistance test, 38
Spalling, 324
Steel, 173
Stone veneer, 279
Strength tests
    E 72: 235
Stress parameters, 403
Striping, 324
Suction, initial rate of (See
    Absorption, Initial rate of)

## T

Tensile stress values, flexural, 235
Tile, clay, sampling and testing
    C 67: 5
Transverse tests, 350

## U

Ultimate strength, 403, 413
United States and Japan
    masonry programs, 378, 403

## W

Walls
    air blast test, 413
    assemblies, 85, 108, 173,
        309, 339, 378
    drainage test, 394
    lateral load, 350
    masonry coupled, 378
    panels, 366
    ties, 173, 279
    veneer, 173, 279
Water penetration, 5, 85, 123, 299
    control, 309
    E 514: 299, 394
Water retention, 85, 123, 164
Wire ties, 279
Workability
    masonry, 108
    ready-mixed mortar, 123

## Z

Zinc, 173